Digital Education and Learning

Series Editor
Michael Thomas
Liverpool John Moores University
Merseyside, UK

Much has been written during the first decade of the new millennium about the potential of digital technologies to produce a transformation of education. Digital technologies are portrayed as tools that will enhance learner collaboration and motivation and develop new multimodal literacy skills. Accompanying this has been the move from understanding literacy on the cognitive level to an appreciation of the sociocultural forces shaping learner development. Responding to these claims, the Digital Education and Learning Series explores the pedagogical potential and realities of digital technologies in a wide range of disciplinary contexts across the educational spectrum both in and outside of class. Focusing on local and global perspectives, the series responds to the shifting landscape of education, the way digital technologies are being used in different educational and cultural contexts, and examines the differences that lie behind the generalizations of the digital age. Incorporating cutting edge volumes with theoretical perspectives and case studies (single authored and edited collections), the series provides an accessible and valuable resource for academic researchers, teacher trainers, administrators and students interested in interdisciplinary studies of education and new and emerging technologies.

Stamatios Papadakis
Georgios Lampropoulos
Editors

Social Robots and Artificial Intelligence in Education

Integrating AI in K-12 and Higher Education

palgrave
macmillan

Editors
Stamatios Papadakis
Department of Preschool Education
University of Crete
Heraklion, Greece

Georgios Lampropoulos
Department of Education
University of Nicosia
Nicosia, Greece

ISSN 2753-0744 ISSN 2753-0752 (electronic)
Digital Education and Learning
ISBN 978-3-032-03976-7 ISBN 978-3-032-03977-4 (eBook)
https://doi.org/10.1007/978-3-032-03977-4

Credit Line: Andriy Onufriyenko/Getty Images

This Palgrave Macmillan imprint is published by the registered company Springer Nature Switzerland AG.
The registered company address is: Gewerbestrasse 11, 6330 Cham, Switzerland

Previous Publications

AI-Assisted Assessment in Education: Transforming Assessment and Measuring Learning
Goran Trajkovski & Heather Hayes

Effective Multimodal Interaction for Online and Hybrid Teaching: Data-Driven Insights
Antonella Giacosa

Vietnamese Higher Education at a Crossroads: Transformation During the Digital Age
Duong Van Thanh, Mark Ashwill & Hoang Auh Tuan

Beyond Digital Distraction: Educating Today's Cyber Student
Kurt C. Schuett

Second Language Teacher Professional Development: Technological Innovations for Post-Emergency Teacher Education
Karim Sadeghi & Michael Thomas

Multimodal Learning Environments in Southern Africa: Embracing Digital Pedagogies
Jako Olivier, Avinash Oojorah & Waaiza Udhin

Researching Virtual Play Experiences: Visual Methods in Education Research
Chris Bailey

Re-imagining Technology Enhanced Learning: Critical Perspectives on Disruptive Innovation
Michael Flavin

Critical Perspectives of Educational Technology in Africa: Design, Implementation, and Evaluation
Bellarmine A. Ezumah

Graduate Skills and Game-Based Learning: Using Video Games for Employability in Higher Education
Matthew Barr

Transmedia Knowledge for Liberal Arts and Community Engagement: A StudioLab Manifesto
Jon McKenzie

Conceptualising the Digital University: The Intersection of Policy, Pedagogy and Practice
Bill Johnston, Sheila MacNeil & Keith Smyth

Education, Narrative Technologies and Digital Learning
Tony Hall

Digital Storytelling in Higher Education: International Perspectives
Grete Jamissen, Pip Hardy, Yngve Nordkvelle & Heather Pleasants

Disruptive Technology Enhanced Learning: The Use and Misuse of Digital Technologies in Higher Education
Michael Flavin

Digital Technology, Schools and Teachers' Workplace Learning: Policy, Practice and Identity
Michael Phillips

Technology and the Politics of University Reform: The Social Shaping of Online Education
Edward C. Hamilton

Augmented Education: Bringing Real and Virtual Learning Together
Kieron Sheehy, Rebecca Ferguson & Gill Clough

Critical Perspectives on Technology and Education
Scott Bulfin, Nicola F. Johnson & Chris Bigum

Digital Skills: Unlocking the Information Society
Jan A. G. M. Dijk & Alexander J. A. M. Deursen

Digital Technologies for School Collaboration
Anastasia Gouseti

Digital Networking for School Reform: The Online Grassroots Efforts of Parent and Teacher Activists
Alison Heron-Hruby & Melanie Landon-Hays

Learning in Real and Virtual Worlds: Commercial Video Games as Educational Tools
Pilar Lacasa

The Politics of Education and Technology: Conflicts, Controversies, and Connections
Neil Selwyn & Keri Facer

Computer Games and Language Learning
Mark Peterson

Rhetoric/Composition/Play Through Video Games: Reshaping Theory and Practice of Writing
Richard Colby, Matthew S. S. Johnson & Rebekah Shultz Colby

Digital Media and Learner Identity: The New Curatorship
John Potter

Digital Education: Opportunities for Social Collaboration
Michael Thomas

Contents

List of Contributors

Durcelina Pimenta Arruda Federal University of Minas Gerais—PPGE, Belo Horizonte, Brazil

Eucidio Pimenta Arruda Federal University of Minas Gerais—PPGE, Belo Horizonte, Brazil

Daniela M. S. Cabella Complete Privacy, Rio de Janeiro, Brazil

Carlo Cattani Unitus, Viterbo, Italy

Christos Chronis Department of Informatics and Telematics, Harokopio University of Athens, Athens, Greece

Pablo Dúo-Terrón University of Granada, Granada, Spain

Todorka Glushkova Department of Computer Technologies, University of Plovdiv, Plovdiv, Bulgaria

Joseph F. Hair Jr. Mitchell College of Business, University of South Alabama, Mobile, AL, USA

Juho Kahila School of Applied Educational Science and Teacher Education, University of Eastern Finland, Joensuu, Finland

Blanka Klimova Faculty of Informatics and Management, University of Hradec Kralove, Hradec Králové, Czech Republic

Georgios Lampropoulos Department of Preschool Education, School of Education, University of Crete, Rethymnon, Greece

Department of Education, School of Education, University of Nicosia, Nicosia, Cyprus

Department of Applied Informatics, School of Information Sciences, University of Macedonia, Thessaloniki, Greece

Anh Nguyen-Nhu Le Ho Chi Minh City University of Education, Ho Chi Minh, Vietnam

Jesús López-Belmonte University of Granada, Granada, Spain

Ela Luria English and Education Department, Levinsky College of Education & Beit Berl College of Education, Tel Aviv-Jaffa, Israel

Damian Maher University of Technology Sydney, Sydney, NSW, Australia

José-Antonio Marín-Marín University of Granada, Granada, Spain

Dioneia Motta Monte-Serrat FFCLRP-USP, Unaerp, Ribeirão Preto, Brazil

Antonio-José Moreno-Guerrero University of Granada, Granada, Spain

Nguyen Hoai Ly Nga University of Economics Ho Chi Minh City, Ho Chi Minh, Vietnam

Stamatios Papadakis Department of Preschool Education, School of Education, University of Crete, Rethymnon, Greece

Eleftheria Papageorgiou Department of Informatics and Telematics, Harokopio University of Athens, Athens, Greece

Marcel Pikhart Faculty of Informatics and Management, University of Hradec Kralove, Hradec Králové, Czech Republic

Nicolas Pope School of Computing, University of Eastern Finland, Joensuu, Finland

Sarah Rana Mitchell College of Business, University of South Alabama, Mobile, AL, USA

Denise Robb Mitchell College of Business, University of South Alabama, Mobile, AL, USA

Chryssa Sofianopoulou Department of Informatics and Telematics, Harokopio University of Athens, Athens, Greece

Ivaylo Staribratov Department of Math, Computer and IT Training, University of Plovdiv, Plovdiv, Bulgaria

Mark Stroud Chazanoff School of Business, College of Staten Island, Staten Island, NY, USA

Matti Tedre School of Computing, University of Eastern Finland, Joensuu, Finland

Iraklis Varlamis Department of Informatics and Telematics, Harokopio University of Athens, Athens, Greece

Henriikka Vartiainen School of Applied Educational Science and Teacher Education, University of Eastern Finland, Joensuu, Finland

Jiaqi Wang Faculty of Education, University of Cambridge, Cambridge, UK

Steven Watson Faculty of Education, University of Cambridge, Cambridge, UK

List of Figures

List of Tables

CHAPTER 1

The Emergence of Artificial Intelligence and Social Robots in Education

Stamatios Papadakis and Georgios Lampropoulos

1.1 Introduction

The rapid advancement of artificial intelligence (AI) and social robots is fundamentally reshaping educational environments globally. These emerging technologies offer transformative capabilities, including personalized learning experiences, enhanced student engagement, and improved

S. Papadakis
Department of Preschool Education, University of Crete,
Heraklion, Greece
e-mail: stpapadakis@uoc.gr

G. Lampropoulos (✉)
Department of Preschool Education, School of Education, University of Crete,
Rethymnon, Greece

Department of Education, School of Education, University of Nicosia,
Nicosia, Cyprus

Department of Applied Informatics, School of Information Sciences,
University of Macedonia, Thessaloniki, Greece
e-mail: glampropoulos@uom.edu.gr

S. Papadakis, G. Lampropoulos (eds.), *Social Robots and Artificial Intelligence in Education*, Digital Education and Learning,
https://doi.org/10.1007/978-3-032-03977-4_1

educational outcomes. The COVID-19 pandemic has further underscored the necessity of leveraging innovative technological solutions to address persistent challenges, such as teacher shortages, educational disparities, and the increasing demands for digital literacy competencies (Zawacki-Richter et al., 2019).

AI-driven educational tools, including intelligent tutoring systems and adaptive learning platforms, are becoming integral parts of contemporary classroom environments. Such systems harness sophisticated algorithms to process large volumes of data and analyze individual student data, allowing the adaptation of customized instructional content and pacing to meet the unique needs and preferences of diverse learners (Khan, 2024; Holmes et al., 2019). Research demonstrates that these adaptive systems can increase learner autonomy and foster self-regulated learning by providing timely feedback and scaffolding (VanLehn, 2011). Moreover, intelligent tutoring systems that integrate affective computing capabilities have the potential to monitor learners' emotional states, thus offering not only cognitive but also socio-emotional support (Lampropoulos & Papadakis, 2025b).

Concurrently, social robots, which engage learners through speech, gestures, and expressive facial cues, are utilized to enhance student motivation and facilitate socio-emotional learning, particularly in early childhood and primary education contexts (Breazeal, 2017; Belpaeme et al., 2018). These robots are increasingly perceived not merely as tools but as active social agents that enhance collaboration and creativity in the classroom (Lampropoulos, 2025). Social robots have also shown promise in supporting children with special educational needs, offering consistent interaction and personalized encouragement that can complement human instruction (Cabibihan et al., 2013). As Wang and Cheung (2024) argue, the integration of generative AI further enhances the autonomy and adaptability of social robots, enabling them to respond dynamically to individual learner profiles.

Despite their evident potential, the integration of AI and social robots into educational settings brings about significant ethical and practical considerations. Critical issues regarding data privacy, security, algorithmic biases, and the potential reduction of teacher autonomy require a deliberate, transparent, and ethically responsible approach to their adoption and deployment (Williamson & Eynon, 2020). Concerns about attachment and emotional dependency, especially in vulnerable populations, highlight the importance of designing socially intelligent robots that adhere to ethical standards and social norms (Kirby et al., 2010; Lampropoulos & Papadakis, 2025a). Furthermore, the successful adoption and integration of AI and robotic technologies rely fundamentally on their thoughtful alignment with pedagogical strategies and clearly defined educational objectives (Luckin et al., 2016). Equally important is the need for ongoing professional development for educators, equipping them with the knowledge and skills to effectively integrate these tools into their teaching practices (Zawacki-Richter et al., 2019). Teachers' perceptions often emphasize the need for clear guidelines and realistic expectations about the capabilities and limitations of social robots (van Ewijk et al., 2020).

As the field advances, interdisciplinary collaboration among educators, technologists, policymakers, and researchers is essential to ensure that AI and social robots serve as enablers of inclusive, equitable, and high-quality education. Hence it is important to continue examining the feasibility, the scalability, and the long-term impacts of deploying these technologies across diverse educational settings (Holmes et al., 2019). There is also a growing consensus on the need for empirical studies that explore the effectiveness of social robots in real-world classroom settings, beyond controlled laboratory environments (LeTendre et al., 2025). Additionally, as noted by Merino-Fernández et al. (2025), integrating educational robotics into STEM curricula offers opportunities for developing students' competencies, including creativity, critical thinking, and digital skills.

This volume explores the multifaceted applications of AI and social robots in education, examining their pedagogical, technological, ethical, and social dimensions. By bringing together interdisciplinary perspectives, it aims to provide a comprehensive understanding of how intelligent systems are transforming learning environments and to offer guidance for their responsible and equitable implementation.

1.2 Academic Rationale and Justification

While AI and robotics have been extensively researched within isolated technical or pedagogical domains, there is still a noticeable gap in integrating interdisciplinary research that holistically examines their educational potential. The majority of existing studies have either emphasized technical functionalities, often at the expense of robust pedagogical grounding, or have presented pedagogical narratives with limited technical insight (Lampropoulos & Papadakis, 2025b). This volume seeks to address these issues by providing a comprehensive synthesis that bridges technological innovation with educational theory, thereby critically assessing how AI-driven tools and social robots influence teaching and learning processes.

The educational application of AI and social robots requires a paradigm that moves beyond technology-centric discourse to one that deeply embeds these innovations within validated pedagogical frameworks. As highlighted by Lampropoulos and Papadakis (2025a), social robots, equipped with advanced AI capabilities, are increasingly recognized as embodied agents capable of fostering both cognitive development and socio-emotional engagement. By integrating adaptive learning pathways and real-time emotional responsiveness, these tools personalize learning experiences and enhance student motivation, engagement, and autonomy (Chen et al., 2020; Leyzberg et al., 2018).

Moreover, this volume builds upon emerging frameworks, providing a critical lens through which to evaluate the educational deployment of AI and social robots (Holmes et al., 2019). Interdisciplinary research is vital for understanding the multifaceted roles these technologies play, not only as tutors or tools but also as social companions and co-creators in learning environments (Breazeal et al., 2016; Lampropoulos, 2025).

Recent empirical studies underscore the potential of AI and social robots to enrich inclusive education, foster collaborative learning, and support differentiated instruction tailored to diverse learner profiles (Johal, 2020; Ouyang et al., 2022). However, challenges remain, particularly concerning ethical considerations, such as data privacy, algorithmic bias, and the socio-emotional implications of human-robot interaction (Arriaga et al., 2025; Edwards et al., 2018). This volume responds to these complexities by advancing theoretical and empirical insights that emphasize ethical design, educational effectiveness, and contextually sensitive implementation strategies.

Furthermore, by synthesizing findings from diverse geographical and socio-economic contexts, including underrepresented regions and special education settings, this volume highlights pathways for equitable technology integration (Merino-Fernández et al., 2025). It advocates for a transformative vision in which AI and social robots serve as integral components of a reimagined educational ecosystem that prioritizes human development and societal well-being (Papadakis & Lampropoulos, 2025).

In conclusion, this volume offers an academically rigorous, interdisciplinary contribution to the evolving discourse on AI and social robots in education. By uniting theoretical depth with empirical breadth, it provides a balanced academic exploration grounded in rigorous theoretical frameworks, such as constructivism, constructionism, social learning theory, and self-determination theory (SDT). It also engages deeply with ethical and critical frameworks in the context of AI literacy. As a result, it provides educators, policymakers, and researchers with critical insights and practical frameworks for harnessing the transformative potential of these technologies in meaningful and ethical ways.

1.3 Objectives and Research Questions

This edited volume is guided by a set of integrated objectives and research questions, which respond to identified theoretical, empirical, and methodological gaps in the field of AI and social robots in education, as well as emerging priorities globally in educational technology research and practice.

The key objectives of this volume are:

- To comprehensively map the state-of-the-art integration of AI and social robots across diverse educational contexts, learner populations, and pedagogical settings: Building on empirical evidence from K-12, higher education, and lifelong learning environments, the volume captures both mainstream applications (such as chatbots and virtual assistants) and pioneering practices (e.g., FOSSBot in STEM education).
- To critically evaluate pedagogical practices, learning outcomes, and instructional strategies associated with AI and social robot-based interventions: The volume extends beyond descriptive case studies to offer analytical perspectives on how these technologies support personalized learning, formative feedback, creative expression, and collaborative knowledge building. Particular emphasis is placed on instructional design, co-learning processes, and teacher facilitation roles.
- To examine the benefits and risks of AI and social robots, with a particular focus on equity, ethics, student well-being, and teacher agency: Recognizing the ethical complexities, the volume examines issues such as algorithmic bias, data privacy, commercialization risks, and the psychological impacts of human-AI interaction. It also addresses socio-technical disparities, including unequal access to resources, technology, and equipment.

- To propose conceptual and practical frameworks for the responsible, ethical, and evidence-informed integration of AI into education: By synthesizing theoretical models and new frameworks, the volume provides educators, policymakers, and developers with practical tools and frameworks to guide ethical adoption and sustainable scaling of AI-enhanced learning environments.
- To prioritize learner and teacher voices in the design and evaluation of AI-driven educational experiences: Throughout the chapters, the role of learners as co-creators of knowledge and the pivotal agency of teachers as facilitators and ethical gatekeepers are consistently highlighted, responding to implementation gaps and reinforcing the importance of human-centered AI integration.

Additionally, the volume aims to address the following research questions to aid in the advancement of scholarly understanding regarding the adoption and integration of AI and social robots in education while offering actionable insights for educational practice and policy:

1. How are AI and social robots currently integrated into educational settings across diverse geographic, cultural, and institutional contexts?
2. What pedagogical theories, models, and instructional strategies underpin the effective use of these technologies in formal and informal learning?
3. What are the potential advantages and limitations of AI and social robots in fostering equitable, inclusive, and learner-centered educational experiences?
4. How can ethical, social, and policy considerations, including data protection, transparency, and teacher professional development, be addressed in the integration of intelligent technologies in education?
5. What conceptual frameworks and empirical evidence can guide the future design, implementation, and evaluation of AI-enhanced learning environments?

1.4 Identification of Research Gaps

A critical analysis of the current literature highlights substantial and multifaceted gaps that hinder the comprehensive understanding and effective implementation of AI and social robots in education. These gaps span across theoretical foundations, methodological rigor, empirical evidence, contextual diversity, practical implementation, and longitudinal evidence. Addressing these gaps is essential to unlock the full pedagogical potential of these technologies. The identified gaps are briefly presented below:

- **Theoretical gaps:** Despite the growing body of research, the integration of AI tools and social robots within established pedagogical paradigms remains limited. Studies frequently adopt a technological determinism approach, focusing on the capabilities of AI systems without embedding them within robust educational theories, such as constructivism, socio-cultural theory, or SDT (Holmes et al., 2019; Lampropoulos & Papadakis, 2025b). Furthermore, there is ambiguity regarding the role of AI and robotic agents in educational settings, for example, whether they should act as tutors, learning companions, or facilitators of peer-like interaction (Breazeal et al., 2016; Leyzberg et al., 2014). Clarifying these roles is crucial to align technological functionalities with pedagogical intentions and learner expectations.
- **Methodological gaps:** Existing research is overly reliant on isolated case studies and short-term interventions, limiting the generalizability and scalability of findings (Gu, 2025; Lampropoulos, 2025). There is a noticeable absence of longitudinal research designs that can capture the evolving dynamics of human-robot interaction and sustained learning outcomes over time (Merino-Fernández et al., 2025). Moreover, mixed-methods studies that combine quantitative learning data with qualitative insights into learner experience, emotional engagement, and contextual variables remain scarce, limiting our ability to fully evaluate their educational impact (Chen et al., 2020).
- **Empirical gaps:** Rigorous empirical evaluations remain limited, especially concerning the impact of AI and social robots on measurable learning outcomes, learner equity, and inclusivity (Lampropoulos & Papadakis, 2025b). While initial findings suggest potential benefits for engagement and motivation, systematic studies assessing emotional, motivational, and socio-relational outcomes are limited (Anzalone et al., 2015; van Ewijk et al., 2020). Further empirical work is needed to explore how AI systems interact with diverse learner profiles, especially in emotionally sensitive or complex social learning scenarios.

- **Contextual gaps:** The research landscape remains disproportionately skewed towards high-income, technologically advanced contexts. There is a critical underrepresentation of studies examining the deployment and efficacy of AI and social robots in low- and middle-income countries, as well as in special education and early childhood education settings (Papakostas et al., 2021; Lorenzo-Lledó et al., 2025). These settings offer unique challenges and opportunities for inclusive and culturally responsive applications; yet, they remain underexplored in the existing literature.
- **Implementation gaps:** Practical implementation barriers persist, notably in the areas of teacher preparedness and institutional readiness. Many educators report feeling insufficiently equipped to integrate AI and social robots into their practice, due to a lack of targeted professional development and accessible pedagogical guidelines (Conti et al., 2025; Edwards et al., 2018). Furthermore, policy frameworks addressing essential issues, such as data privacy, algorithmic transparency, and bias, as well as accountability mechanisms, are inconsistently applied across educational systems (Arriaga et al., 2025; Williamson & Eynon, 2020).
- **Evidence gaps:** There is a lack of evidence concerning the long-term sustainability, scalability, and cost-effectiveness of AI and social robots in education. Although pilot studies seem promising, comprehensive cost-benefit analyses and sustainability assessments are largely absent from the literature (Belpaeme & Tanaka, 2021). Without such data, it remains challenging for educational institutions and policymakers to make informed decisions about the wide-scale adoption and integration of these technologies into formal curricula.

To address the identified gaps, the chapters in this volume provide a systematic and interdisciplinary response to the research, methodological, empirical, contextual, implementation, and theoretical gaps that currently characterize the field of AI and social robots in education. Specifically:

- **Diverse geographic and socio-economic inclusion:** The volume brings together rich studies from 13 countries across 5 continents, that is Asia (Israel, Vietnam), Australia, Europe (Bulgaria, Cyprus, Czech Republic, Finland, Greece, Italy, Spain, United Kingdom),

North America (United States of America), and South America (Brazil).

- **Innovative and varied research methods:** The contributions employ a variety of methodological approaches, including qualitative, quantitative, and mixed-methods studies, review studies, as well as experimental and quasi-experimental designs. Additionally, studies explore the use of AI and social robots by both students and teachers with some of them adopting co-design approaches. The diversity of the approaches and methods used enhances the robustness of the findings and supports triangulated insights into how AI and social robots function across different educational settings and learner demographics.
- **Focus on cognitive, emotional, and social outcomes:** Many chapters go beyond cognitive metrics to address emotional and social aspects of learning, responding to critical gaps in the existing literature. For example, some studies explore the role of chatbots in supporting student well-being and reducing feelings of isolation while others examine how social robots foster emotional engagement and collaboration among learners. Some of the studies also focus on providing empirical insights into the development of creativity, critical thinking, and peer-supported learning in AI-enhanced environments. Besides the emphasis on educational outcomes, the studies also more broadly discuss the societal impact of these technologies from a critical perspective.
- **Ethical, privacy, and AI literacy considerations:** Ethical reflection is a continuous thread throughout the chapters of this volume. The chapters provide a comprehensive examination of AI and social robots' ethical tensions, including algorithmic bias, data privacy, and digital equity. Several chapters also contribute to the emerging discourse on AI literacy, highlighting the need to equip both learners and educators with the competencies to use AI responsibly. Discussions extend to the regulatory landscape and the importance of transparent and accountable AI use in educational decision-making.
- **Theoretical integration for coherent educational outcomes:** The volume bridges technological innovation with pedagogical theory to avoid technological determinism. Foundational theories such as constructionism, social learning theory, universal design for learning, as well as novel frameworks for AI underpin the analyses. This integra-

tive approach ensures that technology use is framed within meaningful educational objectives, respecting the professional role of teachers and the socio-emotional needs of learners.
- **Practical and policy-driven recommendations:** Finally, across multiple chapters, there is a clear commitment to producing actionable recommendations for policymakers, educators, and technology developers. These contributions aim to influence not only academic discourses but also provide practical frameworks for sustainable, ethical, and equitable AI adoption.

Through these multi-dimensional contributions, the volume offers a comprehensive response to the gaps identified in the field, advancing both theoretical knowledge and practical pathways for the ethical and effective integration of AI and social robots in education.

1.5 Volume Structure

The volume is structured into three interrelated sections, each offering critical insights that together provide a comprehensive understanding of the evolving role of AI and social robots in education. Each section builds upon empirical findings and theoretical frameworks discussed across the remaining chapters, ensuring thematic coherence and academic depth. The three sections are briefly presented below:

- **Part I—Artificial Intelligence and Social Robots in Education:** The opening section establishes the conceptual foundations for understanding the multifaceted implications of AI technologies and social robots in education. Specifically, Chaps. 2 through 7 explore AI and social robots in the context of contemporary education focusing on the implementation of these technologies across different educational levels and contexts. Emphasis is placed on the pedagogical practices and application of AI and social robots in classrooms to aid teachers and to provide personalized and inclusive learning experiences. The chapters of this section also explore issues relevant to sustainability in education when using AI-driven solutions and examine the ethical complexities raised by AI, as well as the significance of developing educators' and learners' AI literacy as a precondition for meaningful integration.

- **Paty II—Artificial Intelligence in Student Learning and Cognitive Development:** The second section advances from foundational concepts to explore empirical studies demonstrating the impact of AI on students' learning, cognitive development, emotional engagement, and social learning processes. Drawing from Chaps. 8 through 10, this section highlights the transformative potential of AI-powered tools in fostering adaptive and personalized learning and promoting higher-order thinking skills, such as creativity and critical thinking. These chapters document how robotics, AI-driven tools and platforms, and adaptive learning environments support knowledge acquisition, emotional engagement, and collaborative learning dynamics, especially among learners. Moreover, this section presents participatory approaches, where learners are not passive recipients but co-creators of their learning journeys.
- **Part III—Artificial Intelligence Tools, Implications, and Future Trends in Education:** The final section provides a critical assessment of emerging tools and perspectives on the future of AI in education. Based on Chaps. 11 through 14, this section analyzes the pedagogical affordances and limitations of specific AI technologies, including generative AI models, social robots, and intelligent tutoring systems. It reveals practical implementation strategies, policy implications, professional development needs, and institutional capacities required to successfully embed these technologies within diverse educational systems. Furthermore, this section confronts the challenges of scaling AI applications while safeguarding educational equity and ethical integrity. This section also presents novel frameworks and highlights policy implications, offering actionable recommendations for educators, policymakers, and developers on how to navigate the complex intersection of technological innovation and educational values.

These three sections provide a holistic view of adopting and using AI and social robots in education from theoretical foundations to empirical applications and future-oriented frameworks. Collectively, the volume does not strive to solely be an academic contribution but to also offer practical guidelines for shaping the next generation of educational technologies.

1.6 Conclusions and Future Directions

This volume expands beyond binary narratives of technological optimism or skepticism, offering a mature, evidence-based perspective on the integration of AI and social robots in education. The chapters of this volume provide a balanced analysis that acknowledges both the transformative potential and the inherent risks of these technologies. Drawing on robust empirical studies and interdisciplinary theoretical foundations, the volume emphasizes that the future of AI in education must be carefully crafted and not be passively accepted.

The findings across the chapters underscore the urgency of pursuing sustained and rigorous research that transcends isolated case studies and short-term interventions. The volume collectively advocates for longitudinal, comparative, and participatory research designs that capture the evolving dynamics of AI-enhanced learning environments. Additionally, it recognizes the need for more inclusive research agendas that prioritize underrepresented regions, diverse learner profiles, and the socio-cultural contexts that shape educational technology adoption.

The chapters collectively affirm that while AI and social robots can augment learning experiences, they should not displace the essential human relationships at the core of education. The role of teachers, as ethical guides and facilitators of meaningful learning, remains irreplaceable. Equally, learners should be empowered as active participants and co-designers of their educational journeys, supported by technologies that respect their agency and individuality. Therefore, the volume highlights the need for human-centered AI integration, for AI literacy development among education stakeholders, and for the creation of iterative, design-based educational approaches that are adaptable, context-sensitive, and grounded in ethical considerations.

Furthermore, this volume addresses the importance of embedding policy foresight into educational planning. It offers actionable recommendations for building robust digital governance frameworks that ensure transparency, data privacy, and equity in the deployment of educational AI tools and social robots. Policymakers are urged to foster interdisciplinary collaborations among educators, technologists, and ethicists to co-create solutions that are not only technically effective but socially acceptable and sustainable.

Overall, this volume serves as both a reflective academic contribution and a pragmatic guide for shaping the future of education in an AI-driven world. It provides a roadmap for academics, practitioners, and policymakers to engage critically and constructively with AI and social robots, ensuring that their integration enhances human flourishing rather than diminishes it. Looking ahead to the future, the challenge is not merely to adopt AI technologies but to harness them thoughtfully and inclusively, so that the future of education remains deeply rooted in human values, equity, and purpose-driven innovation.

We hope that the discussions within this volume will spark new ideas and inspire innovative approaches to harness the power of AI and social robots to enrich teaching and learning practices and experiences.

References

Anzalone, S. M., Boucenna, S., Ivaldi, S., & Chetouani, M. (2015). Evaluating the engagement with social robots. *International Journal of Social Robots, 7*(4), 465–478. https://doi.org/10.1007/s12369-015-0298-7

Arriaga, P., Neto, I., Correia, F., & Soares, A. (2025). Ethical considerations in designing and testing robots for children in educational settings. In G. Lampropoulos & S. Papadakis (Eds.), *Social robots in education: How to effectively introduce social robots into classrooms* (pp. 169–204). Springer. https://doi.org/10.1007/978-3-031-82915-4

Belpaeme, T., & Tanaka, F. (2021). Social robots as educators. In *OECD digital education outlook 2021: Pushing the frontiers with artificial intelligence, blockchain and robots* (p. 143). OECD Publishing. https://doi.org/10.1787/589b283f-en

Belpaeme, T., Kennedy, J., Ramachandran, A., Scassellati, B., & Tanaka, F. (2018). Social robots for education: A review. *Science Robotics, 3*(21), eaat5954. https://doi.org/10.1126/scirobotics.aat5954

Breazeal, C. (2017, March 6–9). Social robots: From research to commercialization. In *Proceedings of the 2017 ACM/IEEE International Conference on Human-Robot Interaction* (pp. 1–1). Vienna, Austria.

Breazeal, C., Dautenhahn, K., & Kanda, T. (2016). Social robots. In B. Siciliano & O. Khatib (Eds.), *Springer handbook of robotics* (pp. 1935–1972). Springer. https://doi.org/10.1007/978-3-319-32552-1_72

Cabibihan, J. J., Javed, H., Ang, M., & Aljunied, S. M. (2013). Why robots? A survey on the roles and benefits of social robots in the therapy of children with autism. *International Journal of Social Robots, 5*, 593–618. https://doi.org/10.1007/s12369-013-0202-2

Chen, L., Chen, P., & Lin, Z. (2020). Artificial intelligence in education: A review. *IEEE Access, 8*, 75264–75278. https://doi.org/10.1109/ACCESS.2020.2988510

Conti, D., Cirasa, C., Høgsdal, H., & Di Nuovo, S. F. (2025). The use of social robots in educational settings: Acceptance and usability. In G. Lampropoulos & S. Papadakis (Eds.), *Social robots in education: How to effectively introduce social robots into classrooms* (pp. 205–220). Springer. https://doi.org/10.1007/978-3-031-82915-4

Edwards, C., Edwards, A., Spence, P. R., & Lin, X. (2018). I, teacher: Using artificial intelligence (AI) and social robots in communication and instruction. *Communication Education, 67*(4), 473–480. https://doi.org/10.1080/03634523.2018.1502459

Gu, W. (2025). A meta-analysis of the efficacy of conversational AI use in EFL learning contexts. In G. Lampropoulos & S. Papadakis (Eds.), *Social robots in education: How to effectively introduce social robots into classrooms* (pp. 335–374). Springer. https://doi.org/10.1007/978-3-031-82915-4

Holmes, W., Bialik, M., & Fadel, C. (2019). *Artificial intelligence in education: Promises and implications for teaching and learning.* Center for Curriculum Redesign.

Johal, W. (2020). Research trends in social robots for learning. *Current Robotics Reports, 1*(3), 75–83. https://doi.org/10.1007/s43154-020-00008-3

Khan, S. (2024). *Brave new words: How AI will revolutionize education (and why that's a good thing).* Penguin.

Kirby, R., Forlizzi, J., & Simmons, R. (2010). Affective social robots. *Robotics and Autonomous Systems, 58*(3), 322–332. https://doi.org/10.1016/j.robot.2009.09.015

Lampropoulos, G. (2025). Social robots in education: Current trends and future perspectives. *Information, 16*(1), 29. https://doi.org/10.3390/info16010029

Lampropoulos, G., & Papadakis, S. (2025a). *Social robots in education: How to effectively introduce social robots into classrooms.* Springer Nature. https://doi.org/10.1007/978-3-031-82915-4

Lampropoulos, G., & Papadakis, S. (2025b). The educational value of artificial intelligence and social robots. In G. Lampropoulos & S. Papadakis (Eds.), *Social robots in education: How to effectively introduce social robots into classrooms* (pp. 3–16). Springer. https://doi.org/10.1007/978-3-031-82915-4_1

LeTendre, G., Gray, R., & Toci, M. (2025). Interactions in the wild: Student and teacher perceptions of social robots in public classrooms. In G. Lampropoulos & S. Papadakis (Eds.), *Social robots in education* (pp. 43–62). Springer Nature.

Leyzberg, D., Spaulding, S., & Scassellati, B. (2014). Personalizing robot tutors to individuals' learning differences. In *Proceedings of the 2014 ACM/IEEE international conference on human-robot interaction* (pp. 423–430). ACM. https://doi.org/10.1145/2559636.2559671

Leyzberg, D., Ramachandran, A., & Scassellati, B. (2018). The effect of personalization in longer-term robot tutoring. *ACM Transactions on Human-Robot Interaction, 7*(3), 1–19. https://doi.org/10.1145/3283453

Lorenzo-Lledó, G., Meliá Beigbeder, S., & Gadzhimusieva, D. (2025). The application of emerging technologies in autistic students during the period 1996–2023: A conceptual, social, and intellectual review. In G. Lampropoulos & S. Papadakis (Eds.), *Social robots in education: How to effectively introduce social robots into classrooms* (pp. 309–334). Springer. https://doi.org/10.1007/978-3-031-82915-4

Luckin, R., Holmes, W., Griffiths, M., & Forcier, L. B. (2016). *Intelligence unleashed: An argument for AI in education*. Pearson.

Merino-Fernández, M. Á., Obregón Cuesta, A. I., Alonso-Centeno, A., et al. (2025). Theoretical and practical coherence of integrated STEM education and educational robotics: Review and analysis of good practices in Europe. In G. Lampropoulos & S. Papadakis (Eds.), *Social robots in education: How to effectively introduce social robots into classrooms* (pp. 427–462). Springer. https://doi.org/10.1007/978-3-031-82915-4

Ouyang, F., Zheng, L., & Jiao, P. (2022). Artificial intelligence in online higher education: A systematic review of empirical research from 2011 to 2020. *Education and Information Technologies, 27*(6), 7893–7925. https://doi.org/10.1007/s10639-022-10925-9

Papakostas, G. A., Sidiropoulos, G. K., Papadopoulou, C. I., Vrochidou, E., Kaburlasos, V. G., Papadopoulou, M. T., et al. (2021). Social robots in special education: A systematic review. *Electronics, 10*(12), 1398. https://doi.org/10.3390/electronics10121398

Van Ewijk, G., Smakman, M., & Konijn, E. A. (2020). Teachers' perspectives on social robots in education: An exploratory case study. In *Proceedings of the interaction design and children conference* (pp. 273–280). ACM. https://doi.org/10.1145/3392063.3394397

VanLehn, K. (2011). The relative effectiveness of human tutoring, intelligent tutoring systems, and other tutoring systems. *Educational Psychologist, 46*(4), 197–221. https://doi.org/10.1080/00461520.2011.611369

Wang, F., & Cheung, A. C. (2024). Robots' social behaviors for language learning: A systematic review and meta-analysis. *Review of Educational Research, 00346543231216437.*

Williamson, B., & Eynon, R. (2020). Historical threads, missing links, and future directions in AI in education. *Learning, Media and Technology, 45*(3), 223–235. https://doi.org/10.1080/17439884.2020.1798995

Zawacki-Richter, O., Marín, V. I., Bond, M., & Gouverneur, F. (2019). Systematic review of research on artificial intelligence applications in higher education—Where are the educators? *International Journal of Educational Technology in Higher Education, 16*(1), 39. https://doi.org/10.1186/s41239-019-0171-0

PART I

Artificial Intelligence and Social Robotics in Education

Part I focuses on the integration of AI, educational robotics, and social robots into educational practices, supporting teaching and learning. The chapters explore the foundational theories, practical applications, and ethical considerations regarding AI and social robots in educational settings. This section includes the following six chapters:

- **Chapter 2: Integrating Social Robots into Education** (Georgios Lampropoulos, Stamatios Papadakis): Provides an overview of social robots in classrooms, emphasizing their role in personalized and adaptive learning and their use in formal and informal educational settings. It highlights social robots' socio-emotional intelligence and communication skills, revealing their potential to enhance students' learning outcomes and social skills.
- **Chapter 3: Artificial Intelligence in the Global Educational Context: Prospects for Pedagogical Practice and Application in the Context of Teacher Shortages** (Eucidio Pimenta Arruda, Durcelina Pimenta Arruda): Discusses the potential of AI to address global teacher shortages and improve education quality. It explores how AI-driven systems can be tailored to meet the unique needs of the educational system of different countries.
- **Chapter 4: Revolutionizing English Language Learning through Social Robots and Artificial Intelligence** (Ela Luria): Focuses on English language learning and explores the intersection of AI, social robots, and language education, emphasizing personalized learning,

inclusivity, and equity. It discusses technologies such as natural language processing and intelligent tutoring systems.

- **Chapter 5: Artificial Intelligence and Robotics in School Education: Analysis of Implementation in Bulgarian Education** (Todorka Glushkova, Ivaylo Staribratov): Examines the integration of AI and robotics into Bulgarian schools, highlighting the impact of these technologies on teaching programming, social robotics, and AI topics to students and reflecting on teacher preparation for these innovations.
- **Chapter 6: Ethics and Sustainability in Education that Makes Use of Artificial Intelligence and Robots** (Dioneia Motta Monte-Serrat, Daniela MS Cabella, Carlo Cattani): Addresses the ethical considerations of using AI and robotics in education. The authors propose sustainable and ethical frameworks to handle issues like data privacy and the commercial use of student information.
- **Chapter 7: The Emerging Role of AI and Social Robots in University Student Recruitment** (Joseph F. Hair Jr., Sarah Rana, Denise Robb, Mark Stroud): Explores the influence of AI on high school students' perceptions, its relation to burnout, and its influence in higher education student recruitment process. It proposes future research into the impact of social robots in higher education.

Although there are some key differences among the chapters of this part, such as physical embodiments versus virtual agents, global policy versus local practices, and K-12 pedagogy versus higher education strategies, the following recurrent themes arose:

- Adaptive and personalized learning: Emphasis is placed on the transformative potential of tailoring instruction to each student's needs through social robots or AI agents and chatbots.
- Teacher readiness and institutional capacity: The effective integration of AI and social robots in education depends on both educators' readiness and the available institutional infrastructure.
- Ethical and equitable practices: The imperative need to protect privacy, prevent bias, and ensure inclusive access is highlighted in all chapters.
- Positive learning outcomes: The outcomes of the related chapters highlight the positive influence that the integration of AI and social robots might have but also comment upon the challenges that need to be addressed.

These six chapters present a dynamic landscape in which social robots and AI co-evolve with pedagogical theory, policy imperatives, and ethical norms. As we move into the detailed contributions that follow, readers are invited to reflect on how social robots and intelligent systems can be harnessed not only to optimize learning outcomes, but also to uphold human dignity, equity, and professional agency across all levels of education.

CHAPTER 2

Integrating Social Robots into Education

Georgios Lampropoulos and *Stamatios Papadakis*

2.1 Introduction

Nowadays, students strive to be involved in learning activities that are customized to their needs and involve their active participation (Kolb, 2014; Zhang et al., 2020). The use of information and communication technology (ICT) in education is highly regarded to help meet students' new educational needs through novel approaches and applications (Papert,

G. Lampropoulos (✉)
Department of Preschool Education, School of Education, University of Crete, Rethymnon, Greece

Department of Education, School of Education, University of Nicosia, Nicosia, Cyprus

Department of Applied Informatics, School of Information Sciences, University of Macedonia, Thessaloniki, Greece
e-mail: glampropoulos@uom.edu.gr

S. Papadakis
Department of Preschool Education, University of Crete, Heraklion, Greece
e-mail: stpapadakis@uoc.gr

S. Papadakis, G. Lampropoulos (eds.), *Social Robots and Artificial Intelligence in Education*, Digital Education and Learning,
https://doi.org/10.1007/978-3-032-03977-4_2

1984). Hence, research around various topics of ICT is actively being conducted.

Educational robotics is a field of study that is being increasingly researched (López-Belmonte et al., 2021) due to its ability to enrich the educational process and help meet students' requirements (Benitti, 2012). Although educational robots can bring about several benefits to the educational process, there are challenges that need to be addressed in terms of human-robot interactions (Sheridan, 2016). Specifically, there is a need for educational robots to undertake a more socially responsive role (Dautenhahn, 2007). In this context, there has been an increasing interest in social robots (Leite et al., 2013) and their use in educational settings (Lampropoulos, 2025). Several studies have been conducted that explore social robots from different perspectives and in different domains (Mejia & Kajikawa, 2017). However, education is one of the fields in which the use of social robots can bring several positive changes (Belpaeme et al., 2018). As the need to provide students with personalized learning experiences increases, social robots can be used as intelligent tutors to provide students with tailored learning material and activities. Additionally, social robots have the potential to also support teachers in carrying out their daily activities. As social robots, in contrast to traditional educational robots, can also understand and express emotions, the potentials that their integration in classrooms can bring about is yet to be defined.

The aim of this chapter is to provide an overview regarding the use of social robots in educational settings. In Sect. 2.2, the chapter goes over the use of educational robotics in teaching and learning activities. In Sect. 2.3, the concept of social robots and their use as educational tools is discussed. In Sect. 2.4, the use of social robots as a means to promote personalized learning is showcased and finally in Sect. 2.5, conclusive remarks and suggestions for future research directions are provided.

2.2 Educational Robotics

Educational robotics is being increasingly used in classrooms (López-Belmonte et al., 2021). Educational robotics constitutes an interdisciplinary field of study that strives to develop, integrate, and evaluate specialized designed robots that aim at supporting and enriching teaching and learning activities (Angel-Fernandez & Vincze, 2018; Eguchi, 2010; Scaradozzi et al., 2019).

Studies have showcased its potential to transform the educational domain and help meet the new educational requirements (Ospennikova et al., 2015) as well as its potential to be applied and positively influence learning outcomes in different learning settings (Evripidou et al., 2020) and across educational levels and age groups when integrating robotics-based learning activities in the curriculum (Gunes & Kucuk, 2022). Specifically, studies have revealed the positive influence that educational robotics has on students in pre-kindergarten to primary education (Chiazzese et al., 2019; Ching & Hsu, 2024), in secondary education (Arís & Orcos, 2019) as well as in higher education (Sánchez et al., 2019).

Although the use of educational robotics is mostly associated with computer science and engineering, recent studies have highlighted the educational benefits that can be yielded when introducing educational robotics in various subjects and have commented upon the enrichment of learning and teaching overall (Anwar et al., 2019; Benitti, 2012; Gunes & Kucuk, 2022). Additionally, although there is a clear need to create effective teacher programs to improve teachers' digital competences and knowledge regarding educational robotics (Schina et al., 2021), they mostly express positive opinions about the use of educational robotics in classrooms (Papadakis et al., 2021).

As there are various types of educational robotics that affect the educational process in a different way (Pedersen et al., 2020), learning goals and students' traits and characteristics should be considered to select the optimal type (Atmatzidou & Demetriadis, 2016; Virnes et al., 2008). By integrating educational robotics in classrooms more inclusive learning experiences can be created (Daniela & Lytras, 2019) and students' collaboration and social interactions can be improved (Denis & Hubert, 2001; Screpanti et al., 2021). Educational robots also support and promote maker education and gameful approaches (Lampropoulos & Papadakis, 2024) which can significantly influence students' learning performance and outcomes (Lampropoulos & Sidiropoulos, 2024). Additionally, activities that promote students' active participation (Eguchi, 2015), provide them with feedback (Chevalier et al., 2022), and improve their twenty-first-century skills (Eguchi, 2014), can be created.

2.3 Social Robots in Education

Offering new opportunities and capabilities, social robots are increasingly being used in various domains, including education (Leite et al., 2013). Similarly to educational robotics, there are various types of social robots that can be applied in educational settings affecting both learning and teaching due to their ability to communicate and interact with humans and other robots; thus, they can engage in a social level. Selecting the appropriate for each case social robot is imperative to ensure their effective integration. There are several different types of social robots that each can affect teaching and learning in a different way; hence, it is important to carefully consider which social robots are more suitable for each use case (Mahdi et al., 2022).

Social robots can display emotion and mood changes and are characterized by their social context, functions, and form, their anthropomorphism, and their social interfaces which render them as social and cognitive entities (Hegel et al., 2009; Kirby et al., 2010). As a result, to increase their social recognition and attribution toward them, it is important to comprehend these elements. Since social robots can relate, comprehend, and interact with humans (Breazeal, 2004), it is also important to assess their engagement and communication with humans (Anzalone et al., 2015). Additionally, social robots are characterized by their receptiveness and social evocation, have high levels of expression, relatedness, realism, and intelligence, and have enhanced social, comprehension, and communication skills (Breazeal, 2003). All these traits render them as embodied agents within a heterogeneous group of humans and robots that can socialize with and learn from others (Fong et al., 2003). Hence, their anthropomorphic design puts emphasis on their interpersonal communication and interaction with robots and humans as well as their cognitive and social skills, emotional and social intelligence, and embodiment (Breazeal et al., 2016).

Given their social traits and physical presence, social robots can effectively be used as peer learners or tutors that can increase cognitive and affective outcomes and, in specific tasks, achieve outcomes similar to those of human tutors (Belpaeme et al., 2018). However, more in-depth research needs to be conducted to better understanding the potentials and implications of integrating social robots in educational settings as some studies highlight their positive impact on learning outcomes (Papakostas et al., 2021) while other studies highlight the fact that they are not as effective as human teachers (Woo et al., 2021). Given the nature of social robots and their potential ability to positively influence students' learning performance and motivation, emphasis should be placed on examining their social behaviors within educational settings (Berghe et al., 2019).

In addition to supporting students and their learning, social robots can also support teachers in their day-to-day activities and tasks so that they can focus more on their students (Smakman et al., 2021). Hence, educational stakeholders report positive perspectives regarding the adoption and integration of social robots in education; however, they also express their concerns and quote the need to overcome existing barriers and challenges (Ewijk et al., 2020). Some of these challenges are associated with technological, economic, and logistic factors (Belpaeme & Tanaka, 2021) while others are related to moral and ethical considerations which should be addressed prior to the integration of social robots in teaching and learning activities (Ewijk et al., 2020; Smakman et al., 2021). As social robots capitalize on hardware and software, there is a need to explore their robustness and standardization (Pachidis et al., 2019). There is also a need to create effective guidelines and standards on how to develop, integrate, and use social robots in classrooms (Mahdi et al., 2022; Woo et al., 2021). Therefore, although social robots can constitute effective learning and teaching companions (Johal, 2020), they should comply with rigid human social norms, regulations, and rules (Kirby et al., 2010). Finally, it is vital to further examine how social robots can affect teaching and learning and to identify best practices and approaches to integrate them in classrooms to fully reap the benefits that they can offer (Barakova et al., 2023).

2.4 Social Robots Toward Personalized Learning

The enrichment of educational robotics with artificial intelligence capabilities is becoming more prominent (Chen et al., 2020; Lampropoulos et al., 2025). Simultaneously, the research toward achieving personalized learning is constantly increasing, which, in turn, highlights its significance in the context of education (Bernacki et al., 2021). Personalized learning is closely related to the use of technological applications to tailor the educational material and activities to the goals, needs, requirements, skills, and knowledge of each individual student; resulting, thus, in an adaptive and individualized learning experience (Lampropoulos, 2023; Xie et al., 2019).

There are different approaches and technological interventions that can improve personalized learning (Zhang et al., 2020). Due to their characteristics and their close relationship with the fields of data science and artificial intelligence, social robots can capitalize on educational data and learner-specific data to provide them with meaningful and personalized learning experiences (Lampropoulos & Papadakis, 2025). Personalized robot tutors have proven to effectively support learners and improve their cognitive learning gains (Leyzberg et al., 2012, 2014). Learners who studied through the support of robot tutors in a personalized manner have also been reported to outperform those who were taught using non-personalized lessons (Leyzberg et al., 2018). Additionally, the use of expressive and empathic robots that are characterized by social supportive behaviors, has also resulted in positive learning outcomes (Saerbeck et al., 2010; Serholt et al., 2014). A key aspect toward these outcomes was the interactivity of robots and the ability to form relations with students (Kanda et al., 2004). Hence, it can be inferred that when adding social robots' affective and social skills and their ability to respond to students' actions, expressions, and emotions to the existing capabilities of educational robots to offer personalized learning experiences, these outcomes can be further amplified.

Moreover, digital literacy and artificial intelligence literacy have proven to be important aspects of twenty-first-century education (Dúo-Terrón et al., 2025 [Chap. 9]). Due to their nature, social robots and artificial intelligence can be applied in various educational contexts, levels, and modules (Glushkova & Staribratov, 2025 [Chap. 5]). These technologies can effectively improve teaching and learning processes (Wang & Watson, 2025 [Chap. 14]), increase their learning motivation and intentions to enroll in higher education institutes (Hair et al., 2025 [Chap. 7]), enhance students' involvement, creativity, and nurture their ethical awareness and data agency (Kahila et al., 2025 [Chap. 8]). Additionally, they can promote students' autonomy and enable them to engage with artificial intelligence and STEM professions (Dúo-Terrón et al., 2025 [Chap. 9]) while simultaneously promoting open educational resources and open-source initiatives (Varlamis et al., 2025 [Chap. 11]). Nonetheless, there are several challenges associated with their integration in education, such as cost, ethical and privacy concerns, bias, etc. (Luria, 2025 [Chap. 4]). Another important issue is the gap in teachers' digital and artificial intelligence literacy skills (Arruda & Arruda, 2025 [Chap. 3]) which highlights the need to develop appropriate training programs to support teachers' development (Glushkova & Staribratov, 2025 [Chap. 5]).

Therefore, although artificial intelligence tools can constitute effective supplementary tools, there is a need to carefully balance the integration of artificial intelligence into traditional teaching (Le & Nga, 2025 [Chap. 10]) as well as the fundamental human elements of teaching with technological innovations (Maher, 2025 [Chap. 12]). There is a clear need to create new educational frameworks that account for the integration of emerging technologies in teaching and learning practices, make regulatory improvements focusing on sustainable ethics (Monte-Serrat, et al., 2025 [Chap. 6]) and introduce suitable pedagogical approaches that promote human interaction, human-computer interaction, and human-robot interaction (Klimova & Pikhart, 2025 [Chap. 13]). As a result, the introduction of new technologies, tools, and applications with an emphasis on generative artificial intelligence within the educational domain has revealed the requirement to reform the educational system (Wang & Watson, 2025 [Chap. 14]).

While the outcomes of this study depict a largely positive impact of introducing social robots in education, it is equally important to explore this matter from a broader and more critical perspective. Social robots constitute key aspects in the domain of educational technology. Therefore, it is important to examine their use based not solely on their outcomes and functionalities but also to consider their social impact as they are not neutral but socially influenced and shaped by broader political, economic, and cultural contexts (Selwyn, 2010; Selwyn et al., 2020).

Hence, focusing on critical educational technology, the need to consider the influence of the introduction of social robots in education from the perspectives of sustainability, social justice, and political economy is highlighted. Given the nature of social robots, the involved stakeholders and enterprises, and the interests that are being advanced, the commercialization of social robots in the educational sector is an important aspect that should be carefully examined. In this context, ensuring equal access to educational resources and opportunities is imperative. Due to the physical presence of social robots and their involving both hardware and software components, there is a clear need to examine if and how sustainable it is to adopt and introduce them in classrooms. Therefore, critically examining the use of social robots in education from the continuously evolving economic, technological, societal, cultural, and political landscapes is encouraged.

2.5 Conclusions

This chapter looked into the use of educational robotics in teaching and learning and focused on the role of social robots and their use in education to improve personalized learning. Based on the outcomes, social robots' abilities to communicate, interact, and understand learners is highly regarded. Additionally, social robots can promote embodied learning and provide tailored learning experiences. Although social robots can support students, they can also assist teachers in carrying out their everyday tasks so that they can focus more on their students.

Due to their close relationship with the fields of artificial intelligence and data science, social robots can identify, collect, process, and analyze educational data and student-specific data to provide them with personalized content and activities. Therefore, social robots can act as both supporters for students and intelligent human-like tutors that can support and enrich the educational process, increase students' cognitive learning gains, and enhance learning outcomes. Studies exploring the use of educational robotics have already revealed the benefits that can be yielded when integrating robots in educational settings. Hence, it can be inferred that when adopting and integrating social robots which add social-emotional intelligence, socio-cognitive skills, communication and interaction skills, embodiment, and a sense of attachment and relatedness, these outcomes can be further amplified. Another benefit that arose was the ability of social robots to be used across educational levels and in both formal and informal settings.

As the field of study of social robots is still in its early stages, there are several challenges and issues that need to be addressed before they are more widely adopted in classrooms. Hence, future studies should further examine the moral, ethical, and security-related concerns regarding social robots as well as defining standards and guidelines to develop, introduce, and use social robots in classrooms. Additionally, there is a clear need for more experimental studies to be conducted to better understand the implications of using social robots in different educational settings.

References

Angel-Fernandez, J. M., & Vincze, M. (2018). *Towards a formal definition of educational robotics*. https://doi.org/10.15203/3187-22-1-08

Anwar, S., Bascou, N. A., Menekse, M., & Kardgar, A. (2019). A systematic review of studies on educational robotics. *Journal of Pre-College Engineering Education Research (J-PEER), 9*(2). https://doi.org/10.7771/2157-9288.1223

Anzalone, S. M., Boucenna, S., Ivaldi, S., & Chetouani, M. (2015). Evaluating the engagement with social robots. *International Journal of Social Robotics, 7*(4), 465–478. https://doi.org/10.1007/s12369-015-0298-7

Arís, N., & Orcos, L. (2019). Educational robotics in the stage of secondary education: Empirical study on motivation and STEM skills. *Education Sciences, 9*(2), 73. https://doi.org/10.3390/educsci9020073

Arruda, E. P., & Arruda, D. P. (2025). Artificial intelligence in the global educational context: Prospects for pedagogical practice and application in the context of teacher shortages. In S. Papadakis & G. Lampropoulos (Eds.), *Social Robots and Artificial Intelligence in Education: Integrating AI in K-12 and Higher Education* (pp. 37–60). Palgrave Macmillan.

Atmatzidou, S., & Demetriadis, S. (2016). Advancing students' computational thinking skills through educational robotics: A study on age and gender relevant differences. *Robotics and Autonomous Systems, 75*, 661–670. https://doi.org/10.1016/j.robot.2015.10.008

Barakova, E., Väänänen, K., Kaipainen, K., & Markopoulos, P. (2023). Benefits, challenges and research recommendations for social robots in education and learning: A Meta-Review. *2023 32nd IEEE International Conference on Robot and Human Interactive Communication (RO-MAN)*, 2555–2561. https://doi.org/10.1109/ro-man57019.2023.10309345

Belpaeme, T., & Tanaka, F. (2021). Social robots as educators. In *OECD digital education outlook 2021 pushing the frontiers with artificial intelligence, blockchain and robots: Pushing the frontiers with artificial intelligence, blockchain and robots* (p. 143). OECD Publishing Paris.

Belpaeme, T., Kennedy, J., Ramachandran, A., Scassellati, B., & Tanaka, F. (2018). Social robots for education: A review. *Science Robotics, 3*(21). https://doi.org/10.1126/scirobotics.aat5954

Benitti, F. B. V. (2012). Exploring the educational potential of robotics in schools: A systematic review. *Computers & Education, 58*(3), 978–988. https://doi.org/10.1016/j.compedu.2011.10.006

Berghe, R. van den, Verhagen, J., Oudgenoeg-Paz, O., Ven, S. van der, & Leseman, P. (2019). Social robots for language learning: A review. *Review of Educational Research, 89*(2), 259–295. https://doi.org/10.3102/0034654318821286

Bernacki, M. L., Greene, M. J., & Lobczowski, N. G. (2021). A systematic review of research on personalized learning: Personalized by whom, to what, how, and for what purpose(s)? *Educational Psychology Review, 33*(4), 1675–1715. https://doi.org/10.1007/s10648-021-09615-8

Breazeal, C. (2003). Toward sociable robots. *Robotics and Autonomous Systems, 42*(3–4), 167–175. https://doi.org/10.1016/s0921-8890(02)00373-1

Breazeal, C. (2004). *Designing sociable robots*. MIT Press.

Breazeal, C., Dautenhahn, K., & Kanda, T. (2016). Social robotics. In *Springer handbooks*(pp.1935–1972).https://doi.org/10.1007/978-3-319-32552-1_72
Chen, X., Xie, H., & Hwang, G.-J. (2020). A multi-perspective study on artificial intelligence in education: Grants, conferences, journals, software tools, institutions, and researchers. *Computers and Education: Artificial Intelligence, 1*, 100005. https://doi.org/10.1016/j.caeai.2020.100005
Chevalier, M., Giang, C., El-Hamamsy, L., Bonnet, E., Papaspyros, V., Pellet, J.-P., Audrin, C., Romero, M., Baumberger, B., & Mondada, F. (2022). The role of feedback and guidance as intervention methods to foster computational thinking in educational robotics learning activities for primary school. *Computers & Education, 180*, 104431. https://doi.org/10.1016/j.compedu.2022.104431
Chiazzese, G., Arrigo, M., Chifari, A., Lonati, V., & Tosto, C. (2019). Educational robotics in primary school: Measuring the development of computational thinking skills with the bebras tasks. *Informatics, 6*(4), 43. https://doi.org/10.3390/informatics6040043
Ching, Y.-H., & Hsu, Y.-C. (2024). Educational robotics for developing computational thinking in young learners: A systematic review. *TechTrends, 68*(3), 423–434. https://doi.org/10.1007/s11528-023-00841-1
Daniela, L., & Lytras, M. D. (2019). Educational robotics for inclusive education. *Technology, Knowledge and Learning, 24*(2), 219–225. https://doi.org/10.1007/s10758-018-9397-5
Dautenhahn, K. (2007). Socially intelligent robots: Dimensions of human–robot interaction. *Philosophical Transactions of the Royal Society B: Biological Sciences, 362*(1480), 679–704. https://doi.org/10.1098/rstb.2006.2004
Denis, B., & Hubert, S. (2001). Collaborative learning in an educational robotics environment. *Computers in Human Behavior, 17*(5–6), 465–480. https://doi.org/10.1016/s0747-5632(01)00018-8
Dúo-Terrón, P., López-Belmonte, J., Marín-Marín, J.-A., & Moreno-Guerrero, A.-J. (2025). Exploring horizons: Study on the perceptions of primary students regarding the future of work in robotics and steam. In S. Papadakis & G. Lampropoulos (Eds.), *Social robots and artificial intelligence in education: Integrating AI in K-12 and higher education* (pp. 213–235). Palgrave Macmillan.
Eguchi, A. (2010). What is educational robotics? Theories behind it and practical implementation. *Society for Information Technology & Teacher Education International Conference*, 4006–4014.
Eguchi, A. (2014). Educational robotics for promoting 21st century skills. *Journal of Automation, Mobile Robotics and Intelligent Systems*, 5–11. https://doi.org/10.14313/jamris_1-2014/1
Eguchi, A. (2015). Educational robotics as a learning tool for promoting rich environments for active learning (REALs). In *Handbook of research on educa-*

tional technology integration and active learning (pp. 19–47). https://doi.org/10.4018/978-1-4666-8363-1.ch002

Evripidou, S., Georgiou, K., Doitsidis, L., Amanatiadis, A. A., Zinonos, Z., & Chatzichristofis, S. A. (2020). Educational robotics: Platforms, competitions and expected learning outcomes. *IEEE Access, 8*, 219534–219562. https://doi.org/10.1109/access.2020.3042555

Ewijk, G. van, Smakman, M., & Konijn, E. A. (2020). Teachers' perspectives on social robots in education. *Proceedings of the Interaction Design and Children Conference*. https://doi.org/10.1145/3392063.3394397

Fong, T., Nourbakhsh, I., & Dautenhahn, K. (2003). A survey of socially interactive robots. *Robotics and Autonomous Systems, 42*(3–4), 143–166. https://doi.org/10.1016/s0921-8890(02)00372-x

Glushkova, T., & Staribratov, I. (2025). Artificial intelligence and robotics in school education: Analysis of implementation of AI and robotics in Bulgarian education. In S. Papadakis & G. Lampropoulos (Eds.), *Social robots and artificial intelligence in education: Integrating AI in K-12 and higher education* (pp. 91–120). Palgrave Macmillan.

Gunes, H., & Kucuk, S. (2022). A systematic review of educational robotics studies for the period 2010–2021. *Review of Education, 10*(3). https://doi.org/10.1002/rev3.3381

Hair, J., Rana, S., Robb, D., & Stroud, M. (2025). The emerging role of AI and social robots in university student recruitment. In S. Papadakis & G. Lampropoulos (Eds.), *Social robots and artificial intelligence in education: Integrating AI in K-12 and higher education* (pp. 153–183). Palgrave Macmillan.

Hegel, F., Muhl, C., Wrede, B., Hielscher-Fastabend, M., & Sagerer, G. (2009). Understanding social robots. *Second International Conferences on Advances in Computer-Human Interactions, 2009*, 169–174. https://doi.org/10.1109/achi.2009.51

Johal, W. (2020). Research trends in social robots for learning. *Current Robotics Reports, 1*(3), 75–83. https://doi.org/10.1007/s43154-020-00008-3

Kahila, J., Vartiainen, H., Pope, N., & Tedre, M. (2025). Engaging young minds in AI through co-designing and making their own AI-Driven Apps in the classroom. In S. Papadakis & G. Lampropoulos (Eds.), *Social robots and artificial intelligence in education: Integrating AI in K-12 and higher education* (pp. 189–212). Palgrave Macmillan.

Kanda, T., Hirano, T., Eaton, D., & Ishiguro, H. (2004). Interactive robots as social partners and peer tutors for children: A field trial. *Human-Computer Interaction, 19*(1), 61–84. https://doi.org/10.1207/s15327051hci1901

Kirby, R., Forlizzi, J., & Simmons, R. (2010). Affective social robots. *Robotics and Autonomous Systems, 58*(3), 322–332. https://doi.org/10.1016/j.robot.2009.09.015

Klimova, B., & Pikhart, M. (2025). Use of artificial intelligence in foreign language education in higher education settings. In S. Papadakis & G. Lampropoulos (Eds.), *Social robots and artificial intelligence in education: Integrating AI in K-12 and higher education* (pp. 329–353). Palgrave Macmillan.

Kolb, D. A. (2014). *Experiential learning: Experience as the source of learning and development.* FT Press.

Lampropoulos, G. (2023). Augmented reality and artificial intelligence in education: Toward immersive intelligent tutoring systems. In *Augmented reality and artificial intelligence* (pp. 137–146). https://doi.org/10.1007/978-3-031-27166-3_8

Lampropoulos, G. (2025). Social robots in education: Current trends and future perspectives. *Information, 16*(1), 29. https://doi.org/10.3390/info16010029

Lampropoulos, G., & Papadakis, S. (2024). Maker education and educational robotics in the age of artificial intelligence. In *Intelligent educational robots: Toward personalized learning environments* (pp. 1–8). De Gruyter. https://doi.org/10.1515/9783111352695-001

Lampropoulos, G., & Papadakis, S. (2025). The educational value of artificial intelligence and social robots. In G. Lampropoulos & S. Papadakis (Eds.), *Social robots in education: How to effectively introduce social robots into classrooms, studies in computational intelligence* (pp. 3–15). Springer. https://doi.org/10.1007/978-3-031-82915-4_1

Lampropoulos, G., & Sidiropoulos, A. (2024). Impact of gamification on students' learning outcomes and academic performance: A longitudinal study comparing online, traditional, and gamified learning. *Education Sciences, 14*(4), 367. https://doi.org/10.3390/app14062524

Lampropoulos, G., López-Belmonte, J., & Marín-Marín, J. (2025). Artificial intelligence and educational robotics in maker education. In S. Papadakis & M. Kalogiannakis (Eds.), *Empowering STEM educators with digital tools* (pp. 85–106). IGI Global Scientific Publishing. https://doi.org/10.4018/979-8-3693-9806-7.ch004

Le, A. N.-N., & Nga, N. H. L. (2025). Elevating EFL paraphrasing skills through generative artificial intelligence: An exploration with ChatGPT. In S. Papadakis & G. Lampropoulos (Eds.), *Social robots and artificial intelligence in education: Integrating AI in K-12 and higher education* (pp. 237–269). Palgrave Macmillan.

Leite, I., Martinho, C., & Paiva, A. (2013). Social robots for long-term interaction: A survey. *International Journal of Social Robotics, 5*(2), 291–308. https://doi.org/10.1007/s12369-013-0178-y

Leyzberg, D., Spaulding, S., Toneva, M., & Scassellati, B. (2012). The physical presence of a robot tutor increases cognitive learning gains. *Proceedings of the Annual Meeting of the Cognitive Science Society*, 1882–1887.

Leyzberg, D., Spaulding, S., & Scassellati, B. (2014). Personalizing robot tutors to individuals' learning differences. *Proceedings of the 2014 ACM/IEEE*

International Conference on Human-Robot Interaction. https://doi.org/10.1145/2559636.2559671

Leyzberg, D., Ramachandran, A., & Scassellati, B. (2018). The effect of personalization in Longer-Term robot tutoring. *ACM Transactions on Human-Robot Interaction, 7*(3), 1–19. https://doi.org/10.1145/3283453

López-Belmonte, J., Segura-Robles, A., Moreno-Guerrero, A.-J., & Parra-González, M.-E. (2021). Robotics in education: A scientific mapping of the literature in web of science. *Electronics, 10*(3), 291. https://doi.org/10.3390/electronics10030291

Luria, E. (2025). Revolutionizing English language learning through social robots and artificial intelligence: Towards inclusive and personalized education. In S. Papadakis & G. Lampropoulos (Eds.), *Social robots and artificial intelligence in education: Integrating AI in K-12 and higher education* (pp. 61–90). Palgrave Macmillan.

Mahdi, H., Akgun, S. A., Saleh, S., & Dautenhahn, K. (2022). A survey on the design and evolution of social robots—Past, present and future. *Robotics and Autonomous Systems, 156*, 104193. https://doi.org/10.1016/j.robot.2022.104193

Maher, D. (2025). Chatbots in schools: Opportunities and considerations. In S. Papadakis & G. Lampropoulos (Eds.), *Social robots and artificial intelligence in education: Integrating AI in K-12 and higher education* (pp. 305–328). Palgrave Macmillan.

Mejia, C., & Kajikawa, Y. (2017). Bibliometric analysis of social robotics research: Identifying research trends and knowledgebase. *Applied Sciences, 7*(12), 1316. https://doi.org/10.3390/app7121316

Monte-Serrat, D. M., Cabella, D. M. S., & Cattani, C. (2025). Ethics and sustainability in education that makes use of artificial intelligence and robots. In S. Papadakis & G. Lampropoulos (Eds.), *Social robots and artificial intelligence in education: Integrating AI in K-12 and higher education* (pp. 121–152). Palgrave Macmillan.

Ospennikova, E., Ershov, M., & Iljin, I. (2015). Educational robotics as an inovative educational technology. *Procedia—Social and Behavioral Sciences, 214*, 18–26. https://doi.org/10.1016/j.sbspro.2015.11.588

Pachidis, T., Vrochidou, E., Kaburlasos, V. G., Kostova, S., Bonković, M., & Papić, V. (2019). Social robotics in education: State-of-the-Art and directions. In *Mechanisms and machine science* (pp. 689–700). https://doi.org/10.1007/978-3-030-00232-9_72

Papadakis, S., Vaiopoulou, J., Sifaki, E., Stamovlasis, D., Kalogiannakis, M., & Vassilakis, K. (2021). Factors that hinder in-service teachers from incorporating educational robotics into their daily or future teaching practice. *Proceedings of the 13th International Conference on Computer Supported Education*. https://doi.org/10.5220/0010413900550063

Papakostas, G. A., Sidiropoulos, G. K., Papadopoulou, C. I., Vrochidou, E., Kaburlasos, V. G., Papadopoulou, M. T., Holeva, V., Nikopoulou, V.-A., & Dalivigkas, N. (2021). Social robots in special education: A systematic review. *Electronics, 10*(12), 1398. https://doi.org/10.3390/electronics10121398

Papert, S. A. (1984). New theories for new learnings. *School Psychology Review, 13*(4), 422–428. https://doi.org/10.1080/02796015.1984.12085122

Pedersen, B. K. M. K., Larsen, J. C., & Nielsen, J. (2020). The effect of commercially available educational robotics: A systematic review. In *Robotics in education* (pp. 14–27). https://doi.org/10.1007/978-3-030-26945-6_2

Saerbeck, M., Schut, T., Bartneck, C., & Janse, M. D. (2010). Expressive robots in education. *Proceedings of the SIGCHI Conference on Human Factors in Computing Systems, 1613–1622.* https://doi.org/10.1145/1753326.1753567

Sánchez, H., Martínez, L. S., & González, J. D. (2019). Educational robotics as a teaching tool in higher education institutions: A bibliographical analysis. *Journal of Physics: Conference Series, 1391*(1), 012128. https://doi.org/10.1088/1742-6596/1391/1/012128

Scaradozzi, D., Screpanti, L., & Cesaretti, L. (2019). Towards a definition of educational robotics: A classification of tools, experiences and assessments. In *Smart learning with educational robotics* (pp. 63–92). https://doi.org/10.1007/978-3-030-19913-5_3

Schina, D., Esteve-González, V., & Usart, M. (2021). An overview of teacher training programs in educational robotics: Characteristics, best practices and recommendations. *Education and Information Technologies, 26*(3), 2831–2852. https://doi.org/10.1007/s10639-020-10377-z

Screpanti, L., Cesaretti, L., Storti, M., & Scaradozzi, D. (2021). Educational robotics and social relationships in the classroom. In *Makers at school, educational robotics and innovative learning environments* (pp. 195–201). https://doi.org/10.1007/978-3-030-77040-2_26

Selwyn, N. (2010). Looking beyond learning: Notes towards the critical study of educational technology. *Journal of Computer Assisted Learning, 26*(1), 65–73. https://doi.org/10.1111/j.1365-2729.2009.00338.x

Selwyn, N., Hillman, T., Eynon, R., Ferreira, G., Knox, J., Macgilchrist, F., & Sancho-Gil, J. M. (2020). What's next for Ed-Tech? Critical hopes and concerns for the 2020s. *Learning, Media and Technology, 45*(1), 1–6. https://doi.org/10.1080/17439884.2020.1694945

Serholt, S., Barendregt, W., Leite, I., Hastie, H., Jones, A., Paiva, A., Vasalou, A., & Castellano, G. (2014). Teachers' views on the use of empathic robotic tutors in the classroom. *The 23rd IEEE International Symposium on Robot and Human Interactive Communication*, 955–960. https://doi.org/10.1109/roman.2014.6926376

Sheridan, T. B. (2016). Human-Robot interaction. *Human Factors: The Journal of the Human Factors and Ergonomics Society, 58*(4), 525–532. https://doi.org/10.1177/0018720816644364

Smakman, M., Vogt, P., & Konijn, E. A. (2021). Moral considerations on social robots in education: A multi-stakeholder perspective. *Computers & Education, 174*, 104317. https://doi.org/10.1016/j.compedu.2021.104317

Varlamis, I., Chronis, C., Sofianopoulou, C., & Papageorgiou, E. (2025). An open-source and open-design robot for STEM education in K-12. In S. Papadakis & G. Lampropoulos (Eds.), *Social robots and artificial intelligence in education: Integrating AI in K-12 and higher education* (pp. 273–304). Palgrave Macmillan.

Virnes, M., Sutinen, E., & Kärnä-Lin, E. (2008). How children's individual needs challenge the design of educational robotics. *Proceedings of the 7th International Conference on Interaction Design and Children*. https://doi.org/10.1145/1463689.1463766

Wang, J., & Watson, S. (2025). Integrating generative AI in education: Affordances, issues, and directions. In S. Papadakis & G. Lampropoulos (Eds.), *Social robots and artificial intelligence in education: Integrating AI in K-12 and higher education* (pp. 355–379). Palgrave Macmillan.

Woo, H., LeTendre, G. K., Pham-Shouse, T., & Xiong, Y. (2021). The use of social robots in classrooms: A review of field-based studies. *Educational Research Review, 33*, 100388. https://doi.org/10.1016/j.edurev.2021.100388

Xie, H., Chu, H.-C., Hwang, G.-J., & Wang, C.-C. (2019). Trends and development in technology-enhanced adaptive/personalized learning: A systematic review of journal publications from 2007 to 2017. *Computers & Education, 140*, 103599. https://doi.org/10.1016/j.compedu.2019.103599

Zhang, L., Basham, J. D., & Yang, S. (2020). Understanding the implementation of personalized learning: A research synthesis. *Educational Research Review, 31*, 100339. https://doi.org/10.1016/j.edurev.2020.100339

CHAPTER 3

Artificial Intelligence in the Global Educational Context: Prospects for Pedagogical Practice and Application in the Context of Teacher Shortages

Eucidio Pimenta Arruda and Durcelina Pimenta Arruda

3.1 Introduction

The current economic development perspective presents different dilemmas for countries, in a context of development of generative artificial intelligence (AI). On the one hand, countries such as the United States and China have shown massive investment in this type of technology (Roberts et al., 2021; Bhutoria, 2022), while there is a mismatch, especially in the poorest countries, whose global positioning ends up being framed more in the profile of consumption and purchase of third-party technologies than in the creation of innovation and technology (Mannuru et al., 2023). The development of digital technologies, especially AI, amplifies the pressure

E. P. Arruda (✉) • D. P. Arruda
Federal University of Minas Gerais—PPGE, Belo Horizonte, Brazil
e-mail: eucidio@ufmg.br

S. Papadakis, G. Lampropoulos (eds.), *Social Robots and Artificial Intelligence in Education*, Digital Education and Learning,
https://doi.org/10.1007/978-3-032-03977-4_3

on countries, as it also leads them to problematize the possible impacts that these technologies impose on the education system and teaching work (Mao et al., 2024).

The expansion of AI to its generative format, with multimodal production of human verisimilitude, imposes a great challenge for school systems around the world, since it puts the scientific and technological development of a country in evidence in a race for technological, economic, and cultural dominance provided using AI (Nemorin et al., 2023; Bearman et al., 2023). But, on the other hand, there are still basic difficulties related to the implementation of digital technologies in school curricula around the world, as well as teacher training and approximation between contemporary social demands, which involve the changes that technologies provide in society and the school model proposed by countries that still resembles conventional models of times past (Rane et al., 2023; Lin et al., 2023)

Along with this pressure for technological training, there is an increasing lack of teachers to work in the compulsory and free school systems of all countries. There is currently a lack of more than forty-four million new teachers worldwide to meet the universalization of basic education by 2030, with about thirteen million in primary education and thirty-one million in secondary education (UNESCO, 2023). Even rich countries such as Germany, the United States, and the United Kingdom detect shortages of between 30,000 and 120,000 teachers in all areas (Wiggan et al., 2021; See & Gorard, 2019; Arruda & Kerres, 2024)

The lack of interest in the profession is one of the main factors detected, in which the difficulty in building policies for the professional valorization of teachers is highlighted, which are configured in better salaries, an attractive career that draws the attention of the best professionals, as well as better working conditions, which evolve from the available technologies, number of students, infrastructure, personnel, and teaching materials (Arruda & Arruda, 2024; Durso & Arruda, 2022)

Salary presents itself as an essential element in a scenario to understand the lack of teachers worldwide. Only half of the countries pay for primary school teachers as much as they pay for other professions with similar qualifications (UNESCO, 2023; Akiba et al., 2012).

While teacher shortages are a complex problem with solutions still in development, its negative impacts are already evident. The lack of qualified professionals directly compromises the achievement of the United Nations' (UN's) Sustainable Development Goal 4, which seeks to ensure quality education for all. In addition, this shortage also affects Goal 10, which aims to reduce inequalities. The failure to universalize basic education prevents countries from exploiting their potential to produce scientific and technological knowledge and, therefore, can amplify socioeconomic disparities (Arruda & Kerres, 2024; Shiohira, 2021).

Due to these issues, we consider it very pertinent to expand research that discusses the interlocutions between the development of generative AI, school education, as well as the challenges and problems faced by educational systems in meeting the demands of students and teachers in contemporary society.

Given the gaps in education, we propose to explore the integration of AI into teaching work, in confluence with the construction of pedagogical support innovators. I emphasize that the discussion is not limited to the creation of "teaching robots" (Edwards & Cheok, 2018), but rather to the exploration of the potential of generative AI to assist teachers in optimizing their practices, allowing the reorganization of work, the reduction of the overload of activities and, consequently, more effective dedication to students.

Therefore, the objective of this chapter is to critically examine the potential of generative AI to enhance teaching practices and pedagogical support, thereby promoting increased educational opportunities for secondary school students and addressing the global teacher shortage. In this context, we aim to explore educational alternatives that provide access to school knowledge for students who, under current circumstances, would not have other opportunities in the short term, considering the lengthy process of teacher training, which requires many years of university studies (Koetsier, 1995).

The use of AI as a pedagogical support can reconfigure the teaching work and generate conditions to reduce the impacts that the shortage of teachers may cause and, at the same time, reposition the school and educators as relevant and innovative agents in the training of new generations for the challenges of the contemporary world.

This chapter is organized into the presentation of a Theoretical Framework, the methodological analysis of its development, followed by an analysis and discussion of the Theoretical Framework. We then move on to the discussion of the results and an analysis of pedagogical possibilities for the use of AI in development of teaching materials and training of lay teachers for temporary teaching. Finally, we present our conclusions on the study carried out.

3.2 Theoretical Framework

One of the classic ways of presenting solutions to teacher shortages and improving quality in education are measures such as improving teacher salaries, greater support from the federal governments of countries, and restructuring teachers' careers (Podolsky et al., 2016; Andrés, 2012). Younger people tend not to choose the teaching career as a professional horizon, just as those who experience the profession abandon it, due to the numerous contradictions that make the profession uninteresting (Cudowska, 2023).

Among the strategies proposed for the improvement of teaching conditions are the development of gender-sensitive policies to attract qualified candidates for teaching, the review and improvement of the quality of teacher education (both initial and continuing), and the development of a qualifications framework for teachers (UNESCO, 2015). It is important to note that all strategies related to Goal 4 of Education and sustainable development have direct or indirect dependence on an adequate number of teachers, so that aspects such as the universalization of education, guarantee of basic knowledge in mathematics, gender equality, as well as knowledge necessary to meet all other sustainable development goals can become a reality that depend directly on the existence of a population with minimum levels of educational development.

It also emphasizes the importance of implementing inclusive, equitable, and gender-responsive teacher management policies that cover aspects such as hiring, training, allocation, compensation, career development, and working conditions, with a view to improving the status of teachers and educators in the professional and educational contexts. The vulnerability of an education system triggers a vicious cycle: the less efficient school education is, the more fragile the policies to strengthen public schools become, the less pressure society puts on politicians to improve the conditions of students and teachers, and school failures increase in generations that will grow up and have children who connect in an increasingly fragile way with school education.

The references presented show that most countries have difficulties in dealing with the lack of teachers in primary and secondary schools. One of the major implications of these problems is that, year after year, several generations of students lose the opportunity to receive a quality education that allows them to integrate with more quality into the social, economic, and cultural structures of society.

And, because it is a problem that is being constructed as chronic at a global level, there is no prospect of a resolution in the short term. This is not only due to the difficulties observed, but also because social challenges tend to increase as school education fails society. When considering the sustainability goals proposed for 2030, it can be inferred that educational difficulties may make all other goals unlikely to be achieved, since educational level is an element strongly related to the population's adherence to sustainability policies.

From these gaps, we propose to analyze the possibilities that present themselves of transformations in the teacher's work, in the sense of incorporating AI in a way that creates situations of pedagogical support in their activities. It is not, therefore, a matter of thinking about the construction of "robot teachers", but of thinking about the possibilities of generative AI to support teachers to improve their performance as education professionals, to reorganize their work, reduce the intensity of their activities, especially because they deal with an excess of students due to the lack of teachers.

From these analyses, problems arose, which became necessary questions to think about the possibility of using AI as pedagogical support and reconfiguration of teaching work, with a view to reducing the impact of the lack of teachers.

- RQ1: Is it possible to minimize the problem of teacher shortage, worldwide, with the development of pedagogical support tools based on generative AI?
- RQ2: What types of activities and interactions are possible for the application of AI that, at the same time, maintain the relevance of the teacher's pedagogical practice?
- RQ3: What levels of teacher supervision need to be considered for the incorporation of AI in youth education?

3.3 Method

This chapter has as its method the content analysis and the construction of a theoretical essay on the possibilities of AI being incorporated into the educational policies of the countries, to allow the teaching activity to receive pedagogical and administrative support in the management of data and students.

As this is a topic still under construction, most governments, at the international level, are in the debate phase on the different uses of AI in education (Yang, 2022). Our goal is to discuss ways that open as possible for teachers and educational systems to explore, with a view to improving the quality of education for young people in their respective countries.

From the methodological point of view, we carried out broad research on the conditions of lack of teachers, as well as references that propose to present the possibilities of AI as an element of didactic-administrative support for teachers. It is not, therefore, a discursive analysis about the replacement of humans by machines, but about adjustments of machines so that they can improve teacher performance.

The proposed theoretical review included a broad search of international bibliography on the subject, according to criteria of scientific impact, based on the location of these articles in Scopus and Web of Science databases, as well as the number of citations presented. We also carry out analyses of the main AIs available, to understand their functioning and the relationship that can be built with school education. We analyzed the following AIs: Open AI's ChatGPT, Google's Gemini, Meta's Llama, and Anthropic's Claude.

Finally, based on the theoretical frameworks and analyses of AI, we proposed to discuss paths for appropriation in the school environment, especially in the work with young students of compulsory education.

3.4 Discussion and Theoretical Framework

In one of the classic books that conceptually address AI, Ertel and Ertel (2017) presents us with the dimensions and dilemmas of this technology, especially about what is potentially built around AI, from a perspective of it being able to handle human actions at which we are even better and from the perspective of programming adaptability from which intelligent systems develop themselves to continuously improve their responses to human requests.

Intelligence, in this sense, can be linked more to the machine's ability to offer responses that do not fit into conventional commands "if... then... else" (Flach, 2012; Bucher, 2018), in which the computer responds only to what is described in the programming line, but manages to build connections and relationships, rewrite its own program and incorporate new documents, and offer answers that are not only individualized, but that are continuously complexified based on machine learning (ML) (Luan et al., 2020).

AI involves dimensions based on algorithms and programming and ethical issues of the utmost importance to humanity, especially about machine decisions that directly impact life, health, social coexistence, and state structures. This issue is thoroughly discussed in the study of Monte-Serrat et al. (2025 [Chap. 6]), where the authors address the need to apply what they refer to as social ethics—an approach capable of aligning technological innovation with educational innovation.

Accelerated technological transformations, with emphasis on the use of generative AI, have opened a debate on the future of AI in teaching practices and the changes that may occur in the educational environment (Killian et al., 2023). These changes range from the way students produce their educational content to the ways in which teachers can conduct pedagogical activities with the support of AI.

Yu and Lu (2021) point out that the field of scientific management has already witnessed the benefits of AI and big data technologies, benefits that can also be applied to educational management. Dynamic monitoring of the educational and academic process, as well as quality, can drive the organization, integration, recording, and development of intelligent systems for data mining and deep learning. This approach is a technically reliable tool for managing academic-administrative matters in education.

AI can enable more efficient interactions between students and teachers, changing the way we search and create content. However, it is important to emphasize that the concepts of efficiency and interaction need to be problematized, as we still do not have concrete elements to analyze the strengthening of relationships between teachers and students mediated by AI (Farrokhnia et al., 2023). Such competencies not only bring developments related to the change in the way teachers relate to technology, but also open doors to broader changes in the educational context, related to the assessment of learning, the development of intelligent tutoring, and adaptive teaching (Santos & Arruda, 2019; Zhai, 2021).

Despite the possibilities that open, the development of generative AI is still involved in processes in which there is no transparency regarding its algorithms, which are often protected by copyright (Kasneci et al., 2023), but which at the same time learn through access to content available on the web. There are also opaque business issues, in which commercial interests are not understood (Kousa & Niemi, 2023a), in addition to ethical issues, which emerge as the most addressed about AI.

Mastering the procedures of generative AI in education requires the ability to direct technological progress, minimizing risks and maximizing its benefits for teaching practice. It is crucial to avoid a simplistic view of generative AI as a solution or end of education, as the issue is complex and involves several factors. A key aspect is the concentration of generative AI technologies in a small number of companies, in the United States. This raises questions about the autonomy and subjectivity of individuals in their relationships with digital technologies, requiring caution and critical reflection.

It is essential to avoid the impersonation of AI by attributing to it autonomous desires and interests. Instead, it is necessary to understand those involved in the creation of generative AI, the processes used, and the many ways in which it can be applied in education. Generative AI offers a range of possibilities to improve teaching practices, taking on various roles in the educational environment (Markauskaite et al., 2022). This versatility allows each implementation to adapt to the specific objectives of each educational context, providing a varied range of opportunities.

Generative Artificial Intelligence (GAI's) applications in education range from multifunctional and flexible tools such as ChatGPT and other language models (Yu & Guo, 2023), to specific applications with a design that is entirely focused on education. However, before we explore these applications, it is critical to understand the concept of ML, the basis of most generative AI technologies.

ML consists of a model for collecting and analyzing large volumes of data, allowing a machine to learn and create optimized solutions (Osmanoğlu et al., 2020). Through a training process, algorithms can identify patterns and regularities in the data, mimicking the way humans learn on specific tasks (Nasteski, 2017). This capability allows GAI to identify patterns in data that, to humans, would be considered noise (Masters, 2023).

In the field of school assessment, ML has several applications. Among them, we can highlight the development of instruments that allow verifying the progression of student learning with greater precision in relation to the school curriculum, and with the use of more elaborate constructions. These algorithms are aimed at the formative assessment of students, where they can examine and provide automated individual feedback, useful in the face of the large demand of students for teachers (González-Calatayud et al., 2021).

For such an application to generate concrete results, it is crucial that it is aligned with the right learning objectives (Zhai, 2021). Learning assessment instruments can be used individually or integrated into an Intelligent Tutor System (ITS). Another relevant technology is recommendation algorithms, based on ML. These algorithms are already used in research activities, whether to recommend reading materials, form academic social networks, or promote interdisciplinary collaborations (Samin & Azim, 2019).

In education, they serve to suggest study materials to students. Recommendations can be made in three ways: the first is content-based, recommending materials according to the student's learning level; the second is collaborative, user-based, considering the similarities between a group of students to suggest content; and the third is based on association rules, directly correlating unlearned content with total learning (Zhang et al., 2023). These recommendations are a way of personalizing teaching, often linked to self-learning or e-learning (Hu et al., 2023). However, it is important to note that its use often faces a lack of scalability of the data required for an effective application of ML, which can result in unsatisfactory results (Khanal et al., 2020). Like AI-based assessment instruments, recommendation algorithms can be incorporated into an ITS.

ITS are the result of five decades of development, which began with the application of computer-aided instructions (Tavares et al., 2020). In their most advanced versions, these systems can generate content in an automated way and provide personalized tutoring strategies, based on the analysis of students' learning situations (Yu & Guo, 2023).

The ITS are based on the creation of Intelligent Pedagogical Agents (IPA), which aim to meet the individual needs of each student (Conati et al., 2018). To do this, these systems offer individualized feedback, adapt the learning pace, and provide adaptive assessments that adjust the difficulty according to the student's level of knowledge.

Despite the convergence in the literature on the benefits of using ITS, there are debates about their commercial use and their relationship with teaching practice (Santos & Arruda, 2019). This is because there is overlap of activities between ITS and teachers. The difficulty in changing pedagogical approaches and the complexity of knowledge construction pose challenges for the practical and efficient application of ITS (Semensato et al., 2015).

In addition, although technically possible, ITS technologies still present challenges in recognizing different learning pathways, cultural disparities, and special needs (Kousa & Niemi, 2023b). Other applications of AI also have the potential to change school dynamics and influence teaching activity, such as facial recognition and data mining. Facial recognition is often associated with surveillance discourse to enhance security in schools. However, there are reports of using this technology to monitor student behavior and engagement. This practice, known as "advanced emotional AI" (Bhardwaj et al., 2021), can recognize facial expressions and assign meanings to them.

Data mining, on the other hand, allows you to analyze databases of educational institutions through ML and explore scenarios in an unusual way. This technology can broaden knowledge about students and their motivations in the context of their own learning environment (Tavares et al., 2020). Data mining can be used, for example, to reduce school dropouts.

It is worth noting that generative AI will be dealt with exclusively at another time, due to its relevance to the research object. In the context of teachers, several promises are directed to their activities. Among them is the ability to increase the effectiveness of teaching, in terms of quality and personalization (Maghsudi et al., 2021). At this point, it is important to point out that AI as a tool has technological potential to improve quality and adaptability in teaching. However, as a product conceived according to a market, it will hardly be free from deviations for interests that are hidden from education or for the legitimation of pedagogical values that are already outdated. In this way, it is understood that the technological potential alone cannot be treated as sufficient, which would be resorting to solutionism without guarantee of tangible results (Pfotenhauer et al., 2022), focusing on the teaching activity.

Another issue that refers to the conception of AI is the very limits and capabilities of the technology. As already discussed, ML-based technologies need a large mass of data to gain scalability and ensure greater efficiency (Khanal et al. 2020). Poorly structured databases can lead to a variety of problems. As mass and quality data are not always available, an AI-supported teacher needs to be aware of judgment biases (Zhai, 2021), or solutions coming from incorrect weightings (Masters, 2023). In an environment not moderated by the teacher, the risks can be even greater.

It is worth mentioning that the process of creating AIs, especially Large Language Models, is unclear, considering copyrighted algorithms (Kasneci et al., 2023). In addition, because algorithms are not necessarily written by humans, hidden layers often become inaccessible for understanding or inspection. As a result, it becomes impossible to understand the reasoning that led AI to the development of one or another action (Masters, 2023). Such a condition, on the one hand, has the potential to affect teachers' reliability in technology. On the other hand, it makes clear the risks of exercising blind faith in AI-created responses, another challenge to be faced by teachers (Ferreira et al., 2023).

In addition to the issues addressed, it is crucial to note that the influence of the AI design process on teaching work can vary according to geographical context. It is already known that several nations in the global south have structural difficulties related to education that can be affected differently using AI. This raises uncertainties about the power struggle between the corporations that develop the technology, usually coming from the centers of developed countries, and the consumers of these technologies, among them, the less technologically developed countries (Pinto, 2005). For the former, there is an interest in the propagation of a solutionist view of AI in teaching, incorporated in the personalization of education. For the latter, the need for real solutions to education problems is highlighted, which would be difficult to solve with the use of one or another AI technology, given the portion of the population excluded from the digital world (Filgueiras, 2023). This argument questions the degree of congruence between the development of AI and the real needs of teachers, which needs to be clarified.

3.4.1 Critical Dimensions of AI

It is essential to underscore that the use of AI entails numerous risks. Poorly structured databases can introduce biases in judgment, such as discriminatory or prejudiced responses. Additionally, the lack of understanding regarding the internal workings of AI systems may lead educators to place undue or unwarranted trust in these tools.

Selwyn (2025) argues that AI presents the prospect of saving time and reducing workload. However, this potential is not easily observable, as the nature of the work itself may shift—becoming more demanding, particularly in relation to the handling of increasingly complex student data. Furthermore, much of what is traditionally considered "work" in education can also serve as an opportunity for pedagogical development—an opportunity that is often lost with the adoption of AI.

Another critical factor to consider is the political use of AI. In some cases, the technology is appropriated as a justification for reducing teaching positions. Selwyn (2025) highlights this as a risk, warning that AI could undermine traditional schooling or steer it toward a fully automated model, where the role of educators is minimized or even eliminated.

These concerns are far from hypothetical. From a political standpoint—particularly in countries with fragile political and legislative structures—the economic influence of large tech corporations often targets education systems, aiming to replace human educators with so-called intelligent machines.

It is imperative that both teachers and students are equipped to critically engage with and question the outputs generated by AI systems. A conscious, informed approach to AI is crucial. The field of education must not be excluded from scientific and pedagogical discourse on the implications of AI in learning processes—otherwise, it risks becoming a passive consumer of technologies developed without educational values in mind.

In this context, the need for in-depth learning about AI becomes evident. Only through the development of robust knowledge will educators, researchers, and students be able to assert themselves and take active roles in making informed decisions about whether—and how—AI should be integrated into educational environments.

3.5 Results

As explained, thinking about the introduction of AI in teaching practice and work does not mean the development of "robot teachers" (Kupferman, 2022). It is possible to think of an environment in which there are teachers who perform specific activities and tasks of human teachers, without the need for the physical presence of teachers. The authors discuss the current role of robots in education, which includes acting as classroom assistants, educational technologies, and teaching support systems. They argue that with the continued development of AI, robotics, and ML, robots may eventually surpass human teachers in terms of domain knowledge and maintenance costs.

There are, however, significant challenges to implementing robot teachers. This includes the need for skills such as emotional intelligence, creativity, and communication, which are inherent characteristics of human teachers. Additionally, the authors highlight the need to consider social and demographic factors, such as culture and religion, when developing AI systems for education (Edwards & Cheok, 2018).

To address RQ1, it can be stated that one of the most essential elements to consider is respect for the authority that the teacher exercises in the classroom. It is quite easy to disconnect from an automated environment—just push a button. This contrasts with the dynamics of a classroom with interaction between students and human teachers, where social, cultural, and authority relations discourage the simple absence of the educational environment.

Celik et al. (2022) argue that AI offers teachers numerous opportunities to improve the planning, implementation, and evaluation of their teaching. Additionally, teachers play vital roles in the development of AI technology by serving as models for training algorithms and by participating in the development process through the verification of automated AI assessment systems.

The teacher's role as an AI "trainer" is the one that generates the most discomfort. This is due to a lack of transparency about the codes and programs, as well as the intentions of large AI providers, which raises concerns that teachers' work may be being appropriated by machines for future teacher replacements. While this replacement faces numerous challenges, the concerns are valid.

Political choices are not always based on scientific production, which is why such concerns are solid enough to be considered a risk to work, practices, and teaching itself in formal school contexts. When we look at the sustainability goals and indicators proposed for 2030, we realize that we are less than six years away and still have a lot to achieve. Although goals and commitments do not guarantee full fulfillment, they are collectively constructed from a perspective that seeks the best possible reality for human improvement.

Given the shortage of millions of teachers, as mentioned earlier, it is unlikely that there will be enough time to train and empower so many teachers to ensure universal access to primary and secondary education. This issue is further discussed in the study of Glushkova and Staribratov (2025 [Chap. 5]), which focuses on the training and preparation of teachers and students in topics related to AI education in Bulgaria. The authors present teaching experiences from various schools, covering subjects such as AI and the programming of social robots. The complete training of a teacher for these levels of education requires a minimum of three to six years, according to the legislation of each country.

The data show, however, that fewer and fewer young people want to become teachers, and many drop out of courses in the first few years.

Brazilian data from National Institute for Educational Studies and Research Anísio Teixeira (INEP) (2023) indicate that, even in areas with a greater number of trained teachers than available jobs, many choose not to work as teachers. Gašević et al. (2023) explore the ways we can enhance learning with the support of AI. For the authors, the explainability of AI, educational evaluation, and design help teachers to better perform their teaching activities.

However, beyond these issues that present themselves as the most immediate answers to educational gaps, we need to build pathways that allow us to include AI as an element that will minimize or allow an improvement in educational performance while countries are unable to solve the shortage of teachers (Pedro et al., 2019).

It is, therefore, not the best situation in which we should think of education as an element for reducing inequalities, but what is most quickly available to us to shorten the development time of the poorest countries when compared to the rich countries. The longer it takes to provide educational access, the greater the social and economic inequalities become in local and international contexts.

As pointed out, it takes a long time to train teachers and, at the current rate, it would take a few decades to train teachers who would serve all students. At the same time, it takes an average of 12 years to fully graduate a student in primary and secondary education. The loss of this learning time ends up becoming permanent, since it is difficult for adults with little schooling to return to school. This is due to the survival conditions to which these adults are enrolled, as well as to the specific situations of adults who, due to having numerous social, economic, and family obligations, are unable to set fixed schedules to attend school.

In other words, the lack of teachers has a direct impact on the long future in each country, if we consider that each child without an adequate school will have a whole path as an adult of inequality and sharing this inequality with their descendants. Therefore, we believe that AI may not be the most appropriate way to address the shortage of teachers, but it can reduce the speed of inequality at a time when accelerating access to education can result in improved conditions for millions of people around the world. In this sense, we propose the following paths to be developed so that AI can be incorporated into educational systems to enhance teaching activity and support student education.

3.6 Elaboration of Didactic Materials and Construction of Individual Tutoring Supports

In general, generative AIs, such as OpenAI's ChatGPT and Google's Gemini, have very generic answers with certain inaccuracies. Neural ML is based on the analysis of numerous materials that are openly available on the world wide web. It is possible that this software uses closed databases, but, due to copyright issues and due to their programming being a closed box, there is no way to know openly, because copyrighted knowledge cannot be accessed and distributed without the proper authorization credentials.

Note, for example, the movements of major international newspapers such as the *New York Times* that have changed their rules of use by prohibiting the use of AI software to access their content as a training mechanism (*New York Times*, 2023). It was also noted, in the last strike movement that united actors and screenwriters in the United States, that one of the significant issues of disagreement concerned the indefinite use of AI to reproduce faces, movements, voices, and other characteristic items of actors, actresses, and screenwriters.

These problems lie in the strategy of companies to obtain information and knowledge at the lowest possible cost. This is because there is not (or was) a business model in which the use of AI would be incorporated.

We believe that a possible business model for the use of AI in education consists of specific and closed commercial agreements, in which educational systems, together with textbook production companies, authors and all others involved in the teaching and learning market, can build an educational environment in which AI can base its learning on the available open knowledge and, in the country-specific learning materials.

This would allow according to RQ3, for example, the development of intelligent tutors who could help the teacher in providing pedagogical support to students who have learning difficulties, who need other pedagogical accompaniments, or even students who at certain times do not have the exceptional presence of a teacher.

AI systems designed specifically for the contexts of countries or educational systems could bring numerous benefits to the lack of teachers, as they would allow students to perform tasks, guided by teachers, more autonomously and in various times and spaces than school. It would enable the teacher to carry out a micro and macro analysis of the level, difficulties, and possibilities of student learning, and would also allow educational systems to make more consistent diagnoses of the students' most difficult contents, as well as paths and analytical possibilities for reconfiguration or transformation, not only of teaching practices, but also of teacher training courses. We believe that one of the best elements in this process would be the speed in the processing of data and the response of diagnoses—not necessarily precise, but that could bring analyses in shorter times so that educational systems could adjust to social transformations.

3.7 Training of Lay Teachers for Temporary Teaching

As stated earlier, it takes a few years to train a teacher at the university level to work in basic education. If the data show that there is a lack of teachers to work in schools, this means that a considerable number of teachers do not have the minimum qualifications to work in primary and secondary education. Brazil, according to data from INEP (2023), has more than 30% of its teaching staff active without the minimum training required to be teachers.

In response to RQ2, it is possible to reflect on the development and training of AI that can support the teaching activity of these professionals. It is not a model of teacher training, but a model of AI that can guide the teacher in relation to the contents to be taught and even offer short-term training on the curricular components less known by the teacher.

A context such as this would have the objective of addressing, urgently and in the short term, a teacher training gap, without, however, removing the need to train teachers within conventional perspectives. A lay teacher does not have the time and curricular direction necessary to become a teacher with the same qualifications as those who have the specific training. However, it can, in the short term, improve its performance through the support of AI and, with that, minimize the low educational quality that would be offered to students without any other interventions.

The problem encountered in this model consists more in political choices than in technological and formative possibilities. There is an elevated risk that governments will replace, or stop being interested in, conventional teacher training, because they consider that AI can solve the training gap with specific guidance actions. An introduction of software in this model would require a lot of discussion with all those involved in the educational process, so that it is understood that this is a short-term, emergency action, to solve a problem that may imply the future impoverishment of the country.

It is important to note that the disinvestment in teacher training, for political reasons, can generate the same future problem—the decrease in teacher quality and the weakening of school knowledge—since machines are not efficient in solving educational issues that are specifically human, such as interpersonal relationships, subjectivities that alter our perceptions of teaching and learning, our reading of inequality, of power and its individual implications. And, finally, what makes us most human: the ability we must put ourselves in the place of the other, who teaches or learns, to understand their difficulties, their possibilities, their potentialities, and their singularities. Only humans can understand feelings that directly and indirectly impact the ways of teaching and learning.

3.8 Conclusions

This chapter focuses on content analysis and the construction of a theoretical essay on the possibilities of AI to be incorporated into educational policies to allow teachers to receive pedagogical and administrative support in the management of data and students. From the methodological point of view, we conducted broad research on the conditions of teaching work worldwide, to detect teacher gaps. We analyze from the perspective of the potential of generative AI, actions that can be produced to support AI to the teacher and the educational system with a view to improving school education in a sustainable way.

It is not, therefore, a discursive analysis about the replacement of humans by machines, but about adjustments of machines so that they can improve teacher performance. Generative AI software is still in a kind of first wave of innovation, in the improvement and learning phase. However, the results of using AI to produce text, videos, and images are already impressive and tend to become even better, in terms of accurate responses, in an abbreviated time.

This provides us with important clues about the future that opens for Education and for the construction of ML models that act to alleviate the shortage of teachers in countries—without, however, replacing them. It is a perspective in which AI becomes a source of teacher support for repetitive activities, for contexts of student excess, and for individualized analyses that allow the teacher to make pedagogical decisions that improve student performance.

In addition, it is necessary to consider that it is not just a technical use of AI in teaching. It is a cycle of transformation of the teaching work, still in an unmeasured way, which will bring different actions and activities to the teacher in their daily school life.

We consider the development of AI to be relevant in contexts of teacher shortages, not only for providing emergency assistance to students who find themselves without pedagogical support, but also as a process of pedagogical and technological development in which teachers become the protagonists of a mechanism that allows them to develop activities that improve their performance as teachers.

This leads us to believe that teacher training courses need to pay attention to the new skills that are present in contemporary times, such as understanding the logical bases of computing, software operation, and the development of commands and AI training.

In addition to the teaching activities previously learned in the training courses, new activities emerge, such as learning how to use educational data generated by AI to improve school performance, how to develop individualized student follow-ups based on the learning feedback provided by AI, or even incorporate new strategies of teacher-student interaction in contexts in which AI elaborates certain pedagogical interventions and releases the teacher for other work, such as the deeper analysis of the learning difficulties of a specific student or collective difficulties in certain contents.

Building ML environments in AI is also a necessary activity, which involves not thinking about generic machine support, but rather about building individualized intelligent tutors who can be trained to support specific situations or groups. Adopting this approach can reduce the impersonal perspective that is unrelated to the educational context of a school and allow the development of AI that can respond as closely as possible to a human teacher, especially regarding local needs.

In this text, we can see that one of the biggest challenges of implementing AI in education is the need for emotional skills to deal with students in different contexts. Machines cannot perform this task, which leads to the conclusion that AI can be used as a support element—never as the main factor in the learning process. Teaching work and practice, its monitoring and understanding of students' social, economic, and cultural difficulties are essentially human and necessary. This means that AI cannot be approached under the guise of political propaganda or cost reduction. It should be analyzed exclusively as an option for improving student and teacher performance.

The path forward involves further research to fully comprehend the long-term impact of GAI on student learning outcomes and teacher practices. It also necessitates the development of comprehensive teacher training programs, equipping educators with the skills to effectively and ethically utilize AI in their classrooms.

Therefore, we can observe a completely open future in which we can build, with the participation of teachers, new models of teaching and learning, in which we can appropriate AI models to reduce educational inequalities and, indirectly, reduce social and economic inequalities.

Acknowledgment Research funded by: Minas Gerais State Agency for Research and Development—FAPEMIG; Coordination for the Improvement of Higher Education Personnel (CAPES) and National Council for Scientific and Technological Development (CNPq).

References

Akiba, M., Chiu, Y. L., Shimizu, K., & Liang, G. (2012). Teacher salary and national achievement: A cross-national analysis of 30 countries. *International Journal of Educational Research, 53*, 171–181. https://doi.org/10.1016/j.ijer.2012.03.007

André, M. E. D. A. (2012). Políticas e programas de apoio aos professores iniciantes no Brasil. *Cadernos de Pesquisa, 42*(145), 112–129. https://doi.org/10.1590/S0100-15742012000100008

Arruda, E. P., & Arruda, D. P. (2024). Artificial intelligence for SDG 4 of the 2030 agenda: Transforming education to achieve quality, equality, and inclusion. *Sustainable Economies, 2*(2), 34–34. https://doi.org/10.62617/se.v2i2.34

Arruda, E. P., & Kerres, M. (2024). Education practices mediated by digital technologies: Mobilization and teachers' strategies in primary and secondary schools in Germany. *Education Sciences., 14*(8), 838. https://doi.org/10.3390/educsci14080838

Bearman, M., Ryan, J., & Ajjawi, R. (2023). Discourses of artificial intelligence in higher education: A critical literature review. *Higher Education, 86*(2), 369–385. https://doi.org/10.1007/s10734-022-00937-2

Bhardwaj, P., Gupta, P. K., Panwar, H., Siddiqui, M. K., Morales-Menendez, R., & Bhaik, A. (2021). Application of deep learning on student engagement in e-learning environments. *Computers & Electrical Engineering, 93*. https://doi.org/10.1016/j.compeleceng.2021.107277

Bhutoria, A. (2022). Personalized education and artificial intelligence in the United States, China, and India: A systematic review using a human-in-the-loop model. *Computers and Education: AI*, 3. https://doi.org/10.1016/j.caeai.2022.100068

Bucher, T. (2018). *If... then: Algorithmic power and politics.* Oxford University Press. https://doi.org/10.1093/oso/9780190493028.001.0001

Celik, I., Dindar, M., Muukkonen, H., & Järvelä, S. (2022). The promises and challenges of artificial intelligence for teachers: A systematic review of research. *TechTrends, 66*(4), 616–630. https://doi.org/10.1007/s11528-022-00715-y

Conati, C., Porayska-Pomsta, K., & Mavrikis, M. (2018). AI in Education needs interpretable machine learning: Lessons from Open Learner Modelling. *arXiv preprint.* https://doi.org/10.48550/arXiv.1807.00154

Cudowska, A. (2023). The condition of teachers in a comparative perspective. *Studia z Teorii Wychowania, 14*(3 (44)), 125–136. https://doi.org/10.5604/01.3001.0053.9200

Durso, S., & Arruda, E. P. (2022). Artificial intelligence in distance education: A systematic literature review of Brazilian studies. *Problems of Education in the 21st Century, 80*(5), 679–692. https://www.scientiasocialis.lt/pec/node/1583

Edwards, B. I., & Cheok, A. D. (2018). Why not robot teachers: Artificial intelligence for addressing teacher shortage. *Applied AI, 32*(4), 345–360. https://doi.org/10.1080/08839514.2018.1464286

Ertel, W., & Ertel, W. (2017). Machine learning and data mining. *Introduction to AI*, 175–243. https://doi.org/10.1007/978-3-319-58487-4

Farrokhnia, M., Banihashem, S. K., Noroozi, O., & Wals, A. (2023). A SWOT analysis of ChatGPT: Implications for educational practice and research. *Innovations in Education and Teaching International*, 1–15. https://doi.org/10.1080/14703297.2023.2195846

Ferreira, G. M. D. S., Lemgruber, M. S., & Cabrera, T. L. (2023). From didachography to AI: Metaphors teaching is automated by *Journal of Interactive Media in Education, 2023*(1). https://doi.org/10.5334/jime.798

Filgueiras, F. (2023). Artificial intelligence and education governance. *Education, Citizenship and Social Justice, 17461979231160674.* https://doi.org/10.1177/17461979231160674

Flach, P. (2012). *Machine learning: The art and science of algorithms that make sense of data.* Cambridge University Press.

Gašević, D., Siemens, G., & Sadiq, S. (2023). Empowering learners for the age of artificial intelligence. *Computers and Education: AI, 4.* https://doi.org/10.1016/j.caeai.2023.100130

Glushkova, T., & Staribratov, I. (2025). Artificial intelligence and robotics in school education: Analysis of implementation of AI and robotics in Bulgarian education. In S. Papadakis & G. Lampropoulos (Eds.), *Social robots and artificial intelligence in education: Integrating AI in K-12 and higher education* (pp. 91–120). Palgrave Macmillan.

González-Calatayud, V., Prendes-Espinosa, P., & Roig-Vila, R. (2021). AI for student assessment: A systematic review. *Appl. Sci., 11*, 5467. https://doi.org/10.3390/app11125467

Hu, Y. H., Fu, J. S., & Yeh, H. C. (2023). Developing an early-warning system through robotic process automation: Are intelligent tutoring robots as effective as human teachers? *Interactive Learning Environments*, 1–14. https://doi.org/10.1080/10494820.2022.2160467

INEP. 2023. *Censo da Educação Básica 2022.* Brasil, Ministerio de Educação. https://www.gov.br/inep/pt-br/areas-de-atuacao/pesquisas-estatisticas-e-indicadores/censo-escolar/resultados

Kasneci, E., Seßler, K., Küchemann, S., Bannert, M., Dementieva, D., Fischer, F., et al. (2023). ChatGPT for good? On opportunities and challenges of large language models for education. *Learning and Individual Differences, 103.* https://doi.org/10.1016/j.lindif.2023.102274

Khanal, S. S., Prasad, P. W. C., Alsadoon, A., & Maag, A. (2020). A systematic review: Machine learning based recommendation systems for e-learning. *Education and Information Technologies, 25*(4), 2635–2664. https://doi.org/10.1007/s10639-019-10063-9

Killian, C. M., Marttinen, R., Howley, D., Sargent, J., & Jones, E. M. (2023). Knock, knock ...Who's there? ChatGPT and AI-powered large language models: Reflections on potential impacts within health and physical education teacher education. *The Journal of Teaching in Physical Education, 42*(3), 385–389. https://doi.org/10.1123/jtpe.2023-0058

Koetsier, C. P. (1995). Bridging the gap between initial teacher training and teacher induction. *Journal of Education for teaching, 21*(3), 333–346. https://doi.org/10.1080/02607479550038545

Kousa, P., & Niemi, H. (2023a). AI ethics and learning: EdTech companies' challenges and solutions. *Interactive Learning Environments, 31*(10), 6735–6746. https://doi.org/10.1080/10494820.2022.2043908

Kousa, P., & Niemi, H. (2023b). Artificial intelligence ethics from the perspective of educational technology companies and schools. In *Learning: designing the future* (p. 283). https://doi.org/10.1007/978-3-031-09687-7_17

Kupferman, D. W. (2022). I, robot teacher. *Educational Philosophy and Theory, 54*(10), 1513–1522. https://doi.org/10.1080/00131857.2020.1793534

Lin, C. C., Huang, A. Y., & Lu, O. H. (2023). Artificial intelligence in intelligent tutoring systems toward sustainable education: A systematic review. *Smart Learning Environments, 10*(1), 41. https://doi.org/10.1186/s40561-023-00260-y

Luan, H., Geczy, P., Lai, H., Gobert, J., Yang, S. J., Ogata, H., et al. (2020). Challenges and future directions of big data and artificial intelligence in education. *Frontiers in Psychologie, 11*, 580820. https://doi.org/10.3389/fpsyg.2020.580820

Maghsudi, S., Lan, A., Xu, J., & van Der Schaar, M. (2021). Personalized education in the artificial intelligence era: What to expect next. *IEEE Signal Processing Magazine, 38*(3), 37–50. https://doi.org/10.1109/MSP.2021.3055032

Mannuru, N. R., Shahriar, S., Teel, Z. A., Wang, T., Lund, B. D., Tijani, S., et al. (2023). Artificial intelligence in developing countries: The impact of generative artificial intelligence (AI) technologies for development. *Information Development*. https://doi.org/10.1177/02666669231200628

Mao, J., Chen, B., & Liu, J. C. (2024). Generative AI in education and its implications for assessment. *TechTrends, 68*(1), 58–66. https://doi.org/10.1007/s11528-023-00911-4

Markauskaite, L., Marrone, R., Poquet, O., Knight, S., Martinez-Maldonado, R., Howard, S., et al. (2022). Rethinking the entwinement between artificial intelligence and human learning: What capabilities do learners need for a world with AI? *Computers and Education: AI, 3*, 100056. https://doi.org/10.1016/j.caeai.2022.100056

Masters, K. (2023). Ethical use of artificial intelligence in health professions education: AMEE Guide No. 158. *Medical Teacher, 45*(6), 574–584. https://doi.org/10.1080/0142159x.2023.2186203

Monte-Serrat, D. M., Cabella, D. M. S., & Cattani, C. (2025). Ethics and sustainability in education that makes use of artificial intelligence and robots. In S. Papadakis & G. Lampropoulos (Eds.), *Social robots and artificial intelligence in education: Integrating AI in K-12 and higher education* (pp. 121–152). Palgrave Macmillan.

Nasteski, V. (2017). An overview of the supervised machine learning methods. *Horizons, 4*(51–62), 56.

Nemorin, S., Vlachidis, A., Ayerakwa, H. M., & Andriotis, P. (2023). AI hyped? A horizon scan of discourse on artificial intelligence in education (AIED) and development. *Learning, Media, and Technology, 48*(1), 38–51. https://doi.org/10.1080/17439884.2022.2095568

Osmanoğlu, U. Ö., Atak, O. N., Çağlar, K., Kayhan, H., & Can, T. (2020). Sentiment analysis for distance education course materials: A machine learning approach. *Journal of Educational Technology and Online Learning, 3*(1), 31–48. https://doi.org/10.31681/jetol.663733

Pedro, F., Subosa, M., Rivas, A., & Valverde, P. (2019). Artificial intelligence in education: Challenges and opportunities for sustainable development. https://unesdoc.unesco.org/ark:/48223/pf0000366994

Pfotenhauer, S., Laurent, B., Papageorgiou, K., & Stilgoe, J. (2022). The politics of scaling. *Social Studies of Science, 52*(1), 3–34. https://doi.org/10.1177/03063127211048945

Pinto, Á. V. (2005) O conceito de Tecnología : volume II. Contraponto, Rio de Janeiro.

Podolsky, A., Kini, T., Bishop, J., & Darling-Hammond, L. (2016). Solving the teacher shortage: How to attract and retain excellent educators. *Learning Policy Institute., 10*(54300/262), 960.

Rane, N., Choudhary, S., & Rane, J. (2023). Education 4.0 and 5.0: Integrating AI (AI) for personalized and adaptive *learning*. https://doi.org/10.2139/ssrn.4638365

Roberts, H., Cowls, J., Morley, J., Taddeo, M., Wang, V., & Floridi, L. (2021). The Chinese approach to artificial intelligence: An analysis of policy, ethics, and regulation. In *Ethics, governance, and policies in artificial intelligence* (pp. 47–79). https://doi.org/10.1007/s00146-020-00992-2

Samin, H., & Azim, T. (2019). Knowledge based recommender system for academia using machine learning: A case study on higher education landscape of Pakistan. *IEEE Access, 7*, 67081–67093. https://doi.org/10.1109/ACCESS.2019.2912012

Santos, B. L., & Arruda, E. P. (2019). Dimensions of AI in the context of contemporary education. *Educação UNISINOS, 23*(4), 725–741. https://doi.org/10.4013/edu.2019.234.08

See, B. H., & Gorard, S. (2019). Why don't we have enough teachers?: A reconsideration of the available evidence. *Research Papers in Education, 35*(4), 416–442. https://doi.org/10.1080/02671522.2019.1568535

Selwyn, N. (2025). Critical studies of educatoin & technology. *Neil Selwyn writes about education, technology & Society*. https://criticaledtech.com/

Semensato, M. R., Francelino, L. D. A., & Malta, L. S. (2015). O uso da inteligência artificial na Educação a Distância. *Revista Cesuca Virtual, 2*(4), 36–39. https://api.semanticscholar.org/CorpusID:171049992

Shiohira, K. (2021). *Understanding the impact of AI on skills development*. Education 2030. UNESCO-UNEVOC International Centre for Technical and Vocational Education and Training. https://unevoc.unesco.org/pub/understanding_the_impact_of_ai_on_skills_development.pdf

Tavares, L. A., Meira, M. C., & do Amaral, S. F. (2020). Inteligência Artificial na Educação: Survey. *Brazilian Journal of Development, 6*(7), 48699–48714. https://doi.org/10.34117/bjdv6n7-496

New York Times. (2023). *Terms of service*. https://help.nytimes.com/hc/en-us/articles/115014893428-Terms-of-Service

UNESCO. (2015). *Incheon Declaration and Framework for Action for the Implementation of Sustainable Development Goal 4: Ensure inclusive and equitable quality education and promote lifelong learning opportunities for all*. UNESCO.

UNESCO. (2023). *Global Education Monitoring Report 2023:* Technology in education—A tool on whose terms? https://doi.org/10.54676/UZQV8501

Wiggan, G., Smith, D., & Watson-Vandiver, M. J. (2021). The national teacher shortage, urban education, and the cognitive sociology of labor. *The Urban Review, 53*, 43–75. https://doi.org/10.1007/s11256-020-00565-z

Yang, W. (2022). AI education for young children: Why, what, and how in curriculum design and implementation. *Computers and Education: AI*, 3. https://doi.org/10.1016/j.caeai.2022.100061

Yu, H., & Guo, Y. (2023). Harnessing the potential of Chat GPT in education: Unveiling its value, navigating challenges, and crafting mitigation pathways. In 2023 *5th International Workshop on AI and Education (WAIE)* (pp. 48–52). IEEE. https://doi.org/10.1109/WAIE60568.2023.00016

Yu, S., & Lu, Y. (2021). *An introduction to artificial intelligence in education*. Springer. https://doi.org/10.1007/978-981-16-2770-5

Zhai, X. (2021). Practices and theories: How can machine learning assist in innovative assessment practices in science education. *Journal of Science Education and Technology, 30*(2), 139–149. https://doi.org/10.1007/s10956-021-09901-8

Zhang, B., Zhu, J., & Su, H. (2023). Toward the third-generation artificial intelligence. *Science China Information Sciences, 66*(2). https://doi.org/10.1007/s11432-021-3449-x

CHAPTER 4

Revolutionizing English Language Learning Through Social Robots and Artificial Intelligence: Towards Inclusive and Personalized Education

Ela Luria

4.1 Introduction

In recent years, rapid advancements in artificial intelligence (AI) and robotics have ushered in a new era of possibilities for education, particularly in the domain of language learning (Singha et al., 2024). This study aims to investigate how the integration of social robots and AI technologies can revolutionize English language education by addressing gaps in personalized learning, inclusivity, and real-time language practice. The study contributes to the growing body of research by offering a comprehensive analysis of the potential these technologies hold for enhancing learner engagement, fostering creativity, and developing critical thinking and problem-solving skills (Qawqzeh, 2024; Agasi, 2024). As also highlighted in the study of Arruda and Arruda (2025 [Chap. 3]), which

E. Luria (✉)
English and Education Department, Levinsky College of Education & Beit Berl College of Education, Tel Aviv-Jaffa, Israel

S. Papadakis, G. Lampropoulos (eds.), *Social Robots and Artificial Intelligence in Education*, Digital Education and Learning,
https://doi.org/10.1007/978-3-032-03977-4_4

discusses AI's potential in addressing teacher shortages and educational inequality, this study explores how adaptive technologies can help close access gaps and support long-term educational improvement on a global scale. Additionally, this study fills a crucial gap in the existing literature by focusing not only on the theoretical underpinnings of AI and robotics in education but also on practical implementation strategies that can be applied across diverse educational contexts.

Current research on AI and language education often overlooks the intersection of these technologies with inclusivity and personalized learning pathways, especially for students with diverse linguistic backgrounds and learning needs. By focusing on this gap, this study highlights how AI technologies, such as Natural Language Processing (NLP), Intelligent Tutoring Systems (ITS), and Automated Speech Recognition (ASR), and social robots can create adaptive learning environments that address the unique needs of individual students. This personalized approach ensures that learners not only receive targeted support but also engage in immersive learning experiences that enhance communicative competence and practical language use (Fathi et al., 2024; Kot et al., 2024; Novawan, 2024).

Social robots, in conjunction with AI systems, offer transformative potential by simulating real-world language use scenarios, engaging students in interactive language practice, and providing immediate feedback (Cumbal, 2024). This study explores how these robots can foster authentic language interactions, helping students gain confidence and communicative skills in a low-pressure, supportive environment (Wu & Li, 2024). These technologies enable educators to offer more tailored, inclusive language learning opportunities, which is essential in addressing existing inequities in educational access and outcomes.

The study is grounded in a robust theoretical framework, drawing from constructivism, cognitive load theory, and connectivism (Voskoglou, 2024). These theories support the development of AI-driven learning environments that promote deep understanding, facilitate personalized knowledge construction, and cultivate collaborative learning experiences (Zohuri et al., 2024; Zheng et al., 2024). By leveraging these educational theories, the study provides insights into how AI and social robots can be effectively integrated to support diverse learning styles and foster critical thinking skills essential for language acquisition.

Furthermore, this study offers practical insights into the implementation of AI and social robots in language classrooms. It discusses the challenges of technological infrastructure, teacher training, and ethical considerations (Sharifuddin et al., 2024), while also providing recommendations for scaling these technologies in diverse educational settings. By addressing these practical considerations, the study aims to provide educators with a roadmap for creating inclusive, equitable, and effective language learning environments.

The article is structured as follows. It begins by exploring the theoretical foundations that underpin the integration of AI and social robots in language education, drawing from key learning theories such as constructivism and cognitive load theory. Following this, the paper delves into the key AI-driven technologies currently used in English language learning, including NLP, ITS, Automated Writing Evaluation (AWE), and ASR. The role of social robots in enhancing interaction and engagement is then examined, with a particular focus on personalized learning and inclusivity. The manuscript also outlines practical strategies for implementing AI and social robots in language classrooms, addressing both the potential benefits and challenges of integrating these technologies. Finally, the paper concludes with design considerations and implementation strategies necessary for optimizing the use of AI and social robots in educational settings.

4.2 Theoretical Foundations Supporting AI and Social Robots in Language Education

The integration of AI and social robots in language education is rooted in a variety of educational and cognitive theories that emphasize interaction, personalization, and adaptive learning. These technologies are designed to enhance learning by creating dynamic, responsive environments that reflect how humans naturally acquire and process language. By aligning with established frameworks that focus on active engagement, cognitive efficiency, social interaction, and motivation, AI and social robots provide innovative solutions that optimize language education. These theoretical foundations, such as Constructivism, Cognitive Load Theory, Vygotsky's Social Development Theory, and others, offer valuable insights into how AI-driven tools can facilitate deeper learning experiences tailored to individual needs, supported by insights from neuroscience and neuroeducation.

4.2.1 *Constructivism and AI in Language Education*

Constructivist learning theory, as developed by Jean Piaget (1978) and Jerome Bruner, posits that learners build knowledge through interaction with their environment. AI tools, such as ITS, NLP-based chatbots, and AWE, align with constructivist principles by offering adaptive, interactive environments where learners can experiment with language and receive immediate feedback (Kolegova & Levina, 2024). AI tools like chatbots encourage learners to engage in simulated conversations that mirror real-world scenarios, providing a space for language experimentation and reflection. This application of constructivist theory enables learners to build on prior knowledge and refine their language skills through interaction. AI-powered systems personalize language tasks to challenge learners at the appropriate level, fostering an active, constructivist approach to language acquisition.

4.2.2 *Vygotsky's Social Development Theory and Social Robots*

Vygotsky's Social Development Theory (1978) emphasizes the importance of social interaction in cognitive development. Social robots align with this theory by serving as conversational partners, offering assistance within the learner's Zone of Proximal Development (ZPD), the range of tasks learners can perform with support but not independently. Kanero et al. (2018) notes that social robots, acting as scaffolds within the ZPD, provide interactive, feedback-rich environments that simulate human interaction, promoting language development through guided practice. These robots adapt to each learner's level, offering personalized feedback that allows students to extend their language proficiency in supported, interactive dialogues (Maroto-Gómez et al., 2024). This emphasis on scaffolded interaction through social robots resonates with the work of Lampropoulos and Papadakis (2025 [Chap. 2]), which outlines how social-emotional engagement and embodiment make robots effective facilitators of learning across both formal and informal educational contexts.

4.2.3 *Cognitive Load Theory and Adaptive AI Systems*

Cognitive Load Theory (Sweller, 1991) focuses on managing the cognitive demands placed on learners. AI systems, such as ITS, reduce cognitive overload by adapting the complexity of tasks in real-time, ensuring that learners are appropriately challenged. As Ghafar et al. (2023) suggest, AI tools can break down complex language tasks into manageable steps, reducing extraneous cognitive load while maintaining the intrinsic cognitive load essential for learning. Tools like Grammarly provide step-by-step feedback, allowing learners to focus on specific areas for improvement without becoming overwhelmed (Koltovskaia, 2022). AI's ability to dynamically adjust difficulty levels ensures that learners engage with tasks at an optimal cognitive level, as discussed by Sweller (1991), maximizing both efficiency and language acquisition.

4.2.4 *Multiple Intelligences Theory and Personalized Learning*

Howard Gardner's Multiple Intelligences Theory (1983) posits that individuals possess various types of intelligences, such as linguistic, logical-mathematical, spatial, and interpersonal. AI and social robots support this theory by providing personalized learning experiences tailored to each student's dominant intelligence. AI platforms like Duolingo or personalized robots can adapt content delivery based on a learner's intelligence profile (Yekollu et al., 2024). For example, visual learners may benefit from AI tools that incorporate diagrams, while interpersonal learners may thrive in robot-facilitated group discussions (Maftoon, 2012; Martin, 2019). This personalization aligns with Gardner's theory, ensuring that students receive language instruction suited to their individual strengths (Hickenbottom, 2022)

4.2.5 *Embodied Cognition and Social Robots*

Embodied Cognition suggests that cognitive processes are influenced by the body's interaction with the physical world. Social robots offer a unique means of applying this theory to language learning, enabling learners to engage with verbal and non-verbal communication simultaneously. Robots like Pepper or NAO allow learners to practice language using both verbal and physical cues, such as gestures and facial expressions, aligning with the principles of Embodied Cognition (Ghafar et al., 2023). By offering multimodal interactions, these robots support the idea that language learning is not purely cognitive but also grounded in physical and social experiences (Konijn et al., 2022).

4.2.6 *Schema Theory and AI-Powered Language Learning*

Schema Theory (Bartlett, 1932) posits that learners organize knowledge into mental frameworks, or schemas, which are used to interpret and process new information. AI systems can tailor learning materials to align with learners' existing schemas, enhancing understanding and retention (Gao et al., 2021). AI-powered systems analyze learners' backgrounds and recommend culturally relevant language tasks, helping learners build upon existing knowledge structures (George, 2023). This tailored approach to language learning enhances both motivation and learning outcomes (Bolotta & Dumas, 2022). Robots can simulate interactions that align with the learner's schemas, providing conversational practice that reflects their cultural and linguistic contexts (Tellex, 2020).

4.2.7 *Experiential Learning Theory and Simulated Language Environments*

David Kolb's Experiential Learning Theory (1984) emphasizes learning through experience. AI and robots support this theory by creating simulated environments in which learners can practice language skills in real-world scenarios. AI-driven virtual reality (VR) platforms immerse learners in scenarios like ordering food or giving a presentation, offering hands-on language practice (Kang et al., 2023). Robots can engage learners in role-playing exercises, offering real-time feedback and opportunities for language experimentation (Goldberg et al., 2023).

4.2.8 *Connectivism and Digital Language Networks*

Connectivism, proposed by Siemens (2005), highlights the role of technology and networks in knowledge acquisition. AI and robots facilitate connectivist learning by helping students access digital resources and engage with diverse language communities. AI systems aggregate vast amounts of language learning materials, curating content that matches learners' proficiency levels and interests (Edmett et al., 2023). Robots can act as facilitators, guiding learners to online language exchanges and virtual communities, thus promoting authentic language practice and collaboration (Lin et al., 2022).

4.2.9 Critical Pedagogy and Language Learning through AI

Critical Pedagogy, as articulated by Freire (1970), emphasizes the need for education to address social inequalities. AI and robots can support this approach by fostering multicultural awareness and facilitating discussions on linguistic diversity. AI tools can provide learners with resources that address issues of language discrimination and cultural appropriation, encouraging critical reflection and dialogue (Gupta et al., 2022). Robots create inclusive environments for language learning, ensuring that all students, regardless of background, can participate equally in conversations and language exercises (Mageira et al., 2022).

4.2.10 Motivation Theory: Self-Determination and Achievement Goal Theories

Self-Determination Theory (SDT) and Achievement Goal Theory (AGT) explore what motivates learners. SDT, developed by Deci and Ryan (1985), focuses on autonomy, competence, and relatedness, while AGT (Ames, 1992) examines the impact of mastery and performance goals on motivation. AI systems promote autonomy by allowing learners to choose their learning paths and receive personalized feedback, enhancing motivation (Deci & Ryan, 2012; Rusmiyanto et al., 2023). AI-powered gamification elements, such as leaderboards and badges, align with AGT by motivating learners to achieve language milestones through healthy competition (Kherazi & Bourray, 2024).

4.2.11 Neuroscience and Neuroeducation in Language Learning

Insights from neuroscience and neuroeducation reveal how the brain processes language. AI systems and robots can use these insights to optimize language learning through strategies such as spaced repetition and real-time emotional feedback (Hsu & Yu, 2022). AI platforms implement spaced repetition to reinforce vocabulary and grammar, enhancing memory retention (Muley Vilamu, 2021). Robots equipped with affective computing can respond to learners' emotional states, offering personalized support that enhances motivation and engagement (Devillers, 2021).

4.3 English Language Learning: Key Technologies

Building on the theoretical foundations that support the integration of AI and social robots in language education, specific AI-driven technologies bring these concepts into practice. Tools like NLP, AWE, ITS, and ASR transform language learning by creating personalized, adaptive, and effective environments that reflect the principles of personalized learning, interaction, social engagement, and cognitive efficiency discussed earlier.

4.3.1 Natural Language Processing

NLP plays a central role in AI-powered language education by enabling machines to process and understand human language (Alqahtani et al., 2023). Applications of NLP include chatbots, machine translation, and personalized feedback mechanisms, all of which are used in platforms like Duolingo, Google Translate, and more sophisticated systems designed specifically for language learning. NLP is one of the most frequently used AI technologies in language education, facilitating vocabulary acquisition, grammar checking, and sentence construction (Meurers, 2020).

Duolingo and Google Translate rely heavily on NLP algorithms to tailor content to learners' needs. Additionally, chatbots, such as those embedded in language apps, use NLP to simulate human-like conversations, allowing learners to practice speaking and listening skills. NLP-based learning tools can also analyze user input to provide real-time corrections and suggestions, which greatly enhance both writing and speaking skills (Koç, 2024).

4.3.2 Automated Writing Evaluation

AWE tools are another key AI application in English language learning, providing instant feedback on writing quality, grammar, and structure (Zhang & Hyland, 2025). These systems use machine learning algorithms and NLP to evaluate student submissions and offer corrections. Research has shown that combining AWE with teacher feedback leads to improved writing performance, as it enables students to receive continuous, personalized guidance (Xu et al., 2022). Tools such as Grammarly and WriteLab are examples of AWE systems that provide immediate feedback on writing, helping learners to develop their grammar and writing skills autonomously (Hwang et al., 2023).

According to Huang et al. (2023), AWE systems are increasingly being used in educational contexts to support collaborative writing tasks and offer automatic grading. Platforms such as Writing Pal provide students with strategies to improve their essay writing skills by offering automatic feedback on grammar, organization, and overall coherence.

4.3.3 *Intelligent Tutoring Systems*

ITS are designed to provide individualized instruction by adapting learning materials based on real-time student performance. ITS systems use AI to analyze a learner's interactions, assess their proficiency, and tailor the difficulty of exercises to match their current level of understanding. This makes ITS particularly effective for language learning, where progress in vocabulary, reading, and grammar skills can vary greatly among students.

A study by Alam (2023) points out that ITS technologies enhance learning by offering immediate feedback and personalized pathways for students, allowing them to learn at their own pace. Platforms such as iWrite and Tactical Language Training System (TLTS) demonstrate the potential of ITS to support language learners in real-world tasks (Huang, 2023).

4.3.4 *Automated Speech Recognition*

ASR technologies allow AI systems to interpret and respond to human speech, which has significant implications for language education. ASR tools enable learners to practice pronunciation, fluency, and conversational skills in real-time by providing immediate feedback on their spoken language (Sun, 2023). Tools like English Learning Speech Assistant (ELSA) Speak use ASR to analyze pronunciation and offer corrective feedback tailored to the learner's native accent and proficiency level (Permatasari et al., 2024). These systems are particularly beneficial for shy or less confident learners, as they allow for practice in a low-pressure environment, which fosters fluency and boosts confidence in speaking English (Aybirdi, 2023).

ASR systems are not only beneficial for individual learners but also facilitate classroom-based language instruction (Renandya et al., 2024). Teachers can use ASR tools to assess students' speaking abilities, track progress, and provide personalized feedback without the need for manual evaluation (Sholekhah & Fakhrurriana, 2023).

4.3.5 *Chatbots and Conversational AI*

Chatbots represent a significant advancement in AI-driven language learning. These AI-powered conversational agents simulate natural dialogues with learners, enabling them to practice real-life conversations in English. Chatbots use NLP to understand learner input and generate appropriate responses, creating an interactive learning experience. Chatbots are particularly effective in providing corrective feedback and helping learners improve their grammar and sentence structure in conversations (Hawanti et al., 2023).

Platforms such as Replika and Duolingo Bots leverage chatbots to engage learners in contextualized dialogue, simulating interactions they would encounter in real-world settings. These systems offer learners the opportunity to practice speaking in a low-stakes environment, where they can make mistakes and learn from them without fear of judgment (Kovalyova, 2024).

4.4 Social Robots: Enhancing Interaction and Engagement

Building on the advancements in AI-driven technologies like NLP and ASR, social robots offer a new dimension to language learning by fostering interactive, human-like engagement, providing learners with opportunities to practice conversational skills in real-world scenarios and enhancing motivation through personalized, adaptive interactions.

Their integration into the classroom offers a dynamic, engaging learning experience by simulating real-world conversations, providing immediate feedback, and promoting consistent practice in a low-pressure environment. These AI-driven robots, equipped with NLP and speech recognition technologies, can interact with students in a personalized and adaptive manner, offering valuable opportunities to practice language skills in both formal and informal contexts.

4.4.1 *Fostering Authentic Communication with Social Robots*

One of the significant benefits of social robots is their ability to foster authentic communication, offering learners opportunities to practice speaking in naturalistic, low-stakes settings (Wu & Yu, 2024). Social robots like Pepper and NAO are widely used in educational settings to create interactive language learning experiences (Rossi et al., 2024). These robots simulate human-like conversation, allowing learners to engage in dialogues that mimic real-world situations (Wu & Yu, 2024). The use of robots in language education addresses the challenge of providing sufficient speaking practice, especially in classrooms with large student-teacher ratios.

Research shows that learners often feel less anxious when practicing language skills with robots compared to human teachers or peers. As these robots can adapt to the learners' language levels, they create a more personalized learning environment, helping learners feel more confident in using the language (Lio & Ahuja, 2024). The non-judgmental nature of robots encourages learners to experiment with new vocabulary and structures without fear of making mistakes, which is crucial for language acquisition (Luria, 2024).

4.4.2 *Adaptivity and Personalized Learning*

Recent research highlights the effectiveness of personalized learning approaches using social robots and AI in language education. Studies have shown that adaptive systems can significantly improve learning outcomes, with personalized robot tutoring leading to substantial gains in English language skills for Spanish-speaking children (Leyzberg et al., 2018). The TLTS, which incorporates individualized feedback and simulated practice environments, has demonstrated comparable effectiveness to one-on-one human instruction (Beal et al., 2005). Long-term studies have revealed that affective reinforcement learning can optimize story selection for each child, enhancing engagement and linguistic progress in bilingual and English language learners (Park et al., 2019). These findings align with the experiences reported in Glushkova and Staribratov (2025 [Chap. 5]), which discusses the integration of AI and robotics in Bulgarian schools and emphasizes the importance of adaptive learning tools tailored to local educational environments.

4.4.3 *Social Robots as Conversational Partners*

Social robots like NAO and Pepper have shown promise as educational tools, serving as conversational partners and tutors for language learners (Kanda et al., 2004; Ruan et al., 2021). These robots can engage students in interactive dialogues, providing opportunities for speaking practice and immediate feedback (Ruan et al., 2021). Studies have demonstrated that robot-assisted learning can lead to improvements in language skills, particularly in fluency and engagement (Kanda et al., 2004; Ruan et al., 2021). The physical presence of social robots contributes to their effectiveness, offering advantages over traditional learning technologies (Belpaeme et al., 2018). In public spaces, robots like NAO and Pepper are frequently used in educational scenarios, primarily for providing information (Mubin et al., 2018). While social robots have shown potential in achieving outcomes comparable to human tutoring in specific tasks, ongoing research is needed to address technical challenges and optimize their design for educational purposes (Belpaeme et al., 2018).

4.4.4 *Social Interaction and Cognitive Development*

From a theoretical perspective, social robots align with Vygotsky's Social Development Theory, which emphasizes the role of social interaction in cognitive development and language acquisition (Gebhard, 2008). This theory aligns with the use of social robots in learning, where robots can function as mediators within a learner's ZPD, providing scaffolding for complex tasks (Dautenhahn & Billard, 1999). The embodied nature of social robots offers a multimodal learning experience, engaging learners in both verbal and non-verbal communication (Lyon et al., 2024). This approach is supported by the concept of embodied cognition, which links cognitive processes with physical interaction. Researchers in robotics are increasingly incorporating these psychological theories into cognitive architectures for social robots, aiming to enable more authentic human-robot interactions and enhance robots' ability to learn from biological agents (Kirtay et al., 2021).

4.4.5 *Improving Engagement and Motivation*

Social robots have shown promise in enhancing learner engagement and motivation in language learning contexts. Studies indicate that these robots can create a non-threatening environment, reducing language anxiety and increasing student participation (van Minkelen et al., 2020; Song et al., 2025). By implementing strategies based on SDT and the Attention, Relevance, Confidence and Satisfaction (ARCS) model, robots can support learners' psychological needs for autonomy, competence, and relatedness, leading to increased task engagement (van Minkelen et al., 2020;

Deublein et al., 2018). Social robots can sense and respond to students' affective states and behaviors, providing personalized feedback and motivational support (Bourguet et al., 2020). While some studies have shown improved engagement and positive evaluations of robot tutors, the impact on learning outcomes remains inconclusive (van Minkelen et al., 2020; Deublein et al., 2018)

Furthermore, studies have shown that robot-assisted language learning (RALL) can lead to improved language acquisition across various domains, including vocabulary, reading, speaking, and grammar (van den Berghe et al., 2019). A meta-analysis revealed that RALL has a medium-sized average effect compared to traditional or other technology-mediated learning methods (Lee & Lee, 2021). Social robots can create non-threatening environments for language practice, reducing learners' anxiety and fear of judgment (Hayashi & Sato, 2020). This increased confidence can lead to greater willingness to participate in language activities and engage in more complex tasks (Halim et al., 2023). The effectiveness of social robots in language learning is influenced by factors such as age group, target language, and interaction type (Lee & Lee, 2021). Overall, social robots show promise in enhancing language learning experiences and outcomes for diverse learner populations.

4.4.6 Challenges and Ethical Considerations

Despite the many benefits of using social robots in language education, there are also challenges and ethical considerations that need to be addressed. One of the primary challenges is the cost of implementing robot-assisted learning on a large scale. Social robots, such as NAO and Pepper, can be expensive, which may limit their accessibility in under-resourced educational settings (Singh et al., 2022). Additionally, the effectiveness of social robots depends heavily on their ability to recognize and respond to diverse accents and dialects. While AI technologies like ASR have made significant progress in this area, there are still limitations in the accuracy of speech recognition for non-native speakers (Lee et al., 2024).

Ethically, the use of robots raises questions about data privacy and the extent to which learners' interactions with these devices are monitored and analyzed. This concern is also thoroughly addressed in the study of Monte-Serrat et al. (2025 [Chap. 6]), which advocates for the adoption of a sustainable ethics model to navigate the complex tensions between innovation, student privacy, and responsible data collection in AI-enhanced learning systems. As robots collect data on student performance and behavior, it is essential to ensure that this data is used responsibly and that learners' privacy is protected. Educators and policymakers must consider these ethical implications when integrating social robots into language education. Ethical concerns include data privacy, potential emotional attachment, and bias in information provided by robots (Lee et al., 2024; Buchem et al., 2025).

4.4.7 *Critical Reflections on AI, Social Robots, and the Ethics of Educational Technology*

While the pedagogical potential of AI and social robots in language education is extensive, it is essential to critically reflect on the broader implications of integrating such technologies into classrooms. A growing body of literature in the field of critical educational technology, particularly the work of Selwyn (2016, 2021), urges educators and researchers to move beyond a purely instrumental view of technology and instead consider the political economy, social justice, and environmental sustainability that underpin digital tools in education.

From a political economy perspective, many AI tools and social robots are developed and controlled by powerful technology corporations, raising concerns about commercial influence over educational practices (Selwyn, 2016). When public education systems become dependent on proprietary technologies, questions arise regarding the ownership of educational content, student data, and pedagogical agendas (Williamson & Hogan, 2020). These dynamics risk turning education into a market-driven activity where student learning is shaped by algorithmic design rather than educational values.

Issues of equity and access are also central. While AI and robots can enhance personalization and inclusion, their cost and infrastructural requirements may widen existing digital divides, particularly in under-resourced schools or countries (Dixon-Román et al., 2019). This raises important social justice concerns about who benefits from these technologies and who may be left behind.

Moreover, there are sustainability and ethical considerations tied to the production and use of AI-driven devices. The environmental impact of manufacturing, maintaining, and powering robotic systems is rarely addressed in educational technology discourse. Critical scholars emphasize the need to consider not only the pedagogical efficiency of these tools but also their ecological footprint and long-term sustainability (Selwyn, 2021).

As it is suggested by Monte-Serrat et al. (2025 [Chap. 6]), navigating the ethical dimension of AI in education requires more than regulation it calls for a context-sensitive, dynamic ethical framework that evolves with technological advancement and centers the needs of learners.

Therefore, while this chapter advocates for the use of AI and social robots to support inclusive and adaptive language learning, it also acknowledges the need for ongoing critical scrutiny. Educational stakeholders must ensure that the integration of emerging technologies aligns with democratic, equitable, and sustainable educational practices, rather than inadvertently reinforcing systemic inequities or commercial interests.

4.5 Practical Strategies for Implementing AI and Social Robots in Language Classrooms

The integration of AI and social robots into language classrooms necessitates careful consideration of pedagogical strategies that maximize the potential of these technologies while ensuring that they are effectively embedded within the curriculum. The following strategies offer a research-based framework for implementing AI and social robots in language education, focusing on personalized learning, learner engagement, and inclusivity.

4.5.1 Intelligent Tutoring Systems: Personalized Support

ITS provide personalized learning experiences by adapting content in real-time based on each student's performance. By using machine learning algorithms, ITS can assess learners' progress and adjust the difficulty level of exercises accordingly. These systems are particularly effective in language education, where student needs can vary widely. According to Alam (2023), ITS supports various aspects of language acquisition, including reading comprehension, vocabulary, and grammar improvement. By incorporating ITS into classroom activities, teachers can ensure continuous, personalized learning support that adapts to individual student needs. Tools like Duolingo exemplify the effectiveness of scaffolded, adaptive learning experiences that promote steady progress (Alqahtani et al., 2023).

4.5.2 *Social Robots for Interaction and Engagement*

Social robots have emerged as powerful tools for fostering student engagement and creating interactive learning environments. These robots simulate real-world conversations, providing students with opportunities to practice speaking and listening skills in a low-stakes setting. Social robots, such as NAO and Pepper, are gaining traction in language classrooms as conversation partners and tutors. They can be used in practice stations where students work on pronunciation, fluency, and comprehension with immediate feedback (Belpaeme et al., 2018). Teachers can program robots to adapt to students' language proficiency levels, providing personalized feedback that enhances learners' communicative competence (Ghafar et al., 2023).

4.5.3 *Conversational AI for Language Practice*

Conversational AI, such as chatbots, offers dynamic language practice opportunities by simulating real-life dialogues. Powered by NLP, chatbots allow students to practice conversational skills, including grammar, vocabulary, and sentence structure. Chatbots provide immediate feedback, enabling learners to correct errors and make improvements in real-time (Halim et al., 2023). Incorporating conversational AI into writing and speaking exercises can enhance students' interaction with language material. Platforms like ELSA help students improve pronunciation through AI-powered speech recognition, providing tailored feedback on articulation (Permatasari & Lubis, 2024).

4.5.4 *Gamification and Motivation*

Gamified AI platforms can significantly enhance student motivation by transforming language learning into an engaging and interactive experience. Platforms like Duolingo utilize points, challenges, and rewards to encourage continuous engagement with language tasks. Gamification leverages students' intrinsic and extrinsic motivation by rewarding their progress through immediate feedback and visual indicators of success (Kherazi & Bourray, 2024). Teachers can incorporate these gamified platforms into daily assignments, utilizing built-in tracking features to monitor student progress and maintain consistent engagement with language learning (Huang et al., 2023).

4.5.5 *Personalized Learning through Learner Profiling*

One of AI's key strengths lies in its capacity to tailor educational content to individual learner profiles. AI systems analyze student performance data and generate personalized learning paths, ensuring that content is neither too easy nor too difficult. This ability to adjust content based on real-time analysis fosters deeper engagement and sustained progress (Alqahtani et al., 2023). Tools like iWrite, which analyzes multiple writing samples to identify recurring errors, offer targeted feedback that helps learners address specific areas for improvement (Ghafar et al., 2023).

4.5.6 *Supporting Learners with Disabilities and Special Needs*

AI tools hold significant promise for supporting learners with disabilities and special needs by offering accessible technologies such as Text-to-Speech (TTS) and Speech-to-Text (STT) systems. These tools assist learners with visual or auditory impairments, dyslexia, or other challenges by making language content more accessible (Ghafar et al., 2023). TTS tools allow visually impaired students to access written content, while STT tools support learners with speech impairments by providing real-time feedback on pronunciation and language use. Incorporating these technologies into classroom activities ensures that all students can participate meaningfully in language exercises (Sun, 2023).

4.5.7 *Facilitating Inclusive Education with AI Tools*

AI technologies foster inclusivity by supporting students from diverse linguistic backgrounds. Platforms that enable code-switching between languages help multilingual learners transition between their native language and English, ensuring equitable access to learning materials. These tools help create classrooms where students from different language backgrounds can succeed (Huang et al., 2023). AI-powered tools also facilitate inclusivity by adapting content to various learning needs, thereby helping teachers cultivate an inclusive environment where all students can thrive regardless of background or ability (Pawan et al., 2023).

4.6 Design Considerations and Implementation Strategies for AI and Social Robots in Language Classrooms

To maximize the potential of AI and social robots in language education, it is essential to consider both the design of these systems and the strategies used for their implementation. The design must prioritize usability, accessibility, and cultural sensitivity, while implementation should focus on scalability and flexibility to accommodate diverse educational settings.

4.6.1 User-Friendly Interfaces

The effectiveness of AI and social robots in the classroom depends significantly on the design of user-friendly interfaces. These interfaces should be intuitive for both educators and students, prioritizing ease of navigation, visual appeal, and accessibility. For example, teachers need to be able to quickly access lesson plans, monitor student progress, and adjust AI-driven content with minimal technical expertise (Halim et al., 2023). Accessibility features, such as TTS functionalities and adjustable font sizes, are critical to ensuring that AI tools are usable by learners with varying abilities (Ghafar et al., 2023; Huang et al., 2023). Customizable interfaces that allow educators to modify lesson plans and difficulty levels enhance the adaptability of AI systems, making them more effective for diverse learners (Almelhes, 2023).

4.6.2 Culturally Sensitive Content

Incorporating culturally sensitive content into AI and social robots is essential for creating inclusive learning environments. AI systems must integrate diverse cultural references, idiomatic expressions, and historical contexts to ensure that learning experiences are relatable to students from various backgrounds (Huang et al., 2023). This cultural inclusivity helps motivate students by making content more relevant and engaging. Social robots and AI-driven platforms must also reflect linguistic diversity to accommodate learners from different cultural communities, thus promoting effective communication and intercultural competence (Ghafar et al., 2023).

4.6.3 Adaptive Algorithms for Personalized Learning

Adaptive algorithms form the core of personalized learning systems in AI. By continuously analyzing student performance data, these algorithms dynamically adjust task difficulty, ensuring that each learner is adequately challenged without becoming overwhelmed (Son et al., 2024). This personalization is particularly valuable in language education, where learners may progress at different rates in areas such as vocabulary, grammar, and pronunciation. Tools like Grammarly, which provide personalized feedback on writing, exemplify the potential of adaptive algorithms to enhance language learning outcomes (Ghafar et al., 2023).

4.6.4 Scalability and Flexibility

To ensure that AI and social robots are widely implemented across various educational settings, scalability and flexibility must be considered in their design. AI systems should be capable of functioning effectively in different class sizes and settings, from small group lessons to large lecture halls (Halim et al., 2023). Flexibility is essential for adapting lesson plans and feedback systems to cater to individual student needs. Scalable AI solutions must also maintain their efficiency and performance across different school districts and regions without compromising on the quality of educational delivery (Alqahtani et al., 2023). This scalability is critical for broadening access to AI-enhanced education and promoting educational equity across regions.

4.6.5 Real-Time Feedback and Adaptation

AI's ability to offer real-time feedback and adaptation is one of its most significant advantages in language education. Immediate feedback enables students to correct errors as they progress through lessons, ensuring that they do not repeat mistakes or misunderstand concepts (Halim et al., 2023). For example, AI-powered platforms like Duolingo offer real-time corrections on pronunciation and grammar, allowing learners to make adjustments immediately. This capability ensures a continuous feedback loop that enhances learning outcomes and motivates students to improve (Huang et al., 2023). Additionally, AI's real-time assessment of student performance allows for ongoing adaptation of lesson plans and task difficulty, ensuring that learners remain engaged and challenged appropriately (Alqahtani et al., 2023).

4.6.6 *Enhanced Engagement and Motivation*

AI-driven gamification elements, such as badges, leaderboards, and rewards, enhance student engagement by transforming language learning into an enjoyable and interactive experience. Platforms like Duolingo demonstrate how these elements encourage students to continue practicing language skills by offering instant rewards for progress (Kherazi & Bourray, 2024). The autonomy that AI tools provide, allowing students to control the pace and direction of their learning, further enhances intrinsic motivation. When students can explore topics at their own pace, they are more likely to engage deeply with content and achieve better learning outcomes (Huang et al., 2023).

4.6.7 *Continuous Assessment and Progress Tracking*

Continuous assessment is a vital component of AI systems in education, offering real-time insights into student performance and learning trajectories. These tools enable educators to monitor progress, identify areas of difficulty, and adjust their teaching strategies as needed (Huang et al., 2023). The ability of AI systems to track individual performance over time helps educators provide timely interventions for students who are struggling, ensuring that they receive the support needed to meet their learning goals (Kherazi & Bourray, 2024). Data collected from continuous assessment can also inform future curriculum design, ensuring that instructional practices align with the evolving needs of students (Ghafar et al., 2023).

4.7 Conclusion

The integration of AI and social robots into language education represents a pivotal advancement, offering significant improvements in personalized, adaptive, and interactive learning experiences. Rooted in key educational theories such as constructivism (Kolegova & Levina, 2024), Vygotsky's Social Development Theory (Kanero et al., 2018; Maroto-Gómez et al., 2024), and Cognitive Load Theory (Ghafar et al., 2023), AI systems like ITS, NLP platforms, and AWE tools are transforming how students engage with language learning.

AI and social robots provide a dynamic approach to language acquisition, enabling learners to progress at their own pace with real-time feedback. These tools create personalized learning pathways (Ruan et al., 2021) and foster immersion through simulated conversations in low-pressure environments that reduce language anxiety (Lio & Ahuja, 2024). Social robots, such as NAO and Pepper, offer unique advantages by engaging learners in dialogues that adapt to their language proficiency, helping them to build confidence and improve communicative skills (Ruan et al., 2021). This immersive, adaptive approach to language education allows learners to experiment with language, gain fluency, and develop a deeper understanding of linguistic structures through consistent, scaffolded interactions (Kanda et al., 2004; Ruan et al., 2021).

However, several challenges must be addressed to ensure the successful implementation of AI and social robots in classrooms. One primary concern is the high cost of advanced robotics technology like NAO and Pepper (Singh et al., 2022). This expense can limit access to under-resourced schools, raising issues of equity and fairness in education. Addressing this requires policymakers to find innovative solutions to make AI and social robots more affordable and accessible, ensuring that students from all socio-economic backgrounds can benefit from these technologies.

Moreover, technical limitations, such as the difficulties AI tools like ASR systems face in recognizing diverse accents (Lee et al., 2024), also need to be overcome. As classrooms become more linguistically diverse, it is essential that AI platforms are developed to handle a wide range of dialects and linguistic variations, ensuring inclusivity in language education. Ethical considerations regarding data privacy and security are equally crucial. With AI systems collecting vast amounts of data on student performance, it is imperative to establish clear guidelines around data handling to protect students' privacy (Lee et al., 2024; Buchem et al., 2025).

Furthermore, there are concerns about emotional attachment between students and social robots, especially younger learners (Mubin et al., 2018). While robots can provide valuable support and engagement, educators must ensure that the use of technology does not undermine the importance of human interaction in the learning process. Balancing the roles of AI and social robots with the human element in education is vital to preserving students' social and emotional development.

Looking ahead, the future of AI and social robots in language education will depend on the design and implementation strategies employed.

As highlighted, user-friendly interfaces that prioritize accessibility and ease of use for both students and teachers are essential (Halim et al., 2023). Culturally sensitive content and adaptive algorithms (Ghafar et al., 2023) are also crucial in creating personalized and inclusive learning experiences. Additionally, the scalability of AI systems must be addressed to ensure that these technologies can be effectively integrated across various educational settings, from small classrooms to large lecture halls (Alqahtani et al., 2023). By ensuring that AI platforms can adapt to diverse educational environments, these innovations can help bridge educational divides and promote equity.

The real-time feedback and continuous assessment offered by AI platforms, such as Duolingo (Huang et al., 2023), exemplify one of the key strengths of these technologies. Immediate feedback enables students to correct errors and adjust their learning trajectory in real-time, ensuring that they stay engaged and motivated. Gamification elements, like badges and rewards, further enhance motivation by transforming language learning into an interactive, enjoyable experience (Kherazi & Bourray, 2024). The ability of AI tools to track student progress and provide ongoing insights into their learning paths allows educators to deliver more targeted interventions and tailor instruction to meet individual needs (Kherazi & Bourray, 2024).

In conclusion, AI and social robots have the potential to revolutionize language education by providing personalized, adaptive learning experiences that cater to diverse student needs. However, the future success of these technologies hinges on addressing the challenges of cost, accessibility, technical limitations, and ethical concerns. By prioritizing cultural sensitivity, scalability, and user-friendliness in AI system design, and by ensuring responsible data handling, AI and social robots can become invaluable tools in the classroom. Ultimately, these technologies should complement the role of human educators, providing enhanced opportunities for learning while maintaining the essential human connections that are vital for students' overall development. The collaboration between AI tools and teachers holds the promise of creating more inclusive, engaging, and effective educational environments, empowering learners to achieve their full potential. This echoes key themes from the works of Lampropoulos and Papadakis (2025 [Chap. 2]), Arruda and Arruda (2025 [Chap. 3]), and Glushkova and Staribratov (2025 [Chap. 5]), which collectively highlight the growing global movement toward AI-supported personalized learning, equity in access, and the reimagining of teacher and learner roles in the digital age.

References

Agasi, K. (2024). Improving critical thinking by using technology in the learning process. *Interdisciplinary Journal of Research and Development, 11*(1), 76–76.

Alam, A. (2023). Media Multitasking with M-Learning Technology in Real-Time Classroom Learning: Analysing the Dynamics in Formal Educational Settings for the Future of E-Learning in India. In *2023 2nd International Conference on Smart Technologies and Systems for Next Generation Computing (ICSTSN)*, 1–6. IEEE. https://doi.org/10.1109/ICSTSN57873.2023.10151509

Almelhes, S. A. (2023). A review of Artificial Intelligence adoption in second-language learning. *Theory and Practice in Language Studies, 13*(5), 1259–1269. https://doi.org/10.17507/tpls.1305.23

Alqahtani, T., Badreldin, H. A., Alrashed, M., Alshaya, A. I., Alghamdi, S. S., Bin Saleh, K., ... & Albekairy, A. M. (2023). The Emergent Role of Artificial Intelligence, Natural Learning Processing, and Large Language Models in Higher Education and Research. *Research in Social and Administrative Pharmacy, 19*(8), 1236–1242. https://doi.org/10.1016/j.sapharm.2023.05.016

Ames, C. (1992). Classrooms: Goals, structures, and student motivation. *Journal of Educational Psychology, 84*(3), 261–271. https://doi.org/10.1037/0022-0663.84.3.261

Arruda, E. P., & Arruda, D. P. (2025). Artificial intelligence in the global educational context: Prospects for pedagogical practice and application in the context of teacher shortages. In S. Papadakis & G. Lampropoulos (Eds.), *Social robots and artificial intelligence in education: Integrating AI in K-12 and higher education* (pp. 145–162). Palgrave Macmillan. https://doi.org/10.1007/978-3-031-56789-0_9

Aybirdi N. (2023). Revolutionizing English language teaching: The role of artificial intelligence. In: *Conceptual Change Theories and Practice Examples in Education, 1.*

Bartlett, F. C. (1932). *Remembering: A study in experimental and social psychology.* Cambridge University Press.

Beal, C. R., Johnson, W. L., Dabrowski, R., & Wu, S. (2005). Individualized feedback and simulation-based practice in the Tactical Language Training System: An experimental evaluation. *International Conference on Artificial Intelligence in Education.*

Belpaeme, T., Kennedy, J., Ramachandran, A., Scassellati, B., & Tanaka, F. (2018). Social robots for education: A review. *Science Robotics, 3*(21), eaat5954. https://doi.org/10.1126/scirobotics.aat5954

Bolotta, S., & G. Dumas. (2022). Social neuro AI: Social Interaction as the 'Dark Matter' of AI. *Frontiers in Computer Science*, 4, 846440. https://doi.org/10.3389/fcomp.2022.846440

Bourguet, M., Jin, Y., Shi, Y., Chen, Y., Ardila, L. K., & Venture, G. (2020). Social robots that can sense and improve student engagement. *2020 IEEE International Conference on Teaching, Assessment, and Learning for Engineering (TALE)*, 127–134.

Buchem, I., Balaban, I., Badari, J., Ružić,I., & Bäcker, N. (2025). NAO robot in school education: Insights from the European survey and sesign of teacher training. *In International Conference on Robotics in Education (RiE)*, 12–18. Springer Nature Switzerland. https://doi.org/10.1007/978-3-031-98762-5_2

Cumbal, R. (2024). *Robots beyond borders: The role of social robots in spoken second language practice* (Doctoral dissertation, KTH Royal Institute of Technology).

Dautenhahn, K., & Billard, A. (1999). Studying robot social cognition within a developmental psychology framework. *1999 Third European Workshop on Advanced Mobile Robots (Eurobot'99)*, 187–194.

Deublein, A., Pfeifer, A., Merbach, K., Bruckner, K., Mengelkamp, C., & Lugrin, B. (2018). Scaffolding of motivation in learning using a social robot. *Computers & Education, 125*, 182–190.

Devillers, L. (2021). Human–Robot interactions and affective computing: The ethical implications. In Robotics, AI, and Humanity: Science, Ethics, and Policy. In J. von Braun, M. S. Archer, G. Campagna, & M. Sánchez Sorondo (Eds.), *Science and Religion: Interdisciplinary Studies, 6*, 205–211. Springer. https://doi.org/10.1007/978-3-030-54173-6_17

Dixon-Román, E. J., Semel, E. W., & Gordon, E. W. (2019). *Thinking comprehensively about education: Spaces of educative possibility and their implications for public policy*. Routledge.

Edmett, A., Ichaporia, N., Crompton, H., & Crichton, R. (2023). *Artificial intelligence and English language teaching: Preparing for the future*. British Council.

Fathi, J., Rahimi, M., & Derakhshan, A. (2024). Improving EFL learners' speaking skills and willingness to communicate via artificial intelligence-mediated interactions. *System, 121*, 103254.

Freire, P. (1970). *Pedagogy of the oppressed*. Continuum.

Gao, P., Li, J., & S. Liu. (2021). An introduction to key technology in artificial intelligence and big data driven E-Learning and E-Education. *Mobile Networks and Applications, 26*(5), 2123–2126. https://doi.org/10.1007/s11036-021-01777-7

Gardner, H. (1983). *Frames of mind: The theory of multiple intelligences*. Basic Books.

Gebhard, J. G. (2008). *Teaching English as a Foreign or Second Language: A Teacher Self-Development and Methodology Guide*. University of Michigan Press.

George, A. S. (2023). Preparing students for an AI-driven world: Rethinking curriculum and pedagogy in the age of artificial intelligence. *Partners Universal Innovative Research Publication, 1*(2), 112–136.

Ghafar, Z. N., Salh, H. F., Abdulrahim, M. A., Farxha, S. S., Arf, S. F., & Rahim, R. I. (2023). The role of artificial intelligence technology on English language learning: A literature review. *Canadian Journal of Language and Literature Studies, 3*(2), 17–31.

Glushkova, T., & Staribratov, I. (2025). Artificial intelligence and robotics in school education: Analysis of implementation of AI and robotics in Bulgarian education. In S. Papadakis & G. Lampropoulos (Eds.), *Social robots and artificial intelligence in education: Integrating AI in K-12 and higher education* (pp. 89–106). Palgrave Macmillan. https://doi.org/10.1007/978-3-031-56789-0_6

Goldberg, B. B., Bruff, D. O., Greenler, R. M., Barnicle, K., Green, N. H., Campbell, L. E., ... & Campa III, H. (2023). Preparing future STEM faculty through flexible teaching professional development. *PLOS ONE, 18*(10), e0276349. https://doi.org/10.1371/journal.pone.0276349

Gupta, S., & Chen, Y. (2022). Supporting inclusive learning using chatbots? A chatbot-led interview study. *Journal of Information Systems Education, 33*(1), 98–108.

Halim, E., Muliawan, T. N., & Sundaram, D. (2023). Analysis and implementation of social robots to increase language learning capability. *2023 1st IEEE International Conference on Smart Technology (ICE-SMARTec)*, 83–88.

Hawanti, S., Silitonga, L. M., Aziez, F., Furqon, M., Zain, D. S. M., Anjarani, S., & Wu. T. T. (2023). The impact of AI chatbot-based learning on students' motivation in English writing classroom. *In International Conference on Innovative Technologies and Learning*, 542–549. Cham: Springer Nature Switzerland. https://doi.org/10.1007/978-3-031-40113-8_53

Hayashi, K., & Sato, T. (2020). Intelligent speaker is watching you: Alleviation of L2 learners' social anxiety. *CALL for widening participation: short papers from EUROCALL 2020*.

Hickenbottom, K. G. (2022). *Almost like being there: Embodiment, social presence, and engagement using telepresence robots in blended courses* (Doctoral dissertation, Seattle Pacific University).

Huang, Q. (2023). What does the online chat box tell: A mixed study of student participation indicated by online chat data and other variables in an online English course. *European Journal of Open Education and ELearning Studies, 8*(3). https://doi.org/10.46827/ejoe.v8i3.5072

Huang, X., Zou, D., Cheng, G., Chen, X., & Xie, H. (2023). Trends, research issues and applications of artificial intelligence in language education. *Educational Technology & Society, 26*(1), 112–131.

Hsu, I. C., & J. D. Yu. (2022). A medical chatbot using machine learning and natural language understanding. *Multimedia Tools and Applications, 81*(17), 23777–23799. https://doi.org/10.1007/s11042-022-12820-4

Hwang, K. H., Heywood, D., & Carrier, J. (2023). The implementation of ChatGPT-assisted writing instruction in ESL/EFL classrooms. *새한영어영문학, 65*(3), 83–106.

Kanda, T., Hirano, T., Eaton, D., & Ishiguro, H. (2004). Interactive robots as social partners and peer tutors for children: A field trial. *Human–Computer Interaction, 19*, 61–84.

Kanero, J., Geçkin, V., Oranç, C., Mamus, E., Küntay, A. C., & Göksun, T. (2018). Social robots for early language learning: Current evidence and future directions. *Child Development Perspectives, 12*(3), 146–151.

Kang, J. W., Kim, T. H., & Lee. M. S. (2023). AI Chat Bot-ChatGPT-4: A New Opportunity and Challenges in Complementary and Alternative Medicine. *Integrative Medicine Research, 12*(3), 100977. https://doi.org/10.1016/j.imr.2023.100977

Kherazi, A., & Bourray, M. (2024). Gamification and m-learning: An innovative approach to sustainable language learning. *E3S Web of Conferences*, 477, 00066.

Kirtay, M., Chevalère, J., Lazarides, R., & Hafner, V. V. (2021). Learning in social interaction: Perspectives from psychology and robotics. *2021 IEEE International Conference on Development and Learning (ICDL)*, 1–8.

Koç, F. Ş. (2024). The development of listening and speaking skills in EFL via an artificially intelligent chatbot application: A quasi-experimental design study. *Journal of Language Teaching and Learning Research, 14*(2), 55–72. https://doi.org/10.1234/jltr.2024.1402.5

Kolb, D. A. (1984). *Experiential learning: Experience as the source of learning and development*. Prentice-Hall.

Kolegova, I. A., & Levina, I. A. (2024). Using artificial intelligence as a digital tool in foreign language teaching. *Вестник Южно-Уральского государственного университета Серия: Образование. Педагогические науки, 16*(1), 102–110.

Koltovskaia, S. (2022). *Automated writing evaluation for formative second language assessment: Exploring performance, teacher use, and student engagement* (Doctoral dissertation, Oklahoma State University).

Konijn, E. A., Jansen, B., Mondaca Bustos, V., Hobbelink, V. L., & Preciado Vanegas, D. (2022). Social robots for (second) language learning in (migrant) primary school children. *International Journal of Social Robotics, 14*(3), 827–843.

Kot, S. O., & Nykyporets, S. S. (2024). Utilization of artificial intelligence in enhancing English language proficiency in tertiary education. In *Science and education in the third millennium: Information technology, education, law, psychology, social sphere, management* (Chap. 10, pp. 250–274). Springer Nature. https://doi.org/10.1007/978-3-031-56700-5_10

Kovalyova, A. (2024). Chatbots as conversation partners: Effect on perceived writing ability. In *Generative AI in teaching and learning* (pp. 292–310). IGI Global.

Lampropoulos, G., & Papadakis, S. (2025). Integrating social robots into education. In S. Papadakis & G. Lampropoulos (Eds.), *Social robots and artificial intelligence in education: Integrating AI in K-12 and higher education* (pp. 21–35). Palgrave Macmillan.

Lee, H., & Lee, J. H. (2021). The effects of robot-assisted language learning: A meta-analysis. *Educational Research Review, 35*, 100425.

Lee, S., Yoon, J., Cho, Y., & Chun. J. (2024). A systematic review of Chatbot-Assisted interventions for Substance use. *Frontiers in Psychiatry, 15*, 1456689. https://doi.org/10.3389/fpsyt.2024.1456689

Leyzberg, D., Ramachandran, A., & Scassellati, B. (2018). The effect of personalization in longer-term robot tutoring. *ACM Transactions on Human-Robot Interaction (THRI), 7*, 1–19.

Lin, V., Yeh, H. C., & Chen, N. S. (2022). A systematic review on oral interactions in robot-assisted language learning. *Electronics, 11*(2), 290.

Lio, P., & K. Ahuja. (2024). Measuring empathy in Artificial Intelligence: Insights from psychodermatology and implications for general practice. *The Primary Care Companion for CNS Disorders, 26*(5), 57196. https://doi.org/10.4088/PCC.24lr03782

Luria, E. (2024). exploring the use of alexa as a motivational tool in English Language Learning Among EFL Students in Israel. In *IoT, AI, and ICT for educational applications: Technologies to enable education for all* (pp. 77–98). Springer Nature Switzerland.

Lyons, R. J., Arepalli, S. R., Fromal, S. R., Choi, J. D., & Jain. N. (2024). Artificial intelligence Chatbot performance in triage of ophthalmic conditions. *Canadian Journal of Ophthalmology, 59*(4), e301–e308. https://doi.org/10.1016/j.jcjo.2023.07.016

Mageira, K., Pittou, D., Papasalouros, A., Kotis, K., Zangogianni, P., & Daradoumis, A. (2022). Educational AI chatbots for content and language integrated learning. *Applied Sciences, 12*(7), 3239.

Maroto-Gómez, M., Malfaz, M., Castillo, J. C., Castro-González, Á., & Salichs, M. Á. (2024). Personalizing activity selection in assistive social robots from explicit and implicit user feedback. *International Journal of Social Robotics*, 1–19.

Meürers, D. (2020). Natural language processing and language learning. In *Encyclopedia of Applied Llinguistics*, John Wiley & Sons, Inc, 817–831.

Minkelen, P. V., Gruson, C., Hees, P. V., Willems, M., Wit, J. D., Aarts, R., Denissen, J. J., & Vogt, P. (2020). Using self-determination theory in social robots to increase motivation in L2 word learning. *2020 15th ACM/IEEE International Conference on Human-Robot Interaction (HRI)*, 369–377.

Monte-Serrat, D. M., Cabella, D. M. S., & Cattani, C. (2025). Ethics and sustainability in education that makes use of artificial intelligence and robots. In S. Papadakis & G. Lampropoulos (Eds.), *Social robots and artificial intelligence in education: Integrating AI in K-12 and higher education* (pp. 121–152). Palgrave Macmillan.

Mubin, O., Brock, C., Obaid, J. D., & Zlotowski, T. (2018). Social robots in education: State-of-the-Art and future challenges. *Robotics, 7*(1), 22. https://doi.org/10.3390/robotics7010022

Muley Vilamu, F. (2021). *Duolingo–optimization of the Spaced Repetition System to improve long-term memorization.* Universitat Pompeu Fabra.

Novawan, A., Walker, S. A., & Ikeda, O. (2024). The new face of technology-enhanced language learning (TELL) with artificial intelligence (AI): Teacher perspectives, practices, and challenges. *Journal of English in Academic and Professional Communication, 10*(1), 1–18.

Park, H., Grover, I., Spaulding, S., Gomez, L., & Breazeal, C. L. (2019). A model-free affective reinforcement learning approach to personalization of an autonomous social robot companion for early literacy education. *AAAI Conference on Artificial Intelligence.*

Pawan, F., Li, S., Nijiati, S., Dopwell, M., Harris, A., & Iruoje, T. W. (2023). *Culturally- and linguistically-responsive online teacher learning professional development.* Online Learning.

Permatasari, S., & Lubis, Y. (2024). Enhancing pronunciation skills in EFL students through the ELSA Speak application. *Indonesian EFL Journal, 10*(2), 205–212.

Permatasari, R. I., Artha, D. P., Wiratama, B. S., & Wulandari. H. (2024). Overview of Chatbot Usage on Mental Health: A Scoping Review. In *BIO Web of Conferences, 132*, 05002. EDP Sciences. https://doi.org/10.1051/bioconf/202413205002

Piaget, J. (1978). *The Development of Thought: Equilibration of Cognitive Structures.* Oxford: Blackwell.

Qawqzeh, Y. (2024). Exploring the Influence of Student Interaction with ChatGPT on Critical Thinking, Problem Solving, and Creativity. International *Journal of Information and Education Technology, 14*(4), 596–601. https://doi.org/10.18178/ijiet.2024.14.4.2082

Renandya, W. A., Ivone, F. M., & Hidayati, M. (2024). Harnessing the power of technology in ELT. *Journal of Studies in the English Language, 18*(2), 143–160.

Rossi, L., Orsenigo, E., Bisagno, E., & Cadamuro, A. (2024). From learning machines to teaching robots: Interaction for educational purposes between the Social Robot NAO and children: A systematic review. *Journal of e-Learning and Knowledge Society, 20*(1), 15–26.

Ruan, S. S., Jiang, L., Xu, Q., Liu, Z., Davis, G. M., Brunskill, E., & Landay, J. A. (2021). EnglishBot: An AI-powered conversational system for second language learning. *Proceedings of the 26th International Conference on Intelligent User Interfaces.*

Selwyn, N. (2016). *Is technology good for education?* Polity Press.

Selwyn, N. (2021). *Should robots replace teachers? AI and the future of education.* Polity Press.

Sharifuddin, N. S., & Hashim, H. (2024). Benefits and challenges in implementing artificial intelligence in education (AIED) in ESL classroom: A systematic review (2019–2022). *International Journal of Academic Research in Business and Social Sciences, 14*(1), 146–164.

Sholekhah, M. F., & Fakhrurriana, R. (2023). The use of ELSA speak as a mobile-assisted language learning (MALL) towards EFL students pronunciation. *JELITA: Journal of Education, Language Innovation, and Applied Linguistics, 2*(2), 93–100.

Siemens, G. (2005). Connectivism: A learning theory for the digital age. *International Journal of Instructional Technology and Distance Learning, 2*(1).

Singh, D. K., Kumar, M., Fosch-Villaronga, E., Singh, D., & Shukla, J. (2022). Ethical considerations from child-robot interactions in under-resourced communities. *International Journal of Social Robotics*, 1–17.

Singha, S., Singha, R., & Jasmine, E. (2024). Enhancing language teaching materials through artificial intelligence: Opportunities and challenges. *AI in Language Teaching, Learning, and Assessment*, 22–42.

Son, T., Yeo, S., & Lee, D. (2024). Exploring elementary preservice teachers' responsive teaching in mathematics through an Artificial Intelligence-Based Chatbot. *Teaching and Teacher Education, 146*, 104640. https://doi.org/10.1016/j.tate.2024.104640

Song, Y., Huang, L., Zheng, L., Fan, M., & Liu, Z. (2025). Interactions with generative AI Chatbots: Unveiling dialogic dynamics, students' perceptions, and practical competencies in creative problem-solving. *International Journal of Educational Technology in Higher Education, 22*(1), 12. https://doi.org/10.1186/s41239-025-00508-2

Sun, W. (2023). The impact of automatic speech recognition technology on second language pronunciation and speaking skills of EFL learners: A mixed methods investigation. *Frontiers in Psychology, 14*, 1210187.

Sweller, J. (1991). *Cognitive Load Theory. Learning and Instruction, 4*(4), 295–312. https://doi.org/10.1016/0959-4752(94)90003-5

Tellex, S., Gopalan, N., Kress-Gazit, H., & Matuszek, C. (2020). Robots that use language. *Annual Review of Control, Robotics, and Autonomous Systems, 3*(1), 25–55.

Van den Berghe, R., Verhagen, J., Oudgenoeg-Paz, O., Van der Ven, S. & Leseman, P. (2019). Social Robots for Language Learning: A Review. *Review of Educational Research, 89*(2), 259–295. https://doi.org/10.3102/0034654318821286

Voskoglou, M. G. (2024). The role of computers in education in the era of the fourth industrial revolution. In *IoT, AI, and ICT for Educational Applications: Technologies to Enable Education for All* (pp. 119–134). Springer Nature Switzerland.

Vygotsky, L. S. (1978). *Mind in Society: The Development of Higher Psychological Processes.* Edited by Michael Cole, Vera John-Steiner, Sylvia Scribner, and Ellen Souberman. Harvard University Press.

Williamson, B., & Hogan, A. (2020). Commercialisation and privatisation in/of education in the context of Covid-19. *Education International Research*, 1–26.

Wu, R., & Z. Yu. (2024). Do AI Chatbots Improve Students' Learning Outcomes? Evidence from a Meta-Analysis. *British Journal of Educational Technology, 55*(1), 10–33. https://doi.org/10.1111/bjet.13334

Wu, X., & Li, R. (2024). Effects of robot-assisted language learning on English-as-a-foreign-language skill development. *Journal of Educational Computing Research, 07356331231226171.*

Xu, Y., Zhang, J., & Deng, G. (2022). Enhancing customer satisfaction with chatbots: The influence of communication styles and consumer attachment anxiety. *Frontiers in Psychology, 13*, 902782. https://doi.org/10.3389/fpsyg.2022.902782

Yekollu, R. K., Ghuge, T. B., Biradar, S. S., Haldikar, S. V., & Abdul Kader O. F. M. (2024). AI-Driven personalized learning paths: Enhancing education through adaptive systems. *In International Conference on Smart Data Intelligence*, 507–517. Springer Nature Singapore. https://doi.org/10.1007/978-981-97-3191-6_38

Zhang, Z., & Hyland, K. (2025). The role of digital literacy in student engagement with automated writing evaluation (AWE) feedback on second language writing. *Computer Assisted Language Learning, 38*(5–6), 1060–1085.

Zheng, L., Fan, Y., Gao, L., Huang, Z., Chen, B., & Long, M. (2024). Using AI-empowered assessments and personalized recommendations to promote online collaborative learning performance. *Journal of Research on Technology in Education*, 1–27.

Zohuri, B., & Mossavar-Rahmani, F. (2024). Revolutionizing education: The dynamic synergy of personalized learning and artificial intelligence. *International Journal of Advanced Engineering and Management Research, 9*(1), 143–153.

CHAPTER 5

Artificial Intelligence and Robotics in School Education: Analysis of Implementation of AI and Robotics in Bulgarian Education

Todorka Glushkova and Ivaylo Staribratov

5.1 Introduction

Artificial intelligence (AI) is entering more and more rapidly in all spheres of our life and in education. Curriculum and teaching aids are developed on various AI topics in different subjects, classes, and forms of education; tools are created that, with the help of AI, are able to influence learning and the achievement of learning goals. AIEd (Artificial Intelligence in Education) is a new direction in pedagogical sciences that addresses issues related to the application of AI in school education, exploring and

T. Glushkova (✉)
Department of Computer Technologies, University of Plovdiv, Plovdiv, Bulgaria
e-mail: glushkova@uni-plovdiv.bg

I. Staribratov
Department of Math, Computer and IT Training, University of Plovdiv, Plovdiv, Bulgaria
e-mail: staribratov@uni-plovdiv.bg

S. Papadakis, G. Lampropoulos (eds.), *Social Robots and Artificial Intelligence in Education*, Digital Education and Learning,
https://doi.org/10.1007/978-3-032-03977-4_5

evaluating the possibilities, advantages, and disadvantages of these technologies in achieving global goals in education (Ouyang & Jiao, 2021).

Globally, as well as European and nation-wide packages of frameworks and normative documents have been developed, which to a large extent regulate the possibilities for the application of AI. The White Paper on Artificial Intelligence of the European Community (WPAI-EU, 2020) sets the main guidelines for its development. According to the document, it is necessary to develop skills necessary for work in the field of AI and to adapt the education systems of all European countries. The report of the CULT committee (Culture and Education Committee) to the European Parliament on the application of artificial intelligence in education (Tuomi, 2020) makes an in-depth analysis of the need to train specialists to develop and implement intelligent approaches in various areas of business and services. Based on these documents, Bulgaria has developed a "National Strategy for the Development of Artificial Intelligence in Bulgaria until 2030," as well as a "Concept for the Development of Artificial Intelligence." According to these documents, given the rapid development of artificial intelligence and the accumulated backwardness of the formal education system from the needs of the labor market, it is necessary to increase the quality and efficiency of education, to update its content, and to ensure a connection between the education system and the needs of business and society in the era of digital transformation. Based on these strategic documents, three main directions in the application of AI in school education have been identified:

1. Application of AI for education management.
2. Study of basic AI methods, aspects, and algorithms.
3. Application of AI in the Teaching and Learning Process

The process of managing the education is of great importance for its quality. In secondary education, the documentation and administration is of considerable volume. A large part of the schools in Bulgaria use record keeping systems for incoming and outgoing correspondence. But in addition to this document flow, thousands of orders are also generated for each of the activities in the system. Due to the routine and repeatability of actions, introduction of automation is imperative and absolutely possible. A significant number of documents have been digitized and supporting the process through the use of AI will contribute significantly to the ease of document flow and administration.

Another essential aspect is the implementation of management systems to be used skillfully by directors to optimize time and processes. Social robotics can greatly support the organization and implementation of various activities. Such robots can replace service personnel in performing regular and repetitive activities such as cleaning the material base and creating a secure learning environment. By communicating with students, teachers, administrators, and support staff, socially oriented robots can also assist in the early detection of problems, personality conflicts, crisis situations and, last but not least, prevent burnout processes among school teams (Hashem et al., 2024).

The second direction is related to the need to train students who have the necessary knowledge and skills to create and manage processes and systems with AI. In the past few years, many experiments have been conducted and valuable experience has been gained related to the introduction of AI in different types of schools and at different educational levels. For example, many European countries such as Great Britain, Germany, Spain, Portugal, Netherlands, Greece, etc. implement informal training on some aspects of machine learning for students of basic education level. Training takes place mainly in interest clubs. At the Massachusetts Institute of Technology in the USA, an AI curriculum has been developed that also applies to interest club training. The Indian government has launched several initiatives related to teaching AI as a compulsory subject in high school. The experience in Australia, China, South Korea, Japan, etc. is also noteworthy. Interest clubs in STEAM (Science, Technology, Engineering, the Arts, and Mathematics) centers almost everywhere are a preferred form of learning. In the past years, experimental AI education has also been organized in Bulgaria through various forms of formal and informal education, the results of which will be discussed in the second section of this paper.

The third direction requires accelerating the process of developing modern intelligent educational tools, services, and educational platforms to meet the ever-changing characteristics of the learning environment and meet the growing demands and expectations of the education system. The application of chatbots, personal assistants, software environments with embedded AI, as well as comprehensive cyber-physical intelligent educational platforms (Krasteva et al., 2019) is one of the fastest growing topics in educational research.

Human-robot interaction is an important research topic in education. Many studies show that social robots can help ensure quality interaction and communication between all participants in the learning process, as well as stimulate students' motivation, interest, and activity (Johal, 2020). Robotics, considered in different aspects: as an object of programming training in STEAM disciplines or as a teaching tool, is considered as one of the most promising directions for the realization of didactic goals in the classroom.

The capabilities of various robotic devices for social engagement and support are successfully used in the process of inclusive and motivating learning, including for students with special educational needs (Alam, 2022). In the work of Arruda and Arruda (2025 [Chap. 3]), various aspects of the application of AI in the Global Educational Context are discussed.

All these directions require continuous short-term and more in-depth qualification of pedagogical specialists and administrative school staff. It is an indisputable fact that in order to achieve a qualitative development of these technologies, good training of personnel at different levels is necessary—pupils, students, teachers, specialists. To achieve this goal, international standards and frameworks for digital competences have been developed and validated, both for learners and educators; curricula are structured and textbooks and learning resources are created. The accumulated experience provides opportunities to identify both successful approaches and problems and inconsistencies with good pedagogical practices.

Building key digital competencies is the main goal of computer science education and in particular artificial intelligence and robotics. There are various normative documents that determine the directions of modern development. The European Digital Competence Framework for Citizens—DigComp offers a tool to improve citizens' digital competence. The latest version of the DigComp 2.2 framework from summer 2022 includes an additional set of competencies related to artificial intelligence, the Internet of Things, robotics, the development and use of cyber-physical systems (Vuorikari & Holmes, 2022).

The European Digital Competence Framework for Educators DigCompEdu provides a common reference framework to support the development of specific competences for educators. DigCompEdu is aimed at educators at all levels of education, but for the purposes of this study we will concentrate only on school education. DigCompEdu describes 22 competencies organized into domains. The focus of this framework is not on technical skills, but rather aims to specify how digital technologies can be used to improve and innovate in education and training (Bećirović, 2023).

In our study various aspects of the application of AI and robotics in school education in Bulgaria are discussed, as well as the possibilities for increasing the qualification of pedagogical specialists. The main goal is to establish the level of readiness for the application of AI technologies in STEAM education at school, to analyze the results of the studies and to identify guidelines for optimizing this process.

The structure of this chapter is as follows: The first section examines some aspects of the application of AI and social robotics in school education. The experience in formal and informal learning and use of AI and robotics in STEAM education in Bulgaria is presented in the second section. In the last part, the results of conducted experimental studies related to the level of digital competences of teachers are presented and the results of conducted qualification trainings are discussed. The conclusion summarizes the results of the study.

5.2 AI and Social Robotics in Education, Motivation, and Related Works

Modern realities require a digital transformation of school education and the learning process. It is a fact that education helps in the development of the human personality and prepares students for the challenges of the future. The world is currently facing a "Global Education Crisis" as noted by the United Nations on March 3, 2021. The COVID-19 pandemic has imposed social distancing that has had a negative impact on all walks of life, but perhaps most heavily on education.

In its essence, training is a process of stimulating cognitive development, identifying and developing abilities, skills and competencies, which aims to improve students' behavioral attitudes, activity, motivation, and self-esteem. Innovations based on technological development have always been used to improve teaching and increase learning efficiency. In this context, robots and robotic devices are gradually turning from a curious social phenomenon into an integral part of classroom work. Given the problems associated with alienation and the lack of social skills among some students, social robots occupy an increasingly important place in education, where they can be used to acquire the necessary social skills and to support intellectual development.

Social robots are a type of robotic systems designed to interact and communicate with humans in a way that conforms to social behaviors and norms. These robots increasingly provide capabilities in areas such as understanding and processing spoken languages, facilitating verbal communication with users, recognizing and expressing emotions, and social interaction. The use of AI technologies and NLP algorithms in the development of social robots give them the ability to adapt their responses depending on the emotional state of the user and thus participate in conversations, understand social signals, make decisions, and adapt to new situations. They, like all other types of robotic devices, are equipped with cameras, microphones, touch sensors, and other IoT devices to perceive and interact with the environment (Zhenhua & Feng, 2024). All these characteristics provide opportunities for their wide application in all spheres of social interactions.

In the field of education, social robotics has many and varied applications. For younger students, these robots are used as an additional tool in the learning process that increases students' emotional activity (Papadopoulos et al., 2020); in older ones, interaction is provided that improves cognitive and emotional competences. For all participants in the learning process, this technology increases the level of personalization. Due to its importance, a separate scientific field was created, considering the application of social robotics in education. Socio-Educational Robotics (SER) aims to support students' understanding of concepts in STEAM education (Lin et al., 2021). These robots are designed to support the implementation of learning activities. They have great potential and are carriers of numerous social, cognitive, and moral challenges (Tolksdorf et al., 2021). Practice shows that robots in education are effective in performing repetitive, regular tasks, thus allowing teachers to spend more time working with students and personalizing learning.

Children have been found to learn best with robots while playing. Game-based learning is one of the most effective pedagogical approaches, and when the game situation involves a robot that actively interacts with the student, the effectiveness of learning is significantly increased. The robot's ability to interact and clearly express its emotions is one of the reasons for their preference as educational assistants (Pai et al., 2024). Some of the most used educational robots in recent years are:

- LEGO Mindstorm is the most common and popular kit for building, programming, and using robots. They can be used in the teaching of various academic disciplines such as computer science, programming, linguistics, natural sciences.

- Nao is a humanoid robot developed by SoftBank Robotics that is especially appealing to young students. It is used for real-time training, as well as to increase the motivation and activity of students with learning difficulties. It can be programmed in various popular programming languages such as Java, C++, and Python.
- BeeBot and BlueBot are social robots created especially for the youngest students to teach math and programming concepts in accessible ways.
- Romibo is a remote-controlled robot and has the ability to move, gesture, and speak, trying to improve children's cognitive and educational skills.
- Kaspar is a doll-like robot that helps parents and teachers of children with autism and other communication difficulties. Kaspar is intentionally structured with an expressive face cutout because children with autism have difficulty interpreting aloud and reading expressions (Wainer et al., 2014).
- Robobo—a new generation educational social robot that consists of two parts—a "body" (base) with various sensors and a "head" that is a smartphone on which the robot's behavior is programmed. The robot expresses emotions, changes its "face" expression, and can "talk" to the student. The hardware base of this robot can be continuously improved through the use of a more powerful smartphone.

Robots have proven to be an exciting way to learn about electronics, foreign languages, programming, engineering, and more. They have a certain impact on language competences, knowledge, communication, and interactions between students. Various software platforms have been developed and successfully implemented, providing services and tools for using social robotics. (Auerbach et al., 2018) discusses the capabilities of Robogen, an open source software and hardware platform for robotics and artificial intelligence.

Robotics and its application in the educational space in an international aspect is the subject of numerous pedagogical studies, Kerimbayev et al. (2020), Kafila et al. (2024), Schiavo et al. (2024), etc. Particularly important as a perspective is the topic of the application of robotics as an object and subject of education in STEAM (Science, Technology, Engineering, Art, and Mathematics) education (Ouyang & Xu, 2024; Ayeni et al.,

2024, etc.) On the other hand, the issues and regulatory rules for the application of AI and robotics in education are being addressed by all official institutions worldwide (Airaj, 2024). In the EU, the Council of Europe Standing Conference of Ministers of Education is discussing Regulating Artificial Intelligence in Education (RAIEd, 2023), in which a document examines the main dangers and challenges in the mass application of AI in education. However, when analyzing whether robots can replace human teachers, no study has shown that robots can be more effective than humans. As can be seen in Monte-Serrat et al. (2025 [Chap. 6]), sustainable ethics can greatly adapt modern education to "technological innovation, intertwining the parties involved (students and educational institutions), the object (education, intelligent systems, and robots) and the dynamic relationship between them." Additionally, Hair et al. (2025 [Chap. 7]) discuss a significant issue with relationships between perceptions of artificial intelligence (AI), the three elements of burnout (emotional exhaustion, cynicism, and academic ineffectiveness), and student learning.

5.3 STEAM Education, AI, and Social Robotics in Bulgarian Context

In alignment to the purpose of the present research, we will briefly present some approaches for the application of AI technologies and robotics in the organization and management of the school educational environment, as well as in the educational process. In more detail, we will discuss the application of social robotics in the Bulgarian educational environment in the study of the basic principles and algorithms of AI.

The application of AI—technologies in the process of management and organization of the educational school environment is the subject of numerous studies and developments. For solving individual tasks such as generating flexible timetables of the learning process; monitoring student achievement; accounting and management services; document flow management; interaction with parents and the public, etc. various intelligent software systems are developed and implemented. A comprehensive solution to these tasks can be provided by the application of cyber-physical and social multi-agent systems (Tariq et al., 2024), in which services are realized after analyzing the occurring changes in the physical and virtual worlds. The DeLC Laboratory team at Plovdiv University developed a reference architecture of Virtual Physical Space (Stoyanov et al., 2018) that adapts to the conditions of school education. The developed multi-agent platform uses various AI approaches for semantic information processing, analysis, and management of changes in the surrounding world (including emergency situations), management of interaction between teachers, students, parents, and administrators, etc. In this platform, an electronic school diary is being developed related to the management of factory numbered documents (such as certificates and diplomas) through the implementation of blockchain technologies (Todorov et al., 2019).

The application of AI technologies in various aspects of school education has the potential to completely transform the organization of the modern learning process (Kaswan et al., 2024). The principles and theoretical frameworks defined in AIEd address, on the one hand, the development of educational tools with AI, and on the other hand, the use of AI to organize, manage, and evaluate the educational process. This strategy ensures that the tools created meet technological standards and are aligned with predefined educational goals (Chang et al., 2023). On this basis they are offered specific pedagogical principles for the effective integration of AI-technologies and in particular social robotics in the educational process, which on the one hand can motivate and support students in achieving learning goals, and on the other hand help their self-evaluation and personalized development.

The application of AI technologies in conducting the educational process in different academic disciplines and forms of education in Bulgaria is one of the most current topics in pedagogical practice. Chatbots (ChatGPT, Copilot, BgGPT, etc.) are widely used by teachers in the preliminary preparation of lesson plans, learning tasks, test preparation, and overall planning of learning activities. In the course of training, students and teachers also use various AI tools to support the work of students—in project-based learning, in organizing teamwork, in learning in STEAM interest clubs, etc. (Glushkova & Malinova, 2024). The use of game-based learning in which a social robot can participate is one of the most current topics for pedagogical impact in students with special educational needs and especially in students on the autism spectrum (Toskova & Penchev, 2021).

Although this topic is interesting and relevant, in this development we will focus our attention mostly on teaching students in the main areas of AI and the use of these algorithms and approaches in programming social robots.

AI training has been taking place in our country for several years as: a mandatory subject in innovative classes and schools; as an elective module in profiled and vocational high schools; or as learning in STEAM clubs by interest (Glushkova et al., 2020). Some basic curricula were developed and the first textbooks and teaching aids were issued (Fig. 5.1).

In separate modules, the fundamentals of classical and "modern" AI are studied, including the nature and application of Neural Networks, Large Language Models (LLM), basic state space search algorithms, evolutionary strategies, robotics, and programming of robotic devices. The organization and conduct of such training is already taking place in most secondary schools, profiled in the study of natural and mathematical sciences, as well as in some vocational high schools.

PBL is a didactic and methodical approach that has gained momentum in recent years. It started to be implemented as a pilot in Bulgarian education system 10 years ago. Today, more and more schools are conducting short (up to a week) or longer (about a month) PBL. The main problem is that the curricula and programs are not synchronized in all subjects and there is a serious misunderstanding by the controlling authorities about the usefulness of this kind of organization of the learning process. Regardless of the problems in 5% of schools in Bulgaria, for example, this approach is permanently embedded in their development strategies. In the implementation of the synchronization of curricula and cross-curricular connections, teachers use AI-tools, and this significantly supports the preparation of the topics of projects that are provided to students for PBL work.

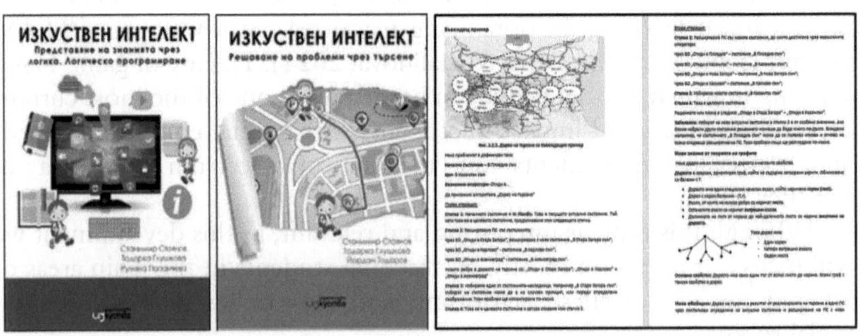

Fig. 5.1 Textbooks for formal AI education (in Bulgarian)

The school STEAM centers created in recent years provide a suitable environment for organizing AI learning in interest clubs involving students of different ages and grades. In these modular clubs, the fundamentals of AI are studied and this knowledge is applied to solving specific practical problems. We found that the highest interest is in programming robotic devices that, based on an assessment of the state of the surrounding world, look for and find a solution to a practical problem—finding a way out of a 3D-labyrinth (through AI search algorithms); assessing the need to water a vegetable bed (through pattern recognition); creating a logic game of the type "guess what animal I am" (through logic programming); development of an educational game where robots react to children's answers, etc. (Fig. 5.2).

The team work on the given problem develops the target social competences, which is also evident from the conducted studies. On the other hand, programming the behavior and emotions of the robotic device during communication with users increases student engagement and increases learning efficiency. In the experimental training conducted, we used various robotic devices that can express emotions as well as react to student behavior. The best results were obtained using the educational robot Robobo, which is able to express emotions such as happiness, fear, surprise, etc. (Fig. 5.3). Through its sensors and the use of a microphone and camera, this robot can recognize the student's behavior and emotional state, reacting and interacting with it.

Initially, a total of 248 students participated in the experimental training, of which:

Fig. 5.2 Robotic programming in STEAM interest clubs

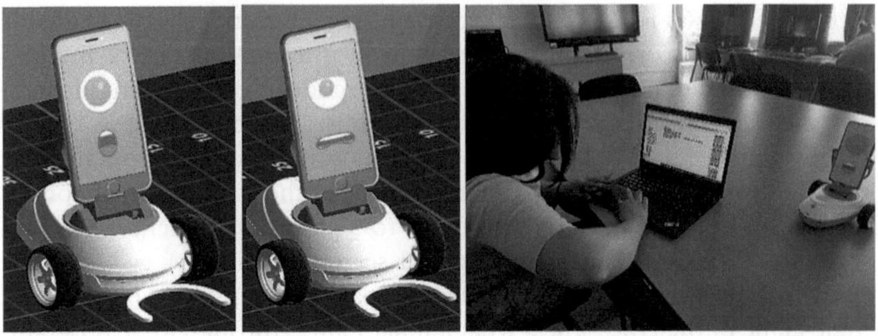

Fig. 5.3 Programming the emotional state and behavior of a Robobo robot

- 104 students studying AI as a compulsory subject in innovative classes in a mathematics high school (Mathematics High School in Plovdiv)
- 92 who study AI and robotics as optional modules in profiled training (Secondary schools in Brezovo, Parvomai, and Asenovgrad)
- 52 students who study AI and robotics in STEAM clubs by interest (Secondary School in Brezovo, Humanitarian Gymnasium in Plovdiv, Vocational High School of Electronics and Electrical Engineering in Plovdiv)

Students from different grades participate in the experimental group—36 in junior high school; 104 in 8th grade; 42 in the 9th grade, and 68 in the 10th and 11th grades. In Fig. 5.4, the results of the conducted training in the different classes and forms of training are visualized in %.

It can be seen from the graph that for high school students who study artificial intelligence within elective modules of their profiled training, the average success varies between 80%–82%, while for students studying AI as a compulsory subject in mathematics high schools this percentage is higher (83%–85%), although the students in this group are younger. The result of learning in STEAM interest groups is indicative, where the results for all age groups are the highest and vary between 85% and 90%. To a large extent, this is due to the fact that in STEAM interest clubs the training in most cases is related to robotics and is at a more applied level, while in the first two groups the training is significantly more theoretical and aims to reach the second level (average) of the target digital competences. The

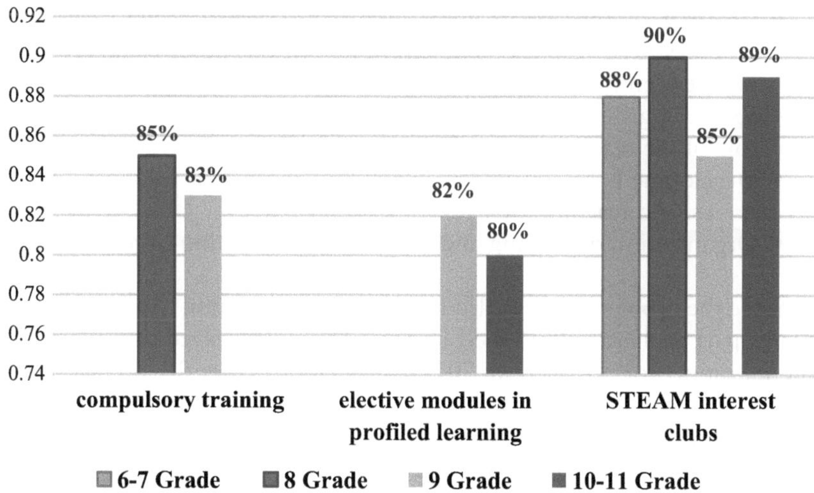

Fig. 5.4 Achievement of students taught in different forms and from different grades

obtained results give reason to deduce an interesting fact—artificial intelligence can be studied in all classes in junior high school and high school, and even in mixed age groups. This also confirms our pre-defined thesis that learning about AI requires more desire, activity, and curiosity from students than basic knowledge.

Although interest in learning about AI and robotics is high, the pedagogical community identifies some significant issues that hinder the learning process. The main ones are:

- Problems with insufficient qualification of pedagogical specialists.
- The lack of well-structured, modularly organized curricula suitable for different groups of learners, specific educational goals, and contexts.
- Insufficient basic knowledge of students.
- Based on the experience gained in recent years, our team from Plovdiv University offers the following approach to solving these problems:

- Creating an appropriate curriculum with clearly defined individual modules that can be studied both within other subjects and independently.
- Expanding the traditional teaching methodology, which defines the role of the teacher as a moderator and facilitator.
- Development of appropriate interactive electronic resources, textbooks, and teaching aids.
- Organization of training and additional qualification of teacher.

Since the problems and goals of education are common to all countries, a large part of these tasks in the European Union are solved through project funding. For the period 2021–2024, more than 20 project lines (https://digital-strategy.ec.europa.eu/en/news/eu-funded-projects-use-artificial-intelligence-technology), related to the application of AI in education have been financed. The implementation of specific projects such as "Guidelines for facilitating the learning of Artificial Intelligence (AI) by School Students of Grades 7-12" (FACILITATE-AI—https://facilitate-ai.eu/), DigiTools (https://digitoolsproject.eu/bg/), STEAME Teacher Academy (https://steame-academy.eu/), or PAIDEA (Preparing teachers for the AI Development in Education as an Innovative Asset) are aimed at the development of basic training courses with appropriate adapted interactive learning resources designed for students of different age groups and forms of learning.

The main task for the development of the processes of the application of AI in the organization and management of education, in the formal training in AI, as well as in the use of AI-tools in school education is the continuous qualification of pedagogical specialists. Plovdiv University is one of the main educational institutions in Bulgaria. In the next part, we will present our experience in this direction and discuss the obtained results.

5.4 Improving the Qualifications of Teachers

The training and continuous qualification of teachers in the field of AI and its applications is one of the most important tasks facing the pedagogical community. The fact is that most teachers did not study AI in their undergraduate or graduate university education. The subject matter is specific and in order to organize and conduct effective training at school, a deeper training of teachers is needed.

Our first task was to ascertain the level of digital competences of teachers related to the teaching and use of AI and robotics. We organized a survey among 377 pedagogical specialists from all over the country who teach various STEAM subjects in profiled, specialized or vocational schools. The selection of the participants is based on their interest, as well as on the invitation of the team. The wide range of participants makes it possible to obtain a clearer picture of the state of the studied problem in Bulgaria. For the purposes of the study, we formed three groups of teachers:

- Group 1: Specialist teachers who teach programming, AI, and robotics.
- Group 2: Teachers of other STEAM subjects who can use AI technologies and robotics in their work.
- Group 3: Other professionals who are trained and retrained as IT teachers.

The studies were conducted in the period 2023–2024 and were carried out in accordance with the EQF for lifelong learning, Digital Competency Framework for Citizens—DigComp 2.2, the teacher-specific framework DigCompEdu. The surveys, although with a single focus, are structured differently with different groups of teachers, but in order to make the self-assessment results comparable, we use the standard DigComp 2.2 framework for the level of relevant digital competences:

- A1—Beginner.
- A2—Researcher.
- B1—Creator.
- B2—Expert.
- C1—Leader.
- C2—Innovator.

The first group included 127 teachers teaching in vocational schools in the professional field of "Computer Science" and profiling in "Software and Computer Technologies." Bearing in mind that all teachers from this group in our country are about 150, it can be said that the sample is representative and the results are indicative. The survey intended for this group is structured in six sections and is aimed at self-assessment of knowledge, skills, and attitudes of teachers related to AI and digitalization in

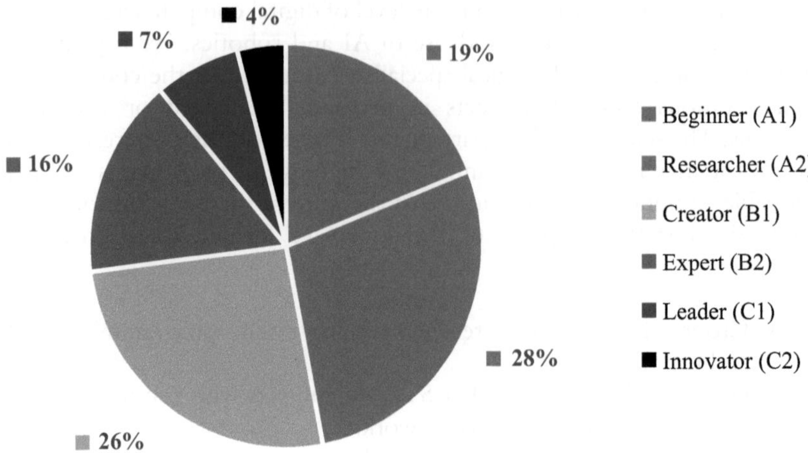

Fig. 5.5 Self-evaluation of the teachers—IT specialists

education. The majority of respondents identified their knowledge as insufficient for both AI and robotics training. Figure 5.5 visualizes the average results of the answers to questions related to the self-assessment of this group of teachers for teaching and using AI technologies in the learning process. It can be seen that around 2/3 of the respondents among this group of the most highly qualified teachers self-rate their digital competences in the field of AI in the low three levels A1, A2, and B1.

The surveys among the second group of teachers were conducted in accordance with the work of our team on several international projects (FACILITATE-AI, STEAME Teacher Academy, DigiTools, PAIDEA) aimed at the application of AI technologies and robotics in the educational process. 216 teachers who teach STEAM subjects and use various digital tools related to AI and robotics in their work are included. The conducted questionnaires also aimed at self-evaluation of teachers and the study of their opinions and attitudes. The results of the summary processing of the questions concerning the present study are presented in Fig. 5.6. It is noteworthy that about 90% of the surveyed teachers self-identify their digital competences related to AI and robotics in the first three levels.

The third group of respondents includes pedagogical specialists who are being trained or retrained as STEAM teachers at Plovdiv University. The study conducted was organized in the initial stage before the start of

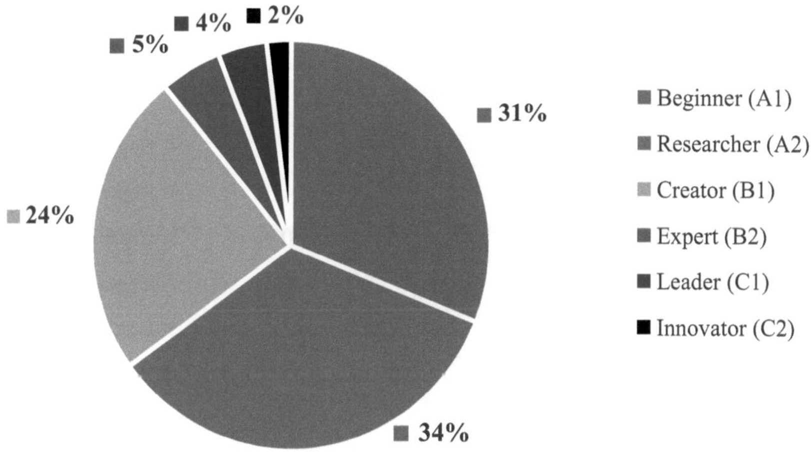

Fig. 5.6 Self-evaluation of the teachers of other STEAM school subjects

their training. This group is quite diverse, but united by their desire to focus their efforts on training and qualification related to the digitization of learning and the application of AI technologies. 34 teachers were included in the study. Figure 5.7 presents the summarized results of the survey. As expected here, over 95% of the respondents self-determine the level of their digital competences at the first three levels A1, A2, and B1.

The comparative analysis of the self-evaluation of teachers from the three observed groups is presented in Fig. 5.8.

Based on our results we can claim that rapidly developing technologies lead to uncertainty among teachers and their attitudes towards the implementation of new technological and methodological approaches related to digitization, learning, and the use of AI and robotics. Logically, the representatives from the first group define a higher level of digital competences compared to the other two groups, but regardless of this, only 27% of them believe that they have knowledge and skills such as Expert (B2), Leader (C1), and Innovator (C2). This group of teachers, however, is responsible for teaching AI and robotics in specialized profiled and vocational schools. The teachers of the second and third groups rather need competences related to the application of AI-technologies and robotics to increase the efficiency and quality of teaching in STEAM subjects. In this sense, it is important for them to have at least intermediate B1–C1

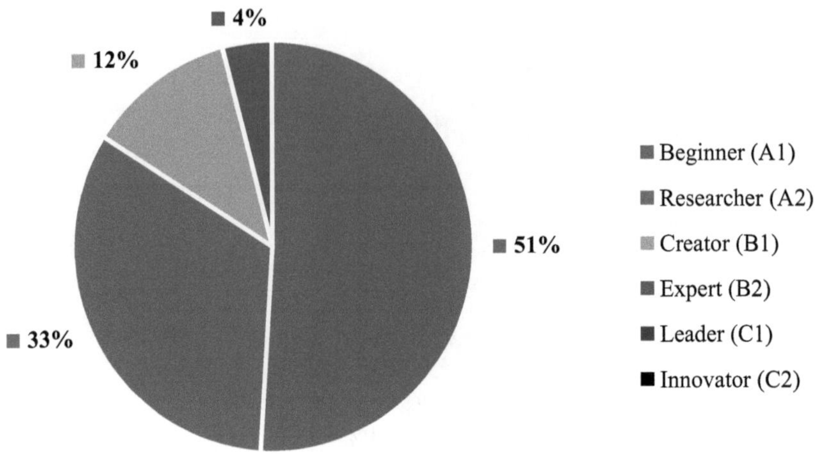

Fig. 5.7 Self-evaluation of other professionals who are trained and retrained as IT teachers

competence levels, but from the above diagram it is clear that they have A1–B1 competence level. All this clearly determines the need to organize and conduct various types of specialized training—in-depth, supporting, and ongoing.

The trainings held in Bulgaria for building key digital competences are a significant number. In the last year, only the Plovdiv University team has participated in the training of over 3,000 teachers to work in a STEAM school environment. Teachers from various subject areas participate in these trainings—natural and mathematical disciplines, humanities, arts, sports, elementary teachers, etc. These trainings are broad-spectrum and have as their main goal familiarization with the possibilities of working in a STEAM environment, giving teachers the opportunity to assess the level of their digital competences and to determine the need for their inclusion in more specialized trainings. For the purposes of this study, we will focus only on qualification courses and specialized training in AI and robotics organized and conducted with the participation of the authors. The observed group of learners consisted of 305 teachers divided into four groups as follows:

- Group 1 is formed by participants in various types of training seminars up to 4 hours. This group includes 175 teachers from all over the country. The trainings are focused on specific topics and aspects of learning and the application of AI in school education. In the majority of cases, the training in this group is conducted online.

- Group 2 is formed by participants in training with a program of 32 hours. The group includes 18 teachers. The training in this group is face-to-face, with topics related to the application of AI and robotics in the learning process in various school subjects.
- Group 3 includes trainees in a course of 60 hours. The number of teachers in this group is 34. The trainings in this group are held in person in certain educational centers in the country. This group features IT teachers mostly from schools specializing in software and hardware sciences, with topics including programming, robotics, and AI algorithms.
- Group 4 unites 78 participants who are preparing and retraining for teachers. Their training is thorough and systematic with a duration of more than 60 hours. This group is the most heterogeneous and includes participants who are studying to be IT teachers at the

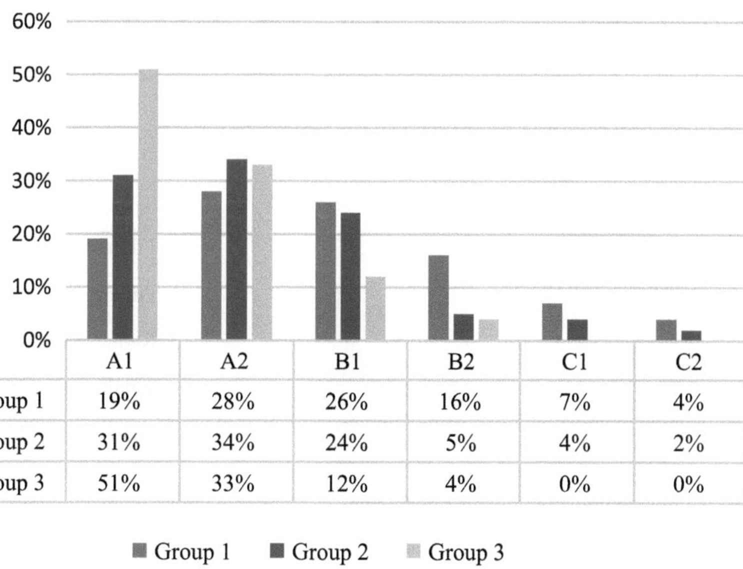

	A1	A2	B1	B2	C1	C2
Group 1	19%	28%	26%	16%	7%	4%
Group 2	31%	34%	24%	5%	4%	2%
Group 3	51%	33%	12%	4%	0%	0%

Fig. 5.8 Comparative analysis of the self-evaluation of the teachers of the three groups

Master's level, teachers from other subject areas who wish to become IT teachers, and non-pedagogical specialists, such as engineers, economists, military, etc., who have an interest in the teaching profession and in particular in AI-technologies.

At the end of the training, we offer the participants a survey, through which we can get feedback on their attitude to training in the field of AI and robotics. According to their opinion, courses for supporting qualification up to 32 hours are extremely insufficient. They share that, despite the developed teaching aids, they do not feel confident enough in their work with students and wish to continue their studies.

However, it is telling that teachers are motivated to teach this subject at school because they believe it is useful and promising for their students (Fig. 5.9).

The motivation and positive attitude of the teachers naturally reflects on their desire to continue their studies with the study of other topics of classical and "modern" artificial intelligence (Fig. 5.10). They share that they wish to pursue robotics, machine learning, and various AI applications.

Because we wanted to understand what the desired topics for further future training were, we analyzed the responses of the learners from the

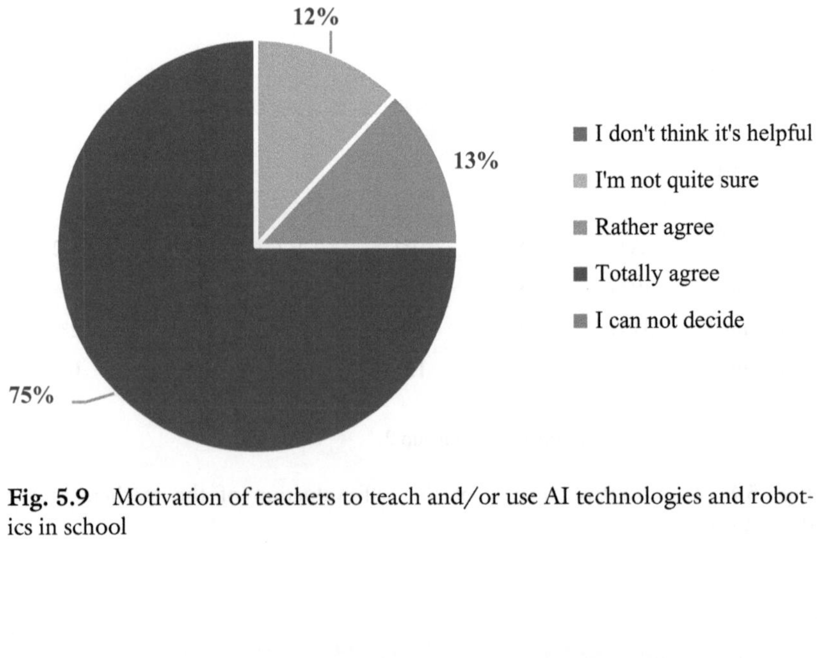

Fig. 5.9 Motivation of teachers to teach and/or use AI technologies and robotics in school

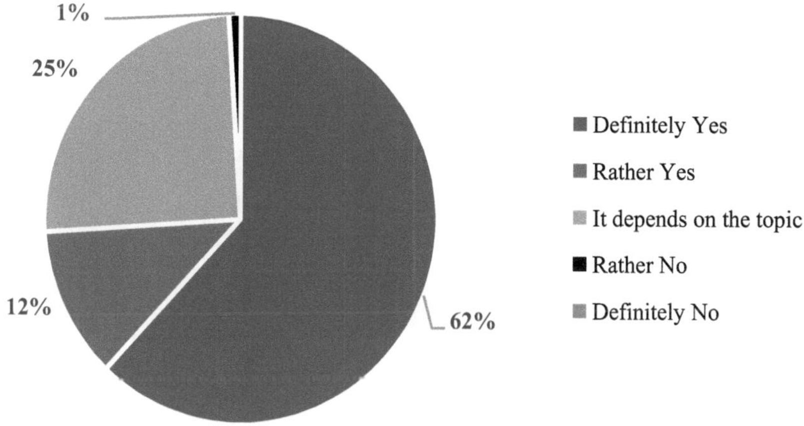

Fig. 5.10 Desire of learners to continue their AI studies

four groups (Fig. 5.11). It is noteworthy that over 90% of participants in all groups are willing to receive information and participate in additional training on topics related to the application of AI and robotics. Logically, teachers from the third and fourth groups have a desire to continue their education on the theoretical topics of AI and programming of robotic devices, while those from the first two groups believe that these topics require much more time and effort. And they are pressed for time to carry out training on topics that are mandatory according to state educational requirements. In the curricula, which have not changed for at least the last 20 years, new technologies are absent. This means that the teacher has to reduce the content of the topics for basic knowledge and conduct practical ones, which significantly worsens the skills and competencies of the students. This balance is very fragile and not every teacher allows himself to violate it for something that he is not sure about and is not convinced that through it he will get a better quality of learning the material (Takeva & Mavrudieva, 2024).

5.5 Results and Discussion

The main findings of the study show that the use of AI and social robotics in education has a positive effect on student motivation and engagement. The impact of programming social robots that express emotions and

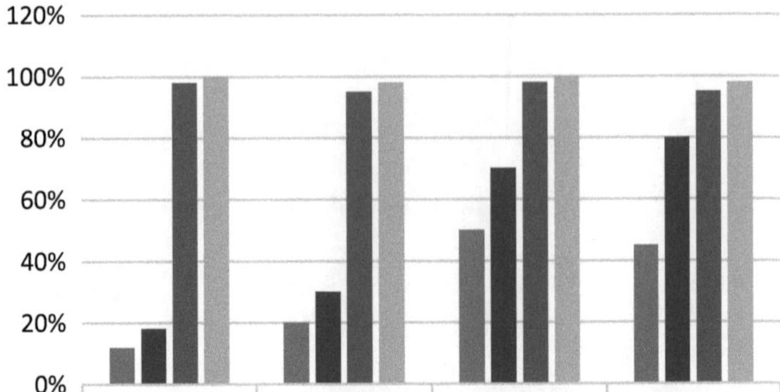

	Group 1	Group 2	Group 3	Group 4
AI algorithms	12%	20%	50%	45%
Program Robots	18%	30%	70%	80%
Using AI tools	98%	95%	98%	95%
Using Robots	100%	98%	100%	98%

Fig. 5.11 Desired learning topics from different groups of teachers

interact with students in real time is particularly strong. This type of technology not only enhances students' interest in the learning process, but also develops their digital and social competencies. It is important to emphasize that practical tasks involving robots have a higher success rate than more theoretical approaches.

The conclusions that emerge as a result of the conducted studies and the practical experience in training and application of AI and the educational process, including through social robots in Bulgaria, can be summarized in the following directions:

- AI-technologies entered life very quickly and educational systems cannot respond to this challenge quickly and adequately.
- Students expect maximum use of AI in the learning process, but curricula and content are not so well adapted.
- From the three directions—use in the administration of school processes; training of students in AI, and the ability of teachers to use AI in their daily activities, the second is developing most systematically

in Bulgaria, although not fast enough to meet the demand of the labor market.

- Teachers experience too much uncertainty in their preparation for using AI and do not find sufficient support from state and administrative bodies for continuous and purposeful improvement of their qualifications in these areas.
- This uncertainty is also based on a certain chaos in the lack of systematicity in curricula and educational content.
- Teachers share that they lack the attitude and willingness to be creators of content responding to the dynamics of development of these technologies.
- The leading entities of the education system—in ministries, regional education administrations, and a significant number of directors hardly accept the new developments, perhaps because their training on these topics is at an insufficient level (Ivanova, 2024; Pinkwart & Liu, 2020; Zhang & Aslan, 2021).
- In order for there to be economic development in a society, there must be well-prepared programmers, including AI and robot management automation. In a significant number of countries, this is a priority, and appropriate strategies are being developed for this. Only in this way is it possible to develop all spheres of the economy. In this aspect, INSAIT was established in Bulgaria in 2022 (INSAIT, 2022).
- The realization of students also requires modern preparation. In a significant part of the companies, the digital transformation is in an advanced phase. Technological development requires knowledge of AI and the use of automation and social robots. A significant number of companies in Bulgaria are experiencing difficulties with personnel and are resorting to increasing their technological development.
- Through good preparation of the teachers, the students will also be able to prepare well, so they will realize themselves well and will be satisfied with their work and will be successful (Saraliev & Antonova, 2024; Zahariev, 2024).

The results of implementing new approaches have broad prospects for the future of AI and robotics education in schools. They suggest that the integration of these technologies can change the traditional approach to teaching by creating a more dynamic, interactive, and motivating learning environment. Also, this research highlights the need for systematic training of educators to effectively use AI and robots in their daily work. In the long term, the application of AI and robotics in education can contribute to the development of skills that are key to the labor market of the future.

The results of the present study are consistent with those of other international studies that highlight the positive effect of using social robots in education (Johal, 2020; Papadopoulos et al., 2020). Studies such as those by Tolksdorf et al. (2021, 2022) show that social robots play a key role in improving the language and social skills of children, especially those with learning disabilities. Although the present study confirms the positive results of the use of social robots, the focus on the development of digital competences of all students offers a wider context and proves that these technologies can be useful for all age groups.

According to studies of Ouyang and Xu (2024) and Manna et al. (2024), it is concluded that educational robotics in STEAM disciplines significantly increases students' interest in science and technology, as well as careers in these fields. These results correspond with the findings of the present study, where social robot programming in STEAM clubs demonstrated higher achievement and motivation among students. This confirms the effectiveness of hands-on learning through robotics.

The studies of He (2024) and Hashem et al. (2024) also highlighted the importance of continuing education and teacher qualification for the successful integration of AI and robotics into school STEAM education. Similarly, in the Bulgarian context, the research shows that the lack of sufficient qualification among teachers is a major obstacle to the implementation of AI technologies. In both cases, the creation of systematic training and support programs for teachers is proposed.

To fully present the problems facing STEAM education, it is worth mentioning that still in Bulgaria (and research shows such a trend in other countries as well), STEAM education is developed based on the desire and capabilities of individual teachers, i.e., there is a lack of systematicity. On the other hand, it is difficult for state educational standards to fit into this type of training. This leads to the fact that project-based STEAM training is largely conducted outside of traditional classes as additional activities and interest clubs. Teachers prefer, in order not to disturb their comfort and if they have ready-made materials, to use them in a traditional learning environment. In addition, each school develops a very different and specific STEAM environment, and each teacher has their own view of this (Dimitrova, 2024). This aspect of the problem, which added to the above-mentioned insufficient and inadequate qualification, is imposed as a major problem for faster and more fruitful STEAM training. State policy is aimed at unity for technical provision, and in a narrow profile of equipment. Furthermore, from a pedagogical perspective, teachers are not provided with the necessary teaching materials and appropriate teaching methodology, as well as assessment methodology in a STEAM environment. All of this definitely worries educators and creates problems and challenges for fast and effective STEAM education.

Kaswan et al. (2024) considered the implementation of personalized AI systems that provide feedback to students and support self-regulated learning. The current research goes in a similar direction, showing that social robots and AI technologies can provide personalized learning, especially in classes with students of mixed age groups and skill levels.

As a result of our analysis and the presented results, it can be said that the implementation of AI development strategies on a national and international level is related to the preparation of future teachers already during their university education and continuous supporting qualification during their pedagogical implementation in school. The obtained experimental results lead to the conclusion that it is imperative to have a well-built system for the qualification of teachers in different areas—to create content; to have more trained to teach high-level programming; and to be able to use and apply AI in practice in every lesson and everything in direct relation to the successful implementation of the students. In addition, it is necessary to significantly improve the digital culture of the leading entities in education. Following the impact processes of digitization, it is good to be much more adaptive and to build the confidence of teachers and managers through quality qualification, and if possible, to carry out anticipatory qualification.

5.6 Conclusions

The present study discusses various aspects of AI application and robotics in school education in Bulgaria as well as the possibilities for increasing the qualifications of pedagogical specialists.

The experimental training in AI and robotics in Bulgaria poses different options for modules and topics which can be studied in different types of schools and in different forms—as a compulsory subject, as an elective module or informally in STEAM clubs based on interests. We found that programming robotic devices—which express emotions, are able to communicate with students, and find a solution to a given practical problem, which in turn increases the interest, activity, and motivation of learners.

On the other hand, the application of AI-technologies and social robotics in the teaching of different academic disciplines and in different contexts provides conditions for increasing the quality of the learning and teaching processes.

A key factor for the success of this training is the additional support and teachers' qualifications. The accumulated experience shows that it is imperative to create a comprehensive strategy for training pedagogical specialists to build a high level of digital competences in the field of AI-technologies and robotics. The future plans of the team are aimed at the development of appropriate curricula for training different groups of teachers, using the possibilities of international projects Facilitate-AI (https://facilitate-ai.eu/) and STEAME TEACHER ACADEMY (https://steame-academy.eu/) for this purpose. In parallel, work is being done on the development of learning materials to be freely available through the Moodle educational platform for continued personalized training of teachers. Success needs only knowledgeable, capable, and motivated teachers who can provide the necessary environment for training and development of successful young people—future artists, scientists, and researchers.

Acknowledgments This study is financed by the European Union—NextGenerationEU, through the National Recovery and Resilience Plan of the Republic of Bulgaria, project № BG-RRP-2.004-0001-C01

References

Airaj, M. (2024). Ethical artificial intelligence for teaching-learning in higher education. *Education and Information Technologies*, 1–23, https://doi.org/10.1007/s10639-024-12545-x

Alam, A. (2022). Social robots in education for long-term human-robot interaction: Socially supportive behaviour of robotic tutor for creating robo-tangible learning environment in a guided discovery learning interaction. *ECS Transactions, 107*(1), 12389. https://doi.org/10.1149/10701.12389ecst

Arruda, E. P., & Arruda, D. P. (2025). Artificial intelligence in the global educational context: Prospects for pedagogical practice and application in the context of teacher shortages. In S. Papadakis & G. Lampropoulos (Eds.), *Social robots and artificial intelligence in education: Integrating AI in K-12 and higher education* (pp. 37–60). Palgrave Macmillan.

Auerbach, J., Concordel, A., Kornatowski, P., & Floreano, D. (2018). Inquiry-based learning with Robogen: An open-source software and hardware platform for robotics and artificial intelligence. *IEEE Transactions on Learning Technologies, 12*(3), 356–369. https://doi.org/10.1109/TLT.2018.2833111

Ayeni, O. O., Unachukwu, C. C., Al Hamad, N. M., Chisom, O. N., & Adewusi, O. E. (2024). The impact of robotics clubs on K-12 students' interest in STEM careers. *Magna Scientia Advanced Research and Reviews, 10*(1), 361–367. https://doi.org/10.30574/msarr.2024.10.1.0027

Bećirović, S. (2023). Fostering digital competence in teachers: A review of existing frameworks. In *Digital pedagogy: The use of digital technologies in contemporary education* (pp. 51–67). Springer. https://doi.org/10.1007/978-981-99-0444-0_5

Chang, D. H., Lin, M. P.-C., Hajian, S., & Wang, Q. Q. (2023). Educational design principles of using AI chatbot that supports self-regulated learning in education: Goal setting, feedback, and personalization. *Sustainability, 15*(17), 12921. https://doi.org/10.3390/su151712921

Dimitrova, N. (2024). Integrirane na obrazovatelen podhod STEM v srednoto uchilishte—predizvikatelstva, vazmozhnosti i perspektivi. Sp. Pedagogika, br.9/2024, str. 1232-1243, 1053656/ped2024-9.04

Głushkova, T., & Malinova, A. (2024). Application of AI technologies in STEAM school education. *International Journal of Research in E-learning*, 1–20. https://doi.org/10.31261/IJREL.2024.10.1.01

Glushkova, T., Stoyanov, S., Tabakova-Komsalova, V., Grancharova-Hristova, M., Krasteva, I. (2020). An approach to teaching artificial intelligence in school, innovative educational technologies, tools and methods for E-learning Scientific Editor Eugenia Smyrnova-Trybulska "E-learning". 12, Katowice–Cieszyn, 257–267. https://doi.org/10.34916/el.2020.12.22

Hair, J., Rana, S., Robb, D., & Stroud, M. (2025). The emerging role of AI and social robots in university student recruitment. In S. Papadakis & G. Lampropoulos (Eds.), *Social robots and artificial intelligence in education: Integrating AI in K-12 and higher education* (pp. 153–183). Palgrave Macmillan.

Hashem, R., Ali, N., El Zein, F., Fidalgo, P., & Khurma, O. A. (2024). AI to the rescue: Exploring the potential of ChatGPT as a teacher ally for workload relief and burnout prevention. In *Research & practice in technology enhanced learning, 19*. https://doi.org/10.58459/rptel.2024.19023

He, L. (2024). *The impact of STEM teacher competence on job performance in the era of artificial intelligence*. 2024 6th International Conference on Computer Science and Technologies in Education (CSTE), Xi'an, China, pp. 337–341, https://doi.org/10.1109/CSTE62025.2024.00070.

INSAIT. (2022). Retrieved August 10, 2024, from https://insait.ai/what-is-insait/

Ivanova, S. (2024). *Artificial intelligence in education*. Second National Scientific-Practical Conference Digital Transformation of Education, pp. 16–20, https://www.conf-dte.bg/docs/2024/p-3.pdf

Johal, W. (2020). Research trends in social robots for learning. *Current Robotics Reports, 1*, 75–83. https://doi.org/10.1007/s43154-020-00008-3

Kafila, K., Rajyalaxmi, M., Rizwana, M., & Sailaja, V. (2024, June). Role of robotics and simulators in teaching learning process. In *AIP Conference Proceedings, 2971* (1), 060031. AIP Publishing. https://doi.org/10.1063/5.0195752

Kaswan, K. S., Dhatterwal, J. S., & Ojha, R. P. (2024). AI in personalized learning. In *Advances in technological innovations in higher education* (pp. 103–117). CRC Press.

Kerimbayev, N., Beisov, N., Kovtun, A., et al. (2020). Robotics in the international educational space: Integration and the experience. *International Journal of Information and Education Technology, 25*, 5835–5851. https://doi.org/10.1007/s10639-020-10257-6

Krasteva, I., Todorov, J., & Stoyanov, S. (2019). Intelligent school education space. *Education and Technologies, 10*(1), 120–126. https://doi.org/10.26883/2010.191.1489

Lin, J., Li, Y., & Yang, G. (2021). FPGAN: Face deidentification method with generative adversarial networks for social robots. *Neural Networks: The Official Journal of the International Neural Network Society, 133*, 132–147. https://doi.org/10.1016/j.neunet.2020.09.001

Manna, S., Azhar, A. & Greace, A. (2024). *A qualitative review of educational robots for STEM: Technical features and impacts.* 2024 47th MIPRO ICT and Electronics Convention (MIPRO), Opatija, Croatia, 2024, pp. 1255–1260. https://doi.org/10.1109/MIPRO60963.2024.10569301

Monte-Serrat, D. M., Cabella, D. M. S., & Cattani, C. (2025). Ethics and sustainability in education that makes use of artificial intelligence and robots. In S. Papadakis & G. Lampropoulos (Eds.), *Social robots and artificial intelligence in education: Integrating AI in K-12 and higher education* (pp. 121–152). Palgrave Macmillan.

Ouyang, F., & Jiao, P. (2021). Artificial intelligence in education: The three paradigms. *Computers & Education: Artificial Intelligence, 2*, 100020. https://doi.org/10.1016/j.caeai.2021.100020

Ouyang, F., & Xu, W. (2024). The effects of educational robotics in STEM education: A multilevel meta-analysis. *IJ STEM Ed, 11*, 7. https://doi.org/10.1186/s40594-024-00469-4

Pai, R. Y., Shetty, A., Dinesh, T. K., Shetty, A. D., & Pillai, N. (2024). Effectiveness of social robots as a tutoring and learning companion: A bibliometric analysis. *Cogent Business & Management, 11*(1). https://doi.org/10.1080/23311975.2023.2299075

Papadopoulos, I., Lazzarino, R., Miah, S., Weaver, T., Thomas, B., & Koulouglioti, C. (2020). A systematic review of the literature regarding socially assistive robots in pre-tertiary education. *Computers and Education, 155*. https://doi.org/10.1016/j.compedu.2020.103924

Pinkwart, N., & Liu, S. (2020). *Artificial intelligence supported educational technologies.* Springer. https://link.springer.com/book/10.1007/978-3-030-41099-5

Regulating Artificial Intelligence in Education (RAIEd). (2023). https://rm.coe.int/regulating-artificial-intelligence-in-education-26th-session-council-o/1680ac9b7c

Saraliev, Z., & Antonova, D. (2024). *The role of AI in digital transformation of the creative industries.* Second National Scientific-Practical Conference Digital Transformation of Education, pp. 476–480. https://www.conf-dte.bg/docs/2024/p-86.pdf

Schiavo, F., Campitiello, L., Todino, M. D., & Di Tore, P. A. (2024). Educational robots, emotion recognition and ASD: New horizon in special education. *Education Sciences, 14*(3), 258. https://doi.org/10.3390/educsci14030258

Stoyanov, S., Glushkova, T., Stoyanova-Doycheva, A., & Doychev, E. (2018). Virtual physical space—An architecture supporting internet of things applications. In *2018 20th International Symposium on Electrical Apparatus and Technologies (SIELA)* (pp. 1–3). IEEE. https://doi.org/10.1109/SIELA.2018.8447156

Takeva, Y., & Mavrudieva, I. (2024). Strategicheski i komunikatsionni aspekti na izpolzvaneto na izkustveniyat intelekt v obrazovanieto, sp. Pedagogika, br.8/2024, str. 1056-1073, 1053656/ped2024-8.02

Tariq, R., Casillas-Muñoz, F. A., Hassan, S. T., & Ramírez-Montoya, M. S. (2024). Synergy of Internet of things and education: Cyber-physical systems contributing towards remote laboratories, improved learning, and school management. *Journal of Social Studies Education Research, 15*(2), 305–352. https://hdl.handle.net/11285/653894

Todorov, J., Krasteva, I., Ivanova, V., & Doychev, E. (2019). BLISS-A CPSS-like application for lifelong learning. In *2019 IEEE International Symposium on INnovations in Intelligent SysTems and Applications (INISTA)* (pp. 1–5). IEEE. https://doi.org/10.1109/INISTA.2019.8778363

Tolksdorf, N. F., Viertel, F. E., & Rohlfing, K. J. (2021). Do shy preschoolers interact differently when learning languagewith a social robot? An analysis of interactional behavior and word learning. *Frontiers in Robotics and AI, 8*, 676123. https://doi.org/10.3389/frobt.2021.676123

Tolksdorf, N. F., Hönemann, D., Viertel, F. E., & Rohlfing, K. J. (2022). Who is that?! Does changing the robot as a learning companion impact preschoolers' language learning? In *2022 17th ACM/IEEE International Conference on Human-Robot Interaction (HRI)*, Sapporo, Japan, pp. 1069–1074, https://doi.org/10.1109/HRI53351.2022.9889420

Toskova, A., & Penchev, G. (2021). Intelligent game recommendation system. In *AIP Conference Proceedings* (Vol. 2333, No. 1). AIP Publishing. https://doi.org/10.1063/5.0042063

Tuomi, I. (2020). Research for CULT Committee—The use of AI in education, 2020, Retrieved June 10, 2024, from https://www.europarl.europa.eu/RegData/etudes/BRIE/2020/629222/IPOL_BRI(2020)629222_EN.pdf

Vuorikari, R., & Holmes, W. (2022). DigComp 2.2. Annex 2. Citizens Interacting with AI Systems. In R. Vuorikari, S. Kluzer, & P. Yves (Eds.), *DigComp 2.2, The Digital Competence framework for citizens: With new examples of knowledge, skills and attitudes* (pp. 77–82). Publication Office of the European Union. https://doi.org/10.2760/490274

Wainer, J., Dautenhahn, K., Robins, B., & Amirabdollahian, F. (2014). A pilot study with a novel setup for collaborativeplay of the humanoid robot KASPAR with children with autism. *International Journal of Social Robotics, 6*(1), 45–65. https://doi.org/10.1007/s12369-013-0195-x

White Paper on the AI of EU (WPAI-EU). (2020). Retrieved June 10, 2024, from https://ec.europa.eu/info/sites/info/files/commission-white-paper-artificial-intelligence-feb2020_bg.pdf

Zahariev, A. (2024). *Artificial intelligence in business and its impact on the labor market.* Second National Scientific-Practical Conference Digital Transformation of Education, pp. 11–15,. https://www.conf-dte.bg/docs/2024/p-2.pdf

Zhang, K., & Aslan, A. (2021). AI technologies for education: Recent research & future directions. *Computers and Education: Artificial Intelligence, 2.* https://www.sciencedirect.com/science/article/pii/S2666920X21000199

Zhenhua, Z., & Feng, G. (2024). Application of social entertainment robots based on machine learning algorithms and the Internet of Things in collaborative art performances. *Entertainment Computing, 100784.* https://doi.org/10.1016/j.entcom.2024.100784

CHAPTER 6

Ethics and Sustainability in Education That Makes Use of Artificial Intelligence and Robots

Dioneia Motta Monte-Serrat, Daniela M. S. Cabella, and Carlo Cattani

6.1 Introduction

This chapter proposes a new way of dealing with ethics in its complex interrelationship between education networks, such as the relationship of students with knowledge or with the school institution, and artificial intelligence, AI, and robots. For some time there has been considerable

D. M. Monte-Serrat (✉)
FFCLRP-USP, Unaerp, Ribeirão Preto, Brazil
e-mail: dserrat@unaerp.br

D. M. S. Cabella
Complete Privacy, Ribeirão Preto, Brazil

C. Cattani
Unitus, Viterbo, Italy
e-mail: cattani@unitus.it

S. Papadakis, G. Lampropoulos (eds.), *Social Robots and Artificial Intelligence in Education*, Digital Education and Learning,
https://doi.org/10.1007/978-3-032-03977-4_6

concern about the overlap between the individual and AI. Rifkin (2010) mentions a platform that unifies the internet of communication and the internet of energy forming the Internet of Things, IoT. This integration has improved communication between machines (computers) and people (environment), forming a global network of objects with intelligent sensors, whose value lies in the amount of data they collect and the impact they have on human life, both negative (among them vulnerability of security and privacy), and positive (such as access to health services, education, and optimization of activities to obtain a better quality of life) (Pinheiro, 2018).

Berners-Lee (1999) mentions the exploitation of information that reaches educational institutions; Saide and Sheng (2021) report the scenario of effective knowledge transfer of the teaching-learning process during the Covid-19 pandemic. AI and robots in education have provided social inclusion and free access to information while the rights of use and exploitation related to people's information leave a business model and go to educational institutions. There is concern about the accumulated data and information about individuality (Berners-Lee, 1999) raising questions about the circumstances in which this data is collected and about the mechanisms to prevent information collection, its use and dissemination outside of limits previously established by laws (Pinheiro, 2018; Magrani, 2018; Cabella & Monte-Serrat, 2022; Monte-Serrat & Cattani, 2023b).

We highlight, in this chapter, that the mediation of intelligent systems to promote learning also permeates individuals, influencing their decisions and behaviors, which implies ethical principles (Monte-Serrat & Cattani, 2023b). This condition of AI, mixed with the individual is little investigated and people take it for granted that there is a separation between the user and the intelligent technological device, which is not true. The use of augmented or virtual reality, or platforms that support classes, making them more didactic and accessible (for example, in the case of free or low-cost applications, instant or asynchronous communication); and the use, by the Educational Institution, of connected devices or robots to teach online classes or turnstiles associated with biometric identification cameras in the control of attendance and student safety, are all current examples that the subjective reality of student users is being experienced under the overlap of disciplines and technologies, without people being aware of it.

Varma et al. (2021) discuss the change in technologies used in the day-to-day operations of the higher education institutes. The adoption of artificial intelligence with educational value changes, according to the authors, the way the institution and students work. The use of intelligent systems in education also opens the door to cyber-attacks in educational institutions, when people steal personal information or confidential one (such as student grades and marks) (Varma et al., 2021). Technological advances have been provoking profound changes in society and education daily, which imply ethical practices that adapt to changes. So far, what we have are ethical principles advocated by norms, which are static and always require new norms aimed at circumventing difficulties that appear over time.

Our proposal in this chapter is that educational institutions adopt the practice of sustainable ethics, which is a new concept of ethics applicable both to present and future facts, ensuring sustainability in the protection of the rights of students using artificial intelligence and robots. To arrive at this concept of sustainable ethics, a qualitative analysis was carried out in education that makes use of technology. Qualitative analysis, based on French discursive theory, is intended to contrast the observed reality with the theory conveyed. In our analysis, the context underlying the collection of data from users of intelligent systems in the educational process is demonstrated: That there is an overlap between the subjectivity of students and the intelligent system. As a result of this overlap, a little noticed aspect arises: the possibility of commercial use of the data collected from the students, since the information collected becomes raw material with property characteristics. It is questioned whether this monetization of data obtained from students violates their privacy as users of AI and robots in education systems. To circumvent these situations, the need to apply ethical principles that effectively accompany the changes in intelligent systems is highlighted, since the existing ethical norms freeze in time, providing inexpressive and obsolete legal protection.

Educational Institutions are the privileged context to address the impacts of technology on the relationship between individual, institution, and education. The knowledge conveyed both in teacher training and in professional relationships in an educational institution must value the impacts of the use of technology in a dynamic context, envisioning ethical aspects, since the robots and AI, which can be used for good, can be used for evil, harming students or education professionals. Institutional decisions as to which technology to adopt can have positive or negative consequences ethically speaking. To avoid damage to individual rights, it is not enough for such decisions to be in accordance with formal legislation, as it is always lagging AI innovation (Cabella & Monte-Serrat, 2022; Magrani, 2018).

It is suggested that educational institutions are attentive to the new practice of sustainable ethics in the treatment of personal data, to avoid damage, present or future, to the relationships between individuals in the educational context that makes use of AI and robots. Ethics and data protection, by going together, is a way that best combine solutions for activities involving data processing, respecting users' rights beyond what is required by law, which gives educational institutions a solid basis to develop education for the long term and without causing harm to the individuals involved.

This new approach on ethics and sustainability in the educational relationship that makes use of artificial intelligence and robots highlights the fact that the mediation of the technological apparatus is mixed with the individual who makes use of it, raising questions about the privacy of the latter and the protection of their personal data. For this, the chapter is divided into eight sections. Section 6.2 presents other related studies. Section 6.3 deals with the explanation that the mediation of artificial intelligence implies the use and exploration of information related to people. As there is still a lack of clarity about the domain of these data, it is important for the reader to be clarified that they reflect the individuality and must be surrounded by care so that laws are not infringed. Section 6.4, besides presenting aspects about AI, robots and ethics, also gives a critical view of the fundamental role of ethics linked to the use of artificial intelligence and robots, since intelligent systems interfere with human behavior. Section 6.5 brings a warning about respect for people's rights (students' subjectivity) related to the use of information technologies like artificial intelligence and robots in education. It is clarified that the data collected by the education platforms represent the person himself, which involves obligations, responsibilities, and the adoption of ethical principles in their treatment. The methodology and literature review on the complex relationship between ethics and digital privacy are the content of Sect. 6.6, which explains the French discourse analysis method used to analyze the theory considering the context underlying the data collection of the users of intelligent systems in education. This method favors the investigation that the use of data permeates the individual himself in the existing relationships in the socio-historical educational process. Therefore, there is interrelation between education, ethics, digital privacy, technology, and norms, so that the technological educational reality mixes the subjectivity of users of the robot or artificial intelligence transforming identities into a product of property nature. Section 6.7 proposes the adoption of sustainable ethics in the educational system. This concept adapts to the changes brought about by new technologies combining security and privacy with a simultaneous assessment of elements at stake such as ethical and legal aspects, security, responsibility, protection of personal data, ownership, and employability of this data. In this way, the parties involved (students and educational institution), the object (education, AI, and robots), and the dynamic relationship between them are intertwined. The conclusion in Sect. 6.8 shows that the internet is borderline in nature, defying regulations on whether databases are proprietary or personal. It also shows that researchers in the area emphasize the need for special treatment and new principles, in line with what is proposed in this chapter: We suggest the adoption of sustainable ethics, which applies in the present and is reflected in the future, contemplating the simultaneous evaluation of laws, ethics, educational context, incessant technological innovation, and protection of access and use of student data.

6.2 Presenting Theoretical Basis Considering Other Related Studies

In this section, it is considered relevant to present other studies that relate AI, ethics, and the educational process. This extensive range of other research is intended to bring elements to, in the following sections, discuss the complexity of applying ethical principles in educational practices that make use of intelligent systems. In Sect. 6.4, we come back to these studies to observe whether these research and development ideas are or are not actually implemented concretely.

Adams et al. (2023) report that AI tools in education need to follow ethical principles due to the “potentially” (emphasis added) harmful social effects of AI. The authors compare existing ethical guidelines and shape their applicability in the educational process: justice, transparency, equity, privacy, responsibility, autonomy, benevolence, and freedom to intensify the collaborative environment between individuals and AI.

Daniela and Strods (2019) suggest that educational robotics can reduce early school leaving and promote sustainable education. The authors argue that the use of robots in education is positive because it increases students’ motivation and helps them build knowledge.

Dieterle et al. (2024) review the literature to discuss the ethical effects of using artificial intelligence in education. The authors show the interrelationship between algorithms and interpretation, citizenship, access, and representation. They investigate how intelligent systems tools could be designed to mitigate harmful effects such as prejudice and seek greater neutrality in results. The authors conclude that there is a vicious cycle between the elements they studied which structurally perpetuates prejudice. They suggest that ways to mitigate prejudice be discussed in the future through a cycle of progressive improvement.

Kamalov et al. (2023) question the sustainability of the educational process that uses ChatGPT, which changes the teaching paradigm. The authors carry out a comprehensive literature review to assess the advantages and challenges of using AI tools in schools and universities. The authors mention studies that suggest that AI has the potential to positively impact teaching but warn that there must be a vigilant eye on its negative effects. They show that AI improves student assessment and speeds up the provision of individualized feedback. Alongside the advantages, the authors highlight several concerns about the disadvantages, claiming that ethical incidents are continually increasing. They claim that there is a lack of policies or guidance on the use of AI in education regarding: ethical challenges, data privacy and security, bias and discrimination, plagiarism and academic integrity, and problems in the teacher-student relationship.

Lin et al. (2014), when discussing the ethics of robots and their social implications, they state that robotics is used to perform many functions and for this reason raises ethical concerns. Bartneck et al. (2021) wrote a book about the foundations of AI and ethics in the interaction between people and intelligent and robotic systems. The authors focus on the tendency for individuals to be impacted, for better or worse, by AI tools, which implies ethical risks.

Meghdari and Alemi (2018) evaluate advances in robotics, highlighting that ethical challenges grow at the same intensity as the growth of technologies and show the need for interdisciplinary research because of systems effects on human interaction. The authors point to possible ethical challenges of technologies in culture, education, nursing, entertainment, and therapy.

Nguyen et al. (2023) wrote an article on ethical principles for artificial intelligence in education in which they state that AI advancement transforms the educational landscape, leading to growing ethical concerns. The authors conducted a topical analysis that includes students, teachers, decision makers, policymakers, and technology developers, to suggest a set of applicable ethical principles. They complete the study with the expectation that interested parties will implement their suggestions.

Pedro et al. (2019) address artificial intelligence in education with a focus on sustainable development. The authors consider AI as something that alters social interactions and, because it depends on innovative ecosystems, causes digital and social exclusion. They mention Sustainable Development Goal 4, which deals with AI in improving the educational process, giving examples of how education systems use data to improve equity and educational quality and for access to education and in tutoring monitoring systems to support teachers. The authors state that educational management information systems improve the state's ability to manage educational systems and endorse the importance of each country preparing the conditions for an AI-driven world. In this same positive direction of AI effects on education, Remian (2019) praises AI services in ameliorating educational results and administrative tasks, and to ensure the mitigation of ethical problems, the author suggests a questionnaire that helps define the emerging ethical concerns of AI in education. We add the study by Serholt and Barendregt (2014), that focuses on students' perceptions regarding robots in education and ethical concerns related to teachers and students. The authors are in favor of the application of robotics in education and conclude that their study helps to shape the future of robotics. Akgun and Greenhow (2022) and Serholt et al. (2017) discuss the case of robots in the classroom from the perspective of ethical tensions. The authors state that although robots facilitate children's learning, there are great ethical concerns regarding children's interactions with robots. They investigate what the role of the robot is, what its effects are on children, and what the implications are for the field of educational robotics.

Sijing and Lan (2018) write about ethical problems and solutions in education that adopt artificial intelligence to improve the personnel training model. While on the one hand the authors evaluate the use of AI as positive in its contribution to improving the cognitive level of students and reducing the workload of teachers, on the other hand, they point out some problems such as: the irrationality of the algorithm; the inaccuracy of the data; the responsibility of the people responsible for education in artificial intelligence (programmers, teachers, and students). The authors offer specific solutions based on the problems they point out. On the other hand, Tarrés-Puertas et al. (2023) address child-robot interaction using educational robots in the Qui-Bot H2O project, which covers social and ethical aspects of robotics applied to chemistry. The authors demonstrate the usefulness of robots in teaching technical and scientific concepts in a playful way.

Some ethical considerations about applying robots in kindergarten environments are studied by Tolksdorf et al. (2021). The impact of technology on children is analyzed from a macro perspective and from an individual perspective, since children are seen as a vulnerable group. In dynamic interaction, it is observed that both the robot and the child can learn from each other and, as children are vulnerable, challenges arise regarding decision-making. To avoid this problem, the authors suggest limitations in the use of robots in early childhood education. Although Tolksdorf et al. (2021) admit the impact of technology on children and suggest limitations on its use to overcome challenges in decision-making, they do not address the fundamental causes of these challenges. Conversely, *The Natural Language for Artificial Intelligence* (Monte-Serrat & Cattani, 2021b) is a book that uses neurolinguistics to describe the dynamics of the biological and logical structure typical of the human cognitive linguistic process to, from it, outline the cognitive structure of intelligent systems under the description of the universal structure of language. In this way, the authors show how cross-fertilization occurs between natural language and AI, an important piece of information that can help in the search for solutions for ethical aspects in the use of AI in education, as we can see in the fourth section of this chapter.

6.3 Mediation Through AI and Robots Implies Rights to Use and Exploit People's Information

August 1, 2024, is the entry into force of the EU Artificial Intelligence Law (European Parliament, 2024). This law establishes promises to guarantee security, transparency, and monitoring of the use of AI systems. However, we note that these promises do not correspond to a real guarantee that intelligent systems and robots will be designed in a way that meets transparency, security, and ethical requirements. Authors question whether there will be access to AI tools, but we show in this chapter that what really matters is "how" intelligent systems were designed in their cognitive core.

The regulation for AI systems by the European Parliament considers intelligent systems and robots that carry out cognitive-behavioral manipulation of people or specific public groups, such as students, a threat to people. Students at institutions can be encouraged to use systems that classify them based on behavior, or their socioeconomic status or personal biometric identification characteristics. This categorization of students can be vexatious for them, affecting their academic life or even their personality qualities, infringing their fundamental rights to freedom, dignity, family, and social coexistence.

It could be argued that perhaps the inability to address ethics and privacy issues in education stems from the fact that digital privacy regulations have not yet reached consensus. Many countries, despite having specific legislation approved and in force, face several discussions about the lack of clarity regarding the data domain. Thus, if a company or institution is not sure about the rights to the data it has, it can be misled by the rules of the digital economy game (Pinheiro, 2018). It is proposed that a comprehensive set of laws be brought together on the rights to access databases collected by sensors or by computer processes (Internet of Things, IoT and AI), but what has been observed so far is that, for the most part, these Laws are still restricted to the financial, health, and public sectors (Pinheiro, 2018).

Some laws and regulations—e.g., Regulation (EU) 2016/679, as known as the General Data Protection Regulation (GDPR) (Krzysztofek, 2018), the Brazilian laws (Brasil, 2014, 2018), and the Brazilian General Data Protection Law, LGPD (2018) still do not adequately protect personal data, increasing the risks to the protection of data collected from the user's life through Internet of Things devices, IoT, and subject to decisions made by Artificial Intelligence. Magrani (2018) says that governance that combines the mass market with security and privacy standards is necessary. This would circumvent the existing tension between innovation and privacy, which reflects new contexts of social relationships that are often not guaranteed by existing norms. Although the laws protect people's rights, there are still gaps in the protection of privacy (which concerns the control that the individual has over their own information) and personal data (which involves guarantees of an instrumental nature, i.e., protection of privacy, and fundamental constitutional guarantees) (Magrani, 2018).

6.4 AI, Robots, and Ethics in the Educational Process

Although Rifkin (2010), Pinheiro (2018), Magrani, (2018), and Varma (2021), among others, have published warnings about the harmful action of AI on individuals' cognition (see Sect. 6.1), research is still found—on the individuals' positive effects of AI—which sails smoothly by negative effects of the use of intelligent systems. It is not difficult to find articles about the changes that artificial intelligence brings to education and schools, changing the role of teachers, educational institutions, and even leaders in education (Gocen & Aydemir, 2020). It is observed that their authors emphasize the benefits and, when mentioning some disadvantages, limit themselves to discussing generalities such as uncertainties in the future of teaching and the need to protect legal aspects. Fiok et al. (2022), for example, in their research on explainable artificial intelligence for education and training, suggest that the benefit of the rapid growth of artificial intelligence should be adopted with caution, as there are limitations and risks associated with making decisions that are not always transparent. However, they leave it open how to resolve the lack of transparency.

We put the focus, in this chapter, on the difficulty and complexity of addressing "how" the interference of artificial intelligence in the educational sector is managed: Either it is seen as something positive, or it is pointed out as something that "possibly" could be dangerous for the educational process, but without clearly identifying the causes of the damage.

There are certainly positive aspects of AI in education highlighted in scientific studies. Given the urgency for researchers to investigate why interaction with artificial intelligence and robots, in fact, interferes with human behavior, in this section we dedicate ourselves to analyzing not only positive elements but also how authors address negative aspects of AI in education. We observe whether they point out concrete solutions that avoid losses or damages to the user of artificial intelligence during its use in the educational process.

6.4.1 Related Studies on AI and/or Robotics in Education and the Complexity of "How" This Interrelationship Is Managed

In Sect. 6.2 of this chapter, we present recent studies related to exploring the role of AI and/or robotics in education. We dedicate this subsection 6.4.1 to critically analyze "how" each of which manages the interference of artificial intelligence in the educational sector. The uncertainties of using intelligent systems over a borderline internet cannot be taken for granted. This brings challenges to assertiveness in protecting the privacy and freedom of individuals and to the effective application of regulations and ethical principles. The search for concrete solutions is aligned with what is called fourth generation qualitative analysis (Campos, 2011), which is not content with studies that cool down and become silent on the shelves (Silva & Brandão, 2011).

Adams et al. (2023) suggest that existing ethical principles be applied to AI tools in education to prevent "potentially" (emphasis added) harmful social effects of AI. Critically analyzed, ethical principles have existed for some time in the form of laws and regulations and have not yet prevented the harm caused by AI. The existence of the law or norm itself does not have the power to prevent losses. What would be missing for AI tools in the educational process to be transparent and benevolent in the collaboration between individuals and AI?

The research by Daniela and Strods (2019), despite reporting positive results for the use of robotics in promoting the educational process, may, over the years, not be consistent in the face of findings that algorithms, through AI, exert influence on the reasoning, and why not, on the learning process, of AI or robot users, thus providing early solutions and preventing them from making free choices (Monte-Serrat & Cattani 2023b). The solution to this would be to integrate ethical principles into the modeling of the intelligent system to protect the individual rights of users (Cabella & Monte-Serrat, 2022; Monte-Serrat & Cattani, 2023b; Bartneck et al., 2021).

The aspects studied by Dieterle et al. (2024) to mitigate the effects of prejudice resulting from the use of artificial intelligence in education did not prove to be adequate. The authors concluded that a vicious cycle forms between the elements they studied, perpetuating prejudice in a structural way. To avoid interpretative biases in facial recognition systems, for example, Monte-Serrat and Cattani (2021a) show the importance of knowing "how" the cognitive linguistic process behaves in the construction of information. The authors consider that the algorithmic core of the AI system must perform interconnection between semantics and statistics to achieve optimization of the AI information classifier, and thus avoid prejudiced bias.

The extensive review of studies on the use of AI in education by Kamalov et al. (2023) brings up some topics that need to be critically analyzed. Although the authors mention that AI has the potential to positively impact teaching and improve student assessment, they paradoxically state that ethical incidents are continually increasing. In our view, although AI in education improves student outcomes, it provides them with "tailored" learning experiences. Who performs and with what criteria the filtering of knowledge? The student does not choose. One chooses "which" knowledge they should learn. Kamalov et al. (2023) also states that AI in education predicts the specific needs of students. Under our critical eye, AI "anticipates" the student by showing "strengths" and "individual limitations" under whose criteria? Who says what is good and what is bad to adapt the students' learning process? Another use of AI in conducting group discussions in the classroom can have AI intervention for "smart triage." Who decides which topics should be discussed? We cannot be naive and take this AI "screening" as neutral. Someone designed its algorithm to direct choices on our behalf. Kamalov et al. (2023) state that AI can help guide teamwork in the "right direction." "Who" determined what the "right direction" would be? This also applies to prompts, which "direct" content creation, with the supposed advantage of "reducing" educators' workload. Furthermore, large-scale experiments on prompt tools run into problems when their strategy does not work for all types of problems, but only some prompts are better in isolation (Rajput, 2024). Who would be behind the algorithms that bias the creation of educational content? It is for these and other reasons that ethical issues are growing in the use of AI in education. Ethical principles need to be applied to the cognitive core of the intelligent system. Kamalov et al. (2023) are right to state that studies on AI-based teaching require a crucial review of the foundations of techniques, applications, and principles.

Lin et al. (2014) claim that robotics used to protect one person can cause harm to others, for example, serving water with parasites, or drowning an individual in a pool, precisely because robots are unaware of risk of harm. The authors describe the complexity of teaching what is right to a robot because if it follows the rules, it can still produce terrible results. There is also the difficulty of determining the "acceptable risk" in the robot's intervention. These are some examples of what can go wrong if there is no ethical intervention in the algorithmic core of the intelligent system, as Cabella and Monte-Serrat (2022) suggest.

Meghdari and Alemi (2018) discuss robotics and ethics under the need to articulate branches of exact sciences with other branches of science such as biology, neuroscience, philosophy, linguistics, and history, in addition to ethical laws. Unlike Cabella and Monte-Serrat (2022) who suggest applying ethics to the robot's cognitive core, the authors (Meghdari & Alemi, 2018) raise questions about which ethical theory to apply; about designing robots that harm people; about using robots to serve as companions for children and the elderly; about ethical implications in the use of robots for intimate relationships; about rights and duties of robots; about the use of robots to perform authority functions, such as police, security guards, or even teachers. Ultimately, all these questions lead to the authors' suggestion of implementing roboethics (ethical standards incorporated into robots (Meghdari & Alemi, 2018).

While addressing benefits of AI in education, Nguyen et al. (2023) show the need to take care of fundamental issues involving human rights. To achieve this, the authors suggest applying a holistic set of ethical principles to AI in the educational context. Thinking critically, one cannot confuse the idealization of principles with their effective application in the complex reality of the interaction between humans and AI. From this, it can be concluded that "suggesting" the application of standards and principles is not the same as "effectively ensuring" that AI in education is truly ethical. The authors come close to the suggestion of Cabella and Monte-Serrat (2022) when they propose that ethical principles should be considered in the development and implementation of AI systems, according to the Privacy by Design, PbD, lesson by Ann Cavoukian (2009).

When warning about potential damage to the human-robot relationship, Bartneck et al. (2021) are not surprised that considerations about AI include the damage it can cause. By stating that addressing negative aspects of AI on humans is part of several national and international policies, the authors express a certain conformity with these negative impacts. On the other hand, they are in line with our research (Cabella & Monte-Serrat, 2022) by recommending the implementation of ethical principles from the conception of the intelligent system, not limited only to regulatory issues. Contrary to these findings of damage to cognition are studies by Remian (2019), who praises AI services in improving educational results, Serholt and Barendregt (2014), who are positive about the application of robotics in education, and Serholt et al. (2017), who understand that robots make learning easier for children. Remian (2019), Serholt and Barendregt (2014), and Serholt et al. (2017) do not analyze the deep effects that interaction with intelligent systems causes in the cognition, emotion, and behavior of their users, as warned by Monte-Serrat and Cattani (2023b), when discussing ethical AI.

6.4.2 *Interface Between Human Cognition and the Cognition of Intelligent Systems*

In recent years, several topics have been discussed in search of theoretical and scientific mastery of the interrelationship between natural language, artificial intelligence and, more recently, cognitive computing. Monte-Serrat and Cattani (2021b) innovate by creating an interface between human cognition and the cognition of intelligent systems. While both are treated as something that does not communicate, Monte-Serrat and Cattani (2021b, 2023a, 2023b) present an innovative perspective in which a universal structure performs a "bridge" between human cognition and artificial intelligence, directing this interface towards the next frontier by linking features of natural language to both artificial intelligence and cognitive computing. In short, the authors (Monte-Serrat & Cattani, 2021b) show, through neurolinguistics, that there is a close relationship between AI and human cognition, which paves the way for AI to interfere in the behavior of the user of an intelligent system or in the behavior of the individual interacting with a robot.

Greater research effort is needed to improve research capacity regarding the application of ethical principles in the design of the artificial intelligence core or robot to be used in an educational institution. From this point of view, there are two possibilities: The intelligent system will interfere with the thinking or behavior of the student interacting with the artificial intelligence, and/or the intelligent system will collect raw data that indicate preferences or choices of a given user. This data or interference in behavior can, unethically, be used, for example, to provide advance solutions that prevent users from freely making their own choices (Monte-Serrat & Cattani, 2023b). The urgency of integrating ethical principles in the artificial intelligence algorithm modeling is due to the pressing need to avoid manipulation of behavior, giving rise to the real protection of individual rights (Cabella & Monte-Serrat, 2022).

Regarding the application of ethical principles to the relationship between humans and AI or robots, the concept of ethics as prior judgment has not been sufficient in the face of innovation. McFarland (1990, 1991) conceives ethics as a practice in which there is a prior judgment intended to establish what must be done to carry out an action. This conception of ethics does not follow the dynamics of innovation in AI, proving to be inconsistent, because, once a system is designed under established ethical principles, new technologies will make ethical application somewhat fragile.

When it comes to ethics and innovation, Cabella and Monte-Serrat (2022) propose strategies for applying ethical principles that are sustainable, i.e., still valid today and looking forward to the future AI innovations. The authors propose that organizations, companies, or any activity related to data processing administer an element of dynamic review at each change in the context of innovation, to guarantee ethical consistency of their products and services. In this way, they (Cabella & Monte-Serrat, 2022) suggest considering not only ethics conceived as practice (McFarland, 1990, 1991), but also ethics conceived as a responsible assessment of the decision making most appropriate to achieve the desired ends (Weber, 1959, 1998). The condensation of the two conceptions of ethics proposed by Cabella and Monte-Serrat (2022) should accompany the regulation and use of AI and robots by students at an educational institution, as we see in the next section.

6.5 Information Technologies in Education: Respect for People's Rights Related to Their Private Life, Personal Data, and Freedom

There is a certain excitement about the advantages of using robots and artificial intelligence in education with the introduction of new tools in the educational environment. These tools simulate human interactions and generate human-like texts with the purpose of enhancing teaching and learning processes. What is most found in published scientific articles are comments on the ethical challenges aimed at the practical and superficial aspects of computational cognition. These publications are aimed at the benefits brought to students, setting aside the harm that the cognition of children and young people may suffer during interaction with intelligent systems. There are authors who advocate a revolution in education with AI, claiming that generative artificial intelligence models and their potential bring advancement to the field of education, improving the efficiency and quality of students' work (Adıgüzel et al. 2023; Kenwright, 2024). Other authors preach that artificial intelligence plays a role in research, enabling independent learning and freeing researchers from the tedious tasks of searching for the information they need. This excitement is accompanied by the affirmation that we live in an era of speed and transparency of technological progress (Kenchakkanavar, 2023).

To this enhanced use of artificial intelligence in education we can add the Covid-19 pandemic. The Brazilian educational system was oriented towards implementation of exceptional standards in education during the state of calamity recognized as Covid-19 pandemic (Brasil, 2020). Since 2020, the tools for the use of computer-mediated communication and computer conferencing to support educational experiences have intensified due to the Covid-19 pandemic. Ministries of Education all over the world have been determining institutions and educators to adopt the use of computer conferencing as a means to facilitate an educational transaction: in the reorganization of learning environments to accommodate technologies, in the performance of online activities synchronous and asynchronous, in online assessment activities, in the use of far-reaching social media (WhatsApp, Facebook, Instagram, Tik Tok etc.) to guide students, in research and projects that integrate curriculum redesign.

Tavares et al. (2023) report on the use of surveillance technologies, especially facial and biometric recognition technologies, in Brazilian schools. The authors restrict themselves to commenting on the importance of surveillance technologies to optimize the management of the school environment by controlling student absences, combating school dropout, and promoting safety. Patriarca (2023) investigates reports that there were failures in the mass installation, authorized by the Department of Education of São Paulo, Brazil in agreement with Google, of the app, without authorization from users. Teachers questioned how their personal data was accessed when the "My School" application was installed on their private cell phone. After the complaints, the Department of Education initiated an administrative process, concluding that a "failure occurred during a test promoted by the department's technical area" and subsequently requested Google process reversal.

Mustać (2024) in turn, suggests the implementation of the Fundamental Rights Impact Assessments FRIAS) for the ethical control of AI over individuals, aimed not only at suppliers of AI systems but also at their implementers, such as public bodies that implement AI systems. Although the author considers the serious consequences for the lives of individuals resulting from the use of AI—by public or private bodies (such as service providers in the areas of education, social services, health, justice, etc.)—the final result in the form of notification, it can turn FRIAs into mere form-filling rather than effective protection of fundamental rights.

Besides those experiences, many innovative and technology-friendly education institutions already adopted other solutions to optimize accountability and day to day activities. For example, parents receive text messages if classes start and their child is not in the classroom, and using the school's smart watch, students can pay for meals at school and have their geolocation monitored (Oliveira, 2021; Revista Segurança Eletrônica, 2022). The initial purpose for adopting cutting-edge technology may be good; however, it is crucial to assess the privacy impact the collection and processing of personal data in such scenarios may have.

Monte-Serrat and Cattani (2023a) stress that education is associated with the political responsibility of specialized institutions, which must mediate the positive and negative aspects of the use of artificial intelligence in education, precisely because of the values at stake. Artificial intelligence and robots cannot be understood as something that is detached from the ethical debate about their effects on the individual's volition and privacy, whether at school or outside of it.

ISO 29134 (2023), the international standard for privacy impact assessment, states that privacy harms can include financial, reputational, and physical harms, as well as embarrassment and invasion of domestic life. To avoid those, it says that the organization responsible for the personal data processing activities should previously assess the potential privacy impact considering different types of privacy such as location, space privacy, and bodily privacy, as well as privacy of communications, privacy of data and image, behavioral privacy, privacy of association, and privacy of feelings and thoughts. It should also consider the privacy principles from ISO/IEC 29100 (2024), which are: data minimization, purpose, consent and choice, legitimacy and specification, collection limitation, use, retention and disclosure limitation, openness, transparency and notice, as well as accountability, accuracy and quality, individual participation and access, information security and privacy compliance.

The privacy risks are closely related to the fact that the use of technology has reached the subjectivity of students (Monte-Serrat & Cattani, 2020) causing discomfort because the body is outside the real context of a classroom, generating an effort on the part of students to project themselves socially and emotionally as people, negotiating with the body, to mold it to an imaginary context, not directly experienced.

Although the issue of data privacy remains extremely critical and worrying, little has been done to make data protection effective. There is a recognition that the protection of data collected by teaching platforms has become a concern, since these data represent the person themselves and involve obligations for those responsible for data processing and rights for their holders (Doneda, 2010). The data represent attributes of the identified person, maintaining a concrete link with the person who is the holder of these data, and their protection is a fundamental right to the protection of human dignity (Doneda, 2010).

The discussion of private rights involves ethical issues. The conception of data as being the person himself recognizes their ethical value, recognizes the right to psychophysical integrity, recognizes the free will of individuals (Magrani, 2018). It is observed that the adherence of individuals to technology in their daily tasks leads to a behavioral intervention, which is not desired by the user (Zuboff, 2019), and to the possible use of data by the wrong hands or by criminal hackers (Pinheiro, 2018).

Cavoukian (2009) proposes that there be balance about practices involving AI, privacy and data protection, suggesting a "win-win," "positive sum" stance in relation to everyone involved. Srour (2016), when referring to business and ethics, clarifies that making ethical decisions is respecting the interest of others, which requires equity and equivalence, suggesting that the parties agree to an exchange in which there is no abuse. It cannot be denied that there is a fine line between the protection of isolated data (information encoded at the syntactic level) and the content of the information obtained through this data (semantic level), providing the collected data with an economic value, which makes more evident the need to protect the ownership of this data (Pinheiro, 2018).

6.5.1 Critical Reflection on the Use of Technology in Education and People's Rights Regarding Their Private Life, Personal Data, and Freedom

The use of intelligent systems in educational practices, both by the institution and by students, is not restricted to the learning of certain subjects, but involves aspects of educational policies on the part of the State, social justice in which the law provides for equality and non-discrimination, and sustainability of actions carried out in the present, so that the future of students is not compromised or neglected. These topics are all implied in the rights of use and exploitation related to the information collected from students of a given educational institution, which must have limits on the use and dissemination of the data it holds, not only within legal parameters, but also within ethical parameters. This combination of law and ethics makes actions sustainable, guaranteeing the protection of students' rights regarding their use of artificial intelligence and robots during the learning process, so that this protection does not become obsolete.

It is essential to raise awareness, on the part of educators and educational institutions, about the guarantees of people's rights in relation to the protection of privacy, personal data, and freedom, to act in accordance with the rules and the ethical principles. And more than that: These legal rules and ethical principles must keep up with the changes that technology imposes to avoid, for example, data processing mechanisms provoke discrimination in facial recognition systems, or that the algorithms of the intelligent system treat underrepresented people differently (Luca, 2020). This awareness is not restricted to legal aspects. Research on the brain-computer interface shows that the actions of individuals can even be directed by the machine, under manipulation of brain mechanisms underlying the emotions, intentions, and decisions of the individual in the interaction with the world, giving rise to discussions about the ethics of neurotechnology and machine intelligence (Yuste et al. 2017).

6.6 Methodology and Discussion on the Complex Relationship Between Ethics and Digital Privacy

The method used in this chapter is qualitative analysis (Pêcheux, 1975; Foucault, 1969) in which theory and context are brought together to observe what underlies the collection of data from users of intelligent education systems. Only content analysis or exposition of ideas on the developed theme is not capable of highlighting the relationship between individual and technological system in the way intended in this chapter. It is understood that content analysis is affiliated with the positivist paradigm, composed of techniques that lead to the partial determination of the conditions of production of a given event, imprisoning the analyst within a theoretical model by privileging what is general (Monte-Serrat, 2014; Serrat, 2013).

To develop the main idea of this chapter, we used the contrast between theory about technological devices (which approach their use, without dealing with subjective aspects) and knowledge of the data that is not very evident (this was done by relating the environment from which the data were collected). This method is called the index paradigm, developed by Ginzburg (2013), which is supported by the knowledge coming from the individual related to the environment of the reality they experience, i.e., the individual under analysis is not detached from the context in which the events were produced (Monte-Serrat, 2014). This methodology is recommended by the French Discourse Analysis theory (Pêcheux, 1975; Foucault, 1969), favoring the investigation of contextual reality since the use of data permeates the individual himself. This methodology meets the content of this chapter precisely in the aspect that a person's data is considered as the person himself (Doneda, 2010).

Analysis of the data related to its environment and the discursive theory (Pêcheux, 1975; Foucault, 1969) go together in such a way that data collection and subjectivity are strictly related to the educational context (Monte-Serrat, 2013). Instead of performing a content analysis focusing only on what authors write about their ideas, a discursive analysis is practiced. In this way it is possible to escape from an idealized or theoretical circularity so that it is possible to relate the facts to their context. The discursive theory emphasizes that meanings emerge from the existing relationship, and this also occurs in the socio-historical educational process that began to adopt technology in teaching activities (Monte-Serrat, 2013).

The link between human cognition and artificial intelligence is shown by Monte-Serrat and Cattani (2021b, 2023b) raising ethical questions about the use of intelligent systems. Language tools and robots must align with data protection and privacy laws because their users may have their choices affected or their ability to make decisions impaired. Despite promoting AI and robots in education as a technological revolution, caution must be exercised because their use cannot be considered a neutral zone in terms of protecting the rights and privacy of AI and robot users.

Considering that the individual merges with the data collected by the technology used in the educational system, it is imperative to discuss ethical practices. Relating ethics—which focus on good or bad human attitudes and moral principles about what is right and what is wrong—to digital privacy—which concerns habits and tools that contain a person's information and must be kept safe—becomes complex when associated with the limits of technology investigated considering existing legal norms. The scope of norms regarding the human-intelligent system relationship is something very complex: Systems are always innovated, moving away from what the norms discipline, and it is to this complexity that organizations must be attentive to guide their own actions (Lawler, 2019), so that user data is collected responsibly, and organizations make correct use of this data. How to properly use the data collected if technological systems are incessantly innovated?

Monte-Serrat and Cattani (2023b) state that, when processing information, AI and robots also collect subjective human experiences, such as affection, sensitivity, morality, and intention. These data collected are part of the human cognitive linguistic process that condenses perception by mixing the subjectivity of the user of the intelligent system with the data collected by AI.

Who owns this data collected by an educational institution? How will they be reused? Is there responsibility on the part of the educational institution for reusing this data? The information collected from students at the educational institution could, for example, give the institution the "power" to interfere in the student's choices without the student realizing it (Monte-Serrat & Cattani, 2023b). The collected data is precious raw material for developers of intelligent systems. Can the institution sell this collected data and earn profits from it? From another perspective, intelligent user-system interaction (like a language tool used for academic work) can lead to the suffocation of human creativity (Monte-Serrat et al., 2020), as these tools offer the user a pre-conceived logic for their decision process, inhibiting new choices (Monte-Serrat and Cattani (2023b).

Ethical principles link human actions to values that guide these actions and the consequent interpretation of phenomena. This is a linguistic-cognitive activity that must balance the application of laws, principles, and practices in the personal data processing scenario (Cabella & Monte-Serrat, 2022). The ethical decision for an action or omission must take into account the facts, who is involved in them, the rule to be applied, the problem and the corresponding solution, the evaluation of what is at stake, the evaluation of costs and benefits, the guarantee balance between the parties involved, the assessment of the context, the extent of responsibility of the parties (McFarland, 1990, 1991; Cabella & Monte-Serrat, 2022).

Our qualitative analysis on ethics in education that makes use of technology raises the need to circumvent the incessant changes to which intelligent systems are subject. The moment a norm or ethical principle is applied to a given situation that violates the user's private rights, the technology changes and challenges these legal principles and norms. To solve this difficulty, we propose a new concept of ethics that is sustainable, applicable in the present, and extendable over time, solving future situations. The new concept of sustainable ethics applied to education that uses technology, in addition to considering ethical aspects, evaluates privacy, conceived as "a set of actions, behaviors, opinions and personal information over which the interested party intends to maintain control" related to the activities and situations of verbal and non-verbal communication that can be translated into information (Rodotà et al., 2008).

If, on the one hand, the Educational Institution promotes teaching through services and devices that promote connectivity, making use of sensors and processors that store data, on the other hand, there is the user of these services and devices, interacting with them and providing elements for this uninterrupted data storage (always recording, Magrani, 2018). This technological reality mixes the subjectivity of the system users, transforming their identities, images, and even their privacy into monetizable data, i.e., into a property nature product. This matter in question leads to investigate the loss of the subjective characteristic of the information obtained by the Educational Institution, since this information is transformed, through processes and systems, into something collective aimed at extracting knowledge, value, and perceptions (Magrani, 2018).

6.7 Sustainable Ethics in the Educational System

It is curious and intriguing, at the same time, that most authors of scientific investigations about artificial intelligence and education only look in the direction of adopting technology as something convenient and useful, leaving aside the issue of protecting data collected from students as users of educational systems. Negi et al. (2011) only mention the changes brought about by new technologies and their interference in daily work, in people's entertainment and in the way education is managed, posing challenges to the traditional teaching and learning process. The authors mention the ease of communication and instantaneous access to a wide range of data as something that challenges content assimilation and evaluation, suggesting the need for constant verification of the learning process itself. Another work, written by Muratuly et al. (2022), in turn, analyzes the practice of biometric authentication of students to control the learning process in online education of students in higher education institutions, without referring to the issue of collecting and protecting user data. The authors focus on research into the use of turnstile associated with facial recognition in Kazakhstan. The process of proving and verifying the identity of the person is associated with the biometric image of the user of a digital scanner. The authors suggest the use of this method in the field of education and in various industries, to base decision making on access to information resources related to factors that significantly affect the probability of correct recognition of students' faces.

Although chapters within this section—on pedagogical practices, language learning, and personal training—(Lampropoulos & Papadakis, 2025 [Chap. 2]; Luria, 2025 [Chap. 4]; Glushkova & Staribratov 2025 [Chap. 5]) as well as many other scientific works, such as those by Negi et al. (2011) and Muratuly et al. (2022), address the use and advantages that technology brings to education, in this chapter we prefer to direct the discussion on education that makes use of technology towards attitudes that combine security standards with privacy, focusing on the tension between innovation and privacy. It is noteworthy that this tension between innovation and privacy is not yet ensured by laws that discipline and protect the rights of individuals and their subjectivity, such as the Brazilian Consumer Code (Brasil, 2017), Civil Code (Brasil, 2002), General Law of Data Protection (Brasil, 2018) and the Brazilian Federal Constitution (Brazil, 1988).

Ethical principles are particularly linked to the subjectivity of the user of artificial intelligence and robots since the cognition of the user of an intelligent system reacts to the images or words shown on the device in use (Monte-Serrat & Cattani, 2023b). The qualities of intelligence, decision-making, and emotions of an individual are not disconnected from intelligent systems and need urgent protection even in educational institutions.

Cabella and Monte-Serrat (2022) propose the new concept of sustainable ethics as a framework that allows the simultaneous assessment of those elements at stake. This is in line with the Pinheiro's (2018) suggestion, about the need to analyze events considering data analysis, the internet of things (IoT), and Artificial Intelligence platforms under ethical and legal aspects in the light of security, liability, protection of personal data, ownership, and employability of that data. Pinheiro (2018) states that freedom and values consistent with respect for human rights must be guaranteed. Concerning how to guarantee the practical application of ethics allied to sustainability, Cabella and Monte-Serrat (2022) assume that:

> *In the organizational activity of judgment or decision making by design about the collection and use of personal data, it is assumed that ethics in privacy and data protection will only be sustainable if the values imposed by norms, rules, and principles are considered together with the context, characterized by the*

> *positive-sum principle [...], so that the organization does not give up its own interests and, at the same time, does not harm the others involved—on the contrary, it creates benefits for them also. Thus, technological development and innovation are made possible, generating value for those involved and mitigating risks in the present and future so that a decision today weighs the effects in the short, medium, and long term. Additionally, there may be changes in the initial project (such as the need to collect new personal data or assign new purposes of use, for example). In that case, the analysis and decision regarding the personal data processing activities must be reviewed to guarantee its sustainability, taking into account the diversity of elements in their dynamic relationships.* (Cabella & Monte-Serrat, 2022, p. 41)

Technology in education involves not only innovation itself, but also appreciation (values) and judgment (subjectivity). Therefore, it is proposed in this chapter that there should be no separation among all these elements. For this reason, it is suggested that the new concept of sustainable ethics is appropriate to the context under discussion, since it considers the topics at stake: the parties (who perform and who suffer the action, as well as the interested parties), the object (education and technology), and the dynamic relationship between all the elements (Cabella & Monte-Serrat, 2022). We demonstrated that there is an intertwining and a dynamic relationship in educational activity so that privacy and data protection go hand in hand with context, legislation, norms and principles, and the element of time (present and future), thus adapting to the concept of sustainability, which guarantees rights in the present and in the future. Cabella and Monte-Serrat (2022) propose that ethical principles will be ensured, and therefore, will acquire the characteristic of sustainability, when all these elements are appreciated simultaneously, under a dynamic relationship between ethics, time, and the parties involved, and not just a relationship with "the" "correct" ethical principle as foreseen in law to be applied to a given situation.

6.8 Conclusions

In this chapter, a new concept was proposed, capable of encompassing, at the same time, the complexity of technological innovation, data protection, and the application of ethical principles in education that makes use of intelligent systems. In particular, we explain the new concept of sustainable ethics to overcome the challenges of technological innovation and their interference in the subjectivity of individuals. Our proposal is for educational institutions to simultaneously appreciate all the elements that are at stake: the parties (who perform and who suffer the action, as well as the interested parties), the object (education and technology), and the dynamic relationship between all the elements.

To achieve ethical sustainability, an in-depth analysis was made of the elements that are present in education carried out based on computers and digital technology. Bibliographical survey with didactic books and scientific chapters was run to clarify that intellectual property and artificial intelligence cannot be taken for granted as separate things. We show that the internet has a border nature that defies regulations (in the sense that regulations are contradictory or absent) on the collected databases considered both as something of a property nature, as well as personal data.

Having presented several recent studies related to exploring the role of AI and/or robotics in education, we made a critical analysis of how the interference of artificial intelligence in the educational sector has been managed. The uncertainties generated by the use of intelligent systems on a border Internet cannot be considered taken for granted, which pose challenges to the certainty of the application of ethical principles, the protection of privacy, and the freedom of individuals. Our search for concrete solutions originates from fourth generation qualitative analysis, which is not satisfied with studies that cool and remain silent on the shelves. Against the current of silencing studies, we show in this chapter how cross-fertilization occurs between natural language and AI. These are theoretical foundations about the behavior of the human cognitive linguistic process, which can help in the search for solutions for ethical aspects in the use of AI in education.

We point out the need for special treatment, for new principles that tackle the dynamics of innovation, and that overcome the gap in static rules so that, in the end, ethical AI keeps up with changes of reality whose structural elements are inseparable. While authors suggest that future regulatory improvements be administered, we advocate a simultaneous assessment of laws, ethical principles, the context of reality, looking to the future due to incessant technological innovation. The concept of sustainable ethics is perfectly suited to the educational context that makes use of technology. In this way, it is possible to promote a balance between the rights of the subjects involved, protection of access and use of the data collected, respect for laws and ethical principles, in the present and in the future. The complex apparatus of sustainable ethics becomes an important agent to overcome the difficulty of lack of uniformity of laws and regulations that generates the loss of control over data security, offering the necessary protection to students who use technology in the educational process.

References

Adams, C., Pente, P., Lemermeyer, G., & Rockwell, G. (2023). Ethical principles for artificial intelligence in K-12 education. *Computers and Education: Artificial Intelligence, 4*, 100131. https://doi.org/10.1016/j.caeai.2023.100131

Adıgüzel, T., Kaya, M., & Cansu, F. (2023). *Revolutionizing education with AI: Exploring the transformative potential of ChatGPT. Contemporary educational technology*. Bastas.

Akgun, S., & Greenhow, C. (2022). Artificial intelligence in education: Addressing ethical challenges in K-12 settings. *AI and Ethics, 2*(3), 431–440. https://doi.org/10.1007/s43681-021-00096-7

Bartneck, C., Lütge, C., Wagner, A., & Welsh, S. (2021). *An introduction to ethics in robotics and AI* (p. 117). Springer Nature. https://doi.org/10.1007/978-3-030-51110-4

Berners-Lee, T. (1999) *Weaving the Web: The original design and ultimate destiny of the World Wide Web by its inventor*. Harper San Francisco.

Brasil. (2002). Civil Code. Brazilian civil code and related legislation.—2nd ed.—Brasilia: Federal Senate, Subsecretariat for Technical Editions, 2008, 616 p. Retrieved August 9, 2024, from https://www2.senado.leg.br/bdsf/bitstream/handle/id/70327/C%C3%B3digo%20Civil%202%20ed.pdf

Brasil. (2014). Law No. 12,965 of 04/23/2014. Civil Rights Framework for the Internet. Retrieved August 9, 2024, from https://legis.senado.leg.br/norma/584776

Brasil. (2017). Consumer protection code and related standards.—2nd ed.—Brasilia: Federal Senate, Coordination of Technical Editions, 2017. 132 p. Retrieved August 9, 2024, from https://www2.senado.leg.br/bdsf/bitstream/handle/id/533814/cdc_e_normas_correlatas_2ed.pdf

Brasil. (2018). Law No. 13,709, of August 14, 2018, provides for the protection of personal data and amends Law No. 12,965, of April 23, 2014 (Marco Civil da Internet). Retrieved August 9, 2024, from https://legis.senado.leg.br/norma/584776

Brasil. (2020). MEC Ministry of Education, National Education Council, Full Council. CNE/CP Resolution n.2, of December 10, 2020. Retrieved August 9, 2024, from http://portal.mec.gov.br/index.php?option=com_docman&view=download&alias=167141-rcp002-20&category_slug=dezembro-2020-pdf&Itemid=30192

Brazil. (1988). Constitution of the Federative Republic of Brazil. Brasília, DF: Presidency of the Republic. Retrieved August 9, 2024, from https://normas.leg.br/?urn=urn:lex:br:federal:constituicao:1988-10-05;1988

Brazil. (2018). Law No. 13,709 of 08/14/2018. General Data Protection Law (2018) (LGPD). Retrieved August 9, 2024, from https://legis.senado.leg.br/norma/27457334

Cabella, D., & Monte-Serrat, D. (2022). *Sustainable ethics in data processing decisions in the context of innovation* [electronic book]. Rio de Janeiro, Brasil. Retrieved August 9, 2024, from https://iapp.org/resources/article/sustainable-ethics-in-data-processing-decisions-in-the-context-of-innovation/

Campos, R. (2011). *Desafios da avaliação de programas e serviços em saúde: novas tendências e questões emergentes.* Campinas: Ed Unicamp 278 p.

Cavoukian, A. (2009). Privacy by design: The 7 foundational principles. *Information and privacy commissioner of Ontario, Canada, 5*, 12. Retrieved August 9, 2024, from https://student.cs.uwaterloo.ca/~cs492/papers/7foundationalprinciples_longer.pdf

Daniela, L., & Strods, R. (2019). Educational robotics for reducing early school leaving from the perspective of sustainable education. In *Smart learning with educational robotics: Using robots to scaffold learning outcomes* (pp. 43–61). https://doi.org/10.1007/978-3-030-19913-5_2

Dieterle, E., Dede, C., & Walker, M. (2024). The cyclical ethical effects of using artificial intelligence in education. *AI and Society, 39*(2), 633–643. https://doi.org/10.1007/s00146-022-01497-w

Doneda, D. (2010). A proteção de dados pessoais nas relações de consumo: para além da informação creditícia. *Escola Nacional de Defesa do Consumidor. Brasília: SDE/DPDC*, 22.

European Parliament. (2024). EU AI Act: first regulation on artificial intelligence. Retrieved August 9, 2024, from https://www.europarl.europa.eu/topics/en/article/20230601STO93804/eu-ai-act-first-regulation-on-artificial-intelligence

Fiok, K., Farahani, F., Karwowski, W., & Ahram, T. (2022). Explainable artificial intelligence for education and training. *The Journal of Defense Modeling and Simulation, 19*(2), 133–144. https://doi.org/10.1177/15485129211028651

Foucault, M. (1969). *L'Archeologie du savoir.* Gallimard.

Ginzburg, C. (2013). Signs: Roots of an evidentiary paradigm. In C. Ginzburg (Ed.), *Clues, myths, and the historical method.* JHU Press.

Glushkova, T., & Staribratov, I. (2025). Artificial intelligence and robotics in school education: Analysis of implementation of AI and robotics in Bulgarian education. In S. Papadakis & G. Lampropoulos (Eds.), *Social robots and artificial intelligence in education: Integrating AI in K-12 and higher education* (pp. 91–120). Palgrave Macmillan.

Gocen, A., & Aydemir, F. (2020). Artificial intelligence in education and schools. *Research on Education and Media, 12*(1), 13–21. https://doi.org/10.2478/rem-2020-0003

ISO/IEC 29100. (2024). Information technology—Security techniques—Privacy framework. Retrieved August 9, 2024, from https://www.iso.org/obp/ui/#iso:std:iso-iec:29100:ed-2:v1:en

ISO/IEC 29134. (2023). Information technology—Security techniques—Guidelines for privacy impact assessment. Retrieved August 9, 2024, from https://www.iso.org/standard/86012.html

Kamalov, F., Santandreu-Calonge, D., & Gurrib, I. (2023). New era of artificial intelligence in education: Towards a sustainable multifaceted revolution. *Sustainability, 15*(16), 12451. https://doi.org/10.3390/su151612451

Kenchakkanavar, A. (2023). Exploring the artificial intelligence tools: Realizing the advantages in education and research. *Journal of Advances in Library and Information Science, 12*(4), 218–224.

Kenwright, B. (2024). Is it the end of undergraduate dissertations? Exploring the advantages and challenges of generative ai models in education. In *Generative AI in teaching and learning* (pp. 46–65). IGI Global. https://doi.org/10.4018/979-8-3693-0074-9.ch003

Krzysztofek, M. (2018). *GDPR: General Data Protection Regulation (EU) 2016/679: Post-reform Personal Data Protection in the European Union*. Kluwer Law International BV.

Lampropoulos, G., & Papadakis, S. (2025). Integrating social robots into education. In S. Papadakis & G. Lampropoulos (Eds.), *Social robots and artificial intelligence in education: Integrating AI in K-12 and higher education* (pp. 21–35). Palgrave Macmillan.

Lawler, B. (2019, Jan 28). Five global trends in data ethics and privacy in 2019. In *Looker Blog*.

Lin, P., Abney, K., & Bekey, G. (Eds.). (2014). *Robot ethics: The ethical and social implications of robotics*. MIT Press.

Luca, C. (2020) Demissão no Google joga luz sobre a IA Ética, 2020. In *UOL Tilt*. Retrieved August 8, 2024, from https://www.uol.com.br/tilt/colunas/cristina-de-luca/2020/12/05/demissao-no-google-joga-luz-sobre-a-ia-etica.htm

Luria, E. (2025). Revolutionizing English language learning through social robots and artificial intelligence: Towards inclusive and personalized education. In S. Papadakis & G. Lampropoulos (Eds.), *Social robots and artificial intelligence in education: Integrating AI in K-12 and higher education* (pp. 61–90). Palgrave Macmillan.

Magrani, E. (2018). *Entre dados e robôs: ética e privacidade na era da hiperconectividade*. Konrad Adenauer Stiftung.

McFarland, M. (1990). Standards-urgency of ethical standards intensifies in computer community. In *Computer*, 23. Jg., Nr. 3, S. 77–81.

McFarland, M. (1991). Standards-ethics and the safety of computer systems. In *Computer*, 24(2), 72-75.

Meghdari, A., & Alemi, M. (2018). Recent advances in social and cognitive robotics and imminent ethical challenges. In *Recent advances in social and cognitive robotics and imminent ethical challenges (August 22, 2018)*. Proceedings of the 10th International RAIS Conference on Social Sciences and Humanities. https://doi.org/10.2991/rais-18.2018.12

Monte-Serrat, D. (2014). *A questão do sujeito: perspectivas da Análise do Discurso, do Letramento e da Psicanálise Lacaniana*. Pedro e João Editores.

Monte-Serrat, D., & Cattani, C. (2020). Teaching in online classes: Social Presence from the Perspective of Michel Foucault and René Magritte. *People*, *7*, 94. Retrieved August 8, 2024, from https://www.naturalspublishing.com/Article.asp?ArtcID=28515

Monte-Serrat, D., & Cattani, C. (2021a). Interpretability in neural networks towards universal consistency. *International Journal of Cognitive Computing in Engineering*, *2*, 30–39.

Monte-Serrat, D., & Cattani, C. (2021b). *The natural language for artificial intelligence* (p. 233). Elsevier-Academic Press.

Monte-Serrat, D., & Cattani, C. (2023a). Artificial intelligence and scientific research: Values at stake in education. In *Artificial intelligence in higher education and scientific research: Future development* (pp. 1–13). Springer Nature Singapore. https://doi.org/10.1007/978-981-19-8641-3_1

Monte-Serrat, D., & Cattani, C. (2023b). Towards ethical AI: Mathematics influences human behavior. *Journal of Humanistic Mathematics, 13*(2), 469–493. https://doi.org/10.5642/jhummath.TGSX1558

Monte-Serrat, D., Cabella, B., & Cattani, C. (2020). The Schrödinger's cat paradox in the mind creative process. *Information Sciences Letters, 9*(3), 1–10. https://doi.org/10.18576/isl/090301

Muratuly, D., Denissova, N., Krak, Y., & Apayev, K. (2022). Biometric authentication of students to control the learning process in online education. *Scientific Journal of Astana IT University, 10*(10), 22–32.

Mustać, T. (2024, August). The AI Act Series: Fundamental rights impact assessments. In *Medium AI advances.* Retrieved August 12, 2024, from https://ai.gopubby.com/the-ai-act-series-fundamental-rights-impact-assessments-b1de3855f8cc

Negi, P., Negi, V., & Pandey, A. (2011). Impact of information technology on learning, teaching and human resource management in educational sector. *International Journal of Computer Science and Telecommunications, 2*(4), 66–72.

Nguyen, A., Ngo, H., Hong, Y., Dang, B., & Nguyen, B. (2023). Ethical principles for artificial intelligence in education. *Education and Information Technologies, 28*(4), 4221–4241. https://doi.org/10.1007/s10639-022-11316-w

Oliveira, M. (2021). Reconhecimento facial em escolas goianas enviará SMS aos pais se aluno faltar aula. Radio Bandeirantes. Retrieved August 8, 2024, from https://radiobandeirantesgoiania.com.br/reconhecimento-facial-em-escolas-goianas-enviara-sms-aos-pais-se-aluno-faltar-aula/

Patriarca, P. (2023). After installing the app en masse without authorization, the SP Education Department says it will investigate the case and cites 'failure'. G1 São Paulo. Published on 08/09/2023 6:14 pm Retrieved August 8, 2024, from https://g1-globo-com.cdn.ampproject.org/c/s/g1.globo.com/google/amp/sp/sao-paulo/noticia/2023/08/09/apos-instalar-app-em-massa-sem-autorizacao-secretaria-da-educacao-de-sp-diz-que-vai-apurar-caso-e-cita-falha.ghtml

Pêcheux, M. (1975). *Les vérités de La Palice. Linguistique, sémantique, philosophie (Théorie)*. Maspero.

Pedro, F., Subosa, M., Rivas, A., Valverde, P. (2019). Artificial intelligence in education: Challenges and opportunities for sustainable development. Peru: Ministerio de Educación. Retrieved August 13, 2024, from https://repositorio.minedu.gob.pe/handle/20.500.12799/6533

Pinheiro, P. (2018). *O direito internacional da propriedade intelectual aplicado à inteligência artificial.* Tese de Doutorado, Programa de Pós-Graduação em Direito, Direito Internacional e Direito Comparado, Universidade de São Paulo, p. 334. https://doi.org/10.11606/T.2.2018.tde-23102020-005146

Rajput, V. (2024). Prompt Engineering Is Dead: DSPy Is New Paradigm For Prompting.Medium May 29, 2024. Retrieved August 12, 2024, from https://medium.com/aiguys/prompt-engineering-is-dead-dspy-is-new-paradigm-for-prompting-c80ba3fc4896

Remian, Dana, (2019). Augmenting education: Ethical considerations for incorporating artificial intelligence in education. Instructional Design Capstones Collection. 52. https://scholarworks.umb.edu/instruction_capstone/52

Revista Segurança Eletrônica. (2022). Pulseira inteligente Schood pode ser aliada para escolas voltarem às aulas presenciais em segurança. Retrieved August 8, 2024, from https://revistasegurancaeletronica.com.br/pulseira-inteligente-schood-pode-ser-aliada-para-escolas-voltarem-as-aulas-presenciais-em-seguranca/

Rifkin, J. (2010). *La civilización empática.* Paid.

Rodotà, S., Moraes, M., Doneda, D., Doneda, L. (2008). A vida na sociedade de vigilância: a privacidade hoje. Rio de Janeiro: Renovar, *381 p.* Monography em Pt | SES-SP, SESSP-ISACERVO | ID: biblio-1081769. Biblioteca responsável: BR1764.1. Localização: BR1764.1; 342.1, R695v.

Saide, S., & Sheng, M. (2021). Knowledge exploration–exploitation and information technology: crisis management of teaching–learning scenario in the COVID-19 outbreak. *Technology Analysis & Strategic Management, 33*(8), 927–942. https://doi.org/10.1080/09537325.2020.1854714

Serholt, S., Barendregt, W. (2014). Students' attitudes towards the possible future of social robots in education. August. In *Workshop proceedings of RO-MAN.*

Serholt, S., Barendregt, W., Vasalou, A., Alves-Oliveira, P., Jones, A., Petisca, S., & Paiva, A. (2017). The case of classroom robots: Teachers' deliberations on the ethical tensions. *AI and Society, 32*, 613–631. https://doi.org/10.1007/s00146-016-0667-2

Serrat, M. M. D. (2013). *Letramento e discurso jurídico* (Doctoral dissertation, Universidade de São Paulo). https://doi.org/10.11606/T.59.2013.tde-14032013-104350

Sijing, L., & Lan, W. (2018). Artificial intelligence education ethical problems and solutions. In *2018 13th International Conference on Computer Science & Education (ICCSE)* (pp. 1–5). IEEE. https://doi.org/10.1109/ICCSE.2018.8468773

Silva, R., & Brandão, D (2011). Nas rodas da avaliação educadora, In Campos (Ed.), *Desafios da avaliação de programas e serviços em saúde: novas tendências e questões emergentes.* Ed. Unicamp.

Srour, R. H. (2016). *Decisões Éticas nas Empresas: como e por que adotar.* Alta Books Editora.

Tarrés-Puertas, M., Costa, V., Pedreira Alvarez, M., Lemkow-Tovias, G., Rossell, J., & Dorado, A. (2023). Child-robot interactions using educational robots: An ethical and inclusive perspective. *Sensors, 23*(3), 1675. https://doi.org/10.3390/s23031675

Tavares, C., Simão, B., & Vilanova, V. (2023). *Surveillance and education technologies: A mapping of facial recognition policies in Brazilian public schools.* InternetLab. https://internetlab.org.br/wp-content/uploads/2023/06/Educacao-na-mira-PT_06.pdf

Tolksdorf, N., Siebert, S., Zorn, I., Horwath, I., & Rohlfing, K. J. (2021). Ethical considerations of applying robots in kindergarten settings: Towards an approach from a macroperspective. *International Journal of Social Robotics, 13*, 129–140. https://doi.org/10.1007/s12369-020-00622-3

Varma, R. R., Umesh, I. M., Nagesh, Y. N., & Kumara Swamy, K. S. (2021). Digital transformation in higher education institutions—An overview. *International Journal of Applied Engineering Research, 16*(4), 278–282.

Weber, M. (1959). *Le savant et le politique* (pp. 166–185). Union Générale d'Éditions.

Weber, M. (1998). *Ciência e política: duas vocações.* Cultrix.

Yuste, R., Goering, S., Bi, G., Carmena, J., Carter, A., Fins, J., et al. (2017). Four ethical priorities for neurotechnologies and AI. *Nature News, 551*(7679), 159. https://doi.org/10.1038/551159a

Zuboff, S. (2019). *The age of surveillance capitalism: The fight for a human future at the new frontier of power: Barack Obama's books of 2019*. Profile Books.

CHAPTER 7

The Emerging Role of AI and Social Robots in University Student Recruitment

Joseph F. Hair Jr, Sarah Rana, Denise Robb, and Mark Stroud

7.1 Introduction

Birth rates have been declining in the United States (U.S.) since approximately 2007 (Osterman et al., 2023), resulting in a smaller pool of eligible college applicants. Confronted with this looming demographic cliff, higher education institutions must consider implementing practical tools to maintain stable enrollment. An unexplored strategy is incorporating the use of artificial intelligence (AI) and social robots into the recruitment processes. Notably, Lampropoulos and Papadakis (2025 [Chap. 2]) provide a detailed account of how social robots, in general, can be integrated

J. F. Hair Jr. • S. Rana (✉) • D. Robb
Mitchell College of Business, University of South Alabama, Mobile, AL, USA
e-mail: jhair@southalabama.edu; Sr2323@jagmail.southalabama.edu; dnr2321@jagmail.southalabama.edu

M. Stroud
Chazanoff School of Business, College of Staten Island, Staten Island, NY, USA
e-mail: mark.stroud@csi.cuny.edu

S. Papadakis, G. Lampropoulos (eds.), *Social Robots and Artificial Intelligence in Education*, Digital Education and Learning,
https://doi.org/10.1007/978-3-032-03977-4_7

into education. The current study describes artificial intelligence as machines that can process, analyze, learn, and interpret information (e.g., Mikalef & Gupta, 2021; Sheikh et al., 2023). In comparison, social robots often maintain a human or animal-like appearance to interact with humans emotionally (Campa, 2016). Members of Generation Z, traditionally the main group of interest for colleges and universities, have only known a world interconnected by the internet and they use technology to actively participate in society (Szymkowiak et al., 2021). Generation Z is particularly well positioned, therefore, to be targeted by the integration of AI into higher education enrollment. Moreover, colleges and universities are likely to benefit from the adoption of AI-supported enrollment strategies (Dempere et al., 2023). With an estimated global economic impact of 13 trillion dollars over the next 10 years (Fountaine et al., 2019), it is advantageous for educational institutions to consider how to implement this technology successfully.

AI has been studied through the lens of various disciplines in the medical and social sciences. For example, scholars have examined AI in the fields of hospitality and travel (Gaur et al., 2021), medicine (Miller & Brown, 2018), and human resources (Tambe et al., 2019). As the preceding chapters in this text demonstrate, the research on AI and education is also on the rise. For example, Arruda and Arruda (2025 [Chap. 3]) provide a pedagogical prospect for using AI to offset teacher shortages. Similarly, Luria (2025 [Chap. 4]) explores the ability to create personalized education programs through the use of social robots and AI technology. Within the context of higher education, most AI research has focused on issues related to academic integrity, such as cheating and plagiarism (Cotton et al., 2024). Yeadon et al. (2023) posits that AI technology threatens to undermine the legitimacy of student essays as a viable assessment mechanism. Rudolph et al. (2023) recommended the integration of AI tools in the curriculum and call for faculty training, as well as the establishment of clear ethical guidelines on the usage of AI in the classroom. Currently, the impact of AI on higher education enrollment has yet to be robustly examined. An opportunity exists, therefore, to explore AI and social robots to improve the recruiting processes at colleges and universities.

Despite the immense practical applications of AI and social robots, research emphasizing theoretical foundations to support these applications is lacking, particularly among high school students. The goal of this investigation is to explore a theoretical perspective on how AI can improve the rapidly changing landscape of higher education. Building on Hobfoll's

(1989) conservation of resources theory (COR), we postulate that individuals value resources, such as knowledge, money, and time. Moreover, we contend that constraints on these resources induce stress and students who encounter excessive demands on these resources may experience emotional exhaustion, cynicism, and reduced efficacy, which are elements of burnout (Maslach & Leiter, 2017). Burnout, traditionally studied within the context of employees, often leads to several negative results, including high turnover and absenteeism (e.g., Maslach, 2003; Wright & Cropanzano, 1998). Examining this paradigm from a higher education viewpoint, we posit that high school students who have limited time and energy will benefit from the use of AI and social robots, which can provide faster and more personalized assistance with the college enrollment process. The proposed theoretical framework explored in our research is displayed on Fig. 7.1.

This research makes several contributions to the literature. First, our work seeks to extend Hobfoll's (1989) conservation of resources theory (COR) to a new context by analyzing resource gains through the adoption of technology. Specifically, we focus on how AI tools can be utilized by colleges and universities to reduce resource constraints, such as time and energy of enrollment support personnel, and therefore increase enrollment intentions among potential students. Second, while the literature is replete with empirical studies on burnout, most research has concentrated on people-oriented occupations where the employee is providing a service to a customer (e.g., social workers and health care professionals; Maslach & Schaufeli, 2018). In contrast, high school students, whose primary

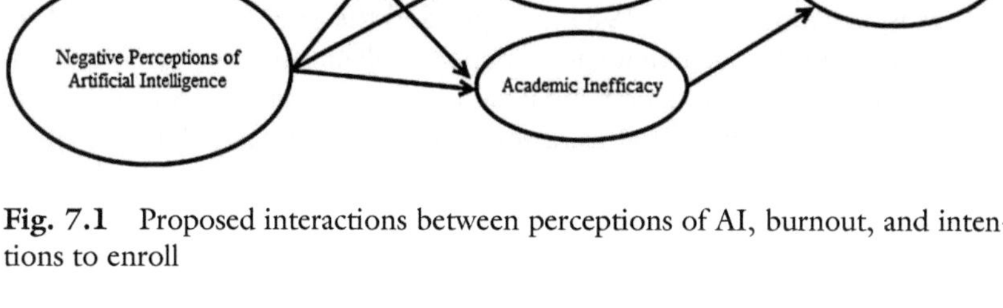

Fig. 7.1 Proposed interactions between perceptions of AI, burnout, and intentions to enroll

occupation is to be a student, have not been the focus of burnout research and little is known about the development or impact of burnout in high school student populations. Consequently, we advance the literature on burnout to include high school students as a focal group. From a business perspective, academic institutions must be concerned with the potential for student burnout contributing to lower enrollment, as this would lead to reduced tuition revenue. Third, while some organizations are making large investments in AI, little is known about high school students' perceptions of the technology and its impact on them. Finally, we will examine high school students' perceptions of the college enrollment process and identify practical implications for colleges and universities so that they can make informed decisions leading to increased enrollment.

The remainder of this chapter will first provide theoretical justifications for our propositions, including a brief discussion on artificial intelligence. Second, we present our methodology for the data collection process and measures. Third, our findings and interpretations are discussed. Finally, we conclude with theoretical and practical implications, limitations, and future research opportunities. Specifically, we focus our future research opportunities on social robots and higher education.

7.2 High School Students Burnout and College Enrollment

7.2.1 *Overview of the Conservation of Resource Theory*

Researchers have long explored the impact of and individual responses to stress from both a personal and organizational outcome perspective (e.g., Boyd et al., 2009; Selye, 1956). Additionally, scholars have focused on the implications of prolonged stress, such as burnout (Freudenberger, 1974). Introduced by Freudenberger in 1974, the term burnout became popularized in the literature as a phenomenon that occurs when an individual experiences excess demands on resources and energies. To measure burnout, Maslach and Jackson (1981) categorized the phenomenon into three symptoms: emotional exhaustion, depersonalization, and reduced personal accomplishment. Depersonalization and personal accomplishment are also known in the literature as cynicism and efficacy (Halbesleben & Buckley, 2004).

COR, a predominant theory utilized in the stress literature to explain burnout (e.g., Halbesleben, 2006), is grounded in the premise that individuals are motivated to obtain, retain, and protect resources (Hobfoll, 1989). Scholars have defined resources as energies (e.g., time, knowledge, money), objects, conditions, and personal characteristics (Hobfoll, 1989). COR has two primary principles: the primacy of resource loss and resource investment. Halbesleben et al. (2014) conclude the first principle of COR, the primacy of resource loss, suggests individuals experience more psychological harm from resource loss. Moreover, while individuals can regain lost resources, regaining these lost resources is less challenging and impactful than the initial acquisition of these resources.

Furthermore, resource depletion is associated with negative implications such as physical or behavioral symptoms, such as sleeplessness and irritability (Freudenberger, 1974). The second principle of COR, resource investment, proposes individuals invest resources to safeguard against resource loss, recover from losses, and gain resources (Hobfoll, 2001). From these two primary principles, four related corollaries have emerged: (1) individuals with more resources are better able to gain resources, and those with fewer resources are more likely to experience resource loss; (2) initial resource loss results in future loss; (3) initial resource gains lead to future gains; (4) diminished resources lead individuals to enact defensive attempts to preserve remaining resources (Halbesleben et al., 2014).

7.2.2 *Resource Loss Implications*

In the current study, we rely heavily on the first principle of COR, which is the primacy of resource loss. As mentioned earlier, this principle suggests resource loss has a more substantial impact than resource gain (Halbesleben et al., 2014). For example, high school students in the U.S. invest a significant amount of time and energy into college placement exams (e.g., ACT, SAT), extracurricular activities, taking advanced placement (AP) courses, and navigating the college application process. These activities would be considered a resource loss (time), as they deplete that individual's finite resources. Existing research on the impact of resource loss is widespread, but it primarily centers on the experiences of individuals above the age of 18 (e.g., employees facing resource loss). Multiple negative potential implications have been identified in organizational settings, such as burnout and turnover intentions (e.g., Alarcon, 2011). There is limited research, however, on the impact of resource loss in populations below the age of 18. We believe this is primarily due to the difficulty in accessing these individuals to conduct research.

Previous empirical studies have found resource loss contributes to emotional exhaustion. For instance, Halbesleben et al. (2013) found that furloughs, symbolizing resource loss, contributed to emotional exhaustion. Similarly, we believe high school students are experiencing resource loss through the demands of the college application process, particularly those applying to multiple schools, and are likely to report feelings of emotional exhaustion. As one of the three symptoms of burnout, emotional exhaustion occurs when an individual's emotional resources are depleted. Subsequently, an individual will develop cynicism or negative attitudes towards others (Maslach & Jackson, 1981). Consequently, students can develop a reduced sense of personal accomplishment (Maslach & Jackson, 1981) or the perception they cannot perform well (Halbesleben & Buckley, 2004). Notably, Fiorilli et al. (2020) highlight studies, like Bask and Salmela-Aro (2013), which have investigated students' burnout levels and their risk for dropping out of school. However, there remains a general lack of research focusing specifically on the high school population and the impact of perceptions of AI on burnout. Therefore, we propose two overarching research questions to frame the current study:

Research Question 1: Is there a relationship between perceptions of AI use and burnout in high school students?

Research Question 2: Is there a relationship between burnout in high school students and the likelihood of enrolling in college?

7.2.3 The Role of Artificial Intelligence in Education

In conjunction with seeking to understand the prevalence of burnout in students under the age of 18 in the U.S., Leiter and Maslach (2018) argue that, despite the various recommendations for reducing burnout, few studies have evaluated the effectiveness of strategies to reduce burnout. According to the COR theory (Hobfoll, 1989), stress variability is related to resource accessibility. Given the multitude of AI-powered educational tools becoming more readily accessible to students (e.g., Grammarly—writing assistant, ChatGPT—chatbot), research investigating how AI could minimize resource loss (e.g., time or energy) is justified. Furthermore, AI perceptions could potentially serve as a predictor of student burnout. The current study measures high school students' perceptions of AI. To do so, we apply established technology acceptance models (e.g., Davis's 1989 Technology Acceptance Model), and examine attitudes toward technology as antecedents of potential use.

Despite AI becoming more accessible and assimilated into our daily lives, some individuals remain fearful of this technology (Cugurullo & Acheampong, 2024). In fact, according to popular press reports (Leffer, 2023) the fear of AI is rising. We contend, however, that current high school students, also known as Generation Z or "digital natives" (Bhalla et al., 2021), are considerably more acclimated to technological advances, like AI, given their exposure to technology at a younger age. Notably, many individuals of Generation Z, born between 1997 and 2012, do not remember a time before the use of AI. For example, Siri, an AI-powered voice assistant feature on the iPhone, was incorporated into iOS in 2011 (Hoy, 2018). Thus, current high school students would have had access to tools that leveraged AI as adolescents and are likely to view educational AI tools as resources that can reduce academic stress instead of exhibiting fearful perceptions towards AI.

There is limited knowledge on the perceptions of AI and students experiencing burnout. For the purpose of this study, we examine negative and positive perceptions of AI separately to determine if there is a relationship between the type of perceptions held and experiencing burnout. Therefore, the present study seeks to investigate an overarching research question regarding whether perceptions of AI, whether positive or negative, are related to burnout symptoms. Additionally, we seek to explore whether there is a relationship between each of three components of burnout (emotional exhaustion, cynicism, and academic inefficacy) and intentions to enroll at colleges and universities.

7.2.4 *What Is Artificial Intelligence*

Since Alan Turing posed the legendary 1950's question, "Can machines think?" (Turing, 2009, p. 433), AI has rapidly advanced and increased in popularity (Bahoo et al., 2023; Yigitcanlar & Cugurullo, 2020). Despite a growing body of literature, scholars indicate no formal definition of AI exists (e.g., Duan et al., 2019; Sheikh et al., 2023). Sheikh et al. (2023) note, however, that the definition provided by the High-Level Expert Group on Artificial Intelligence (AI HLEG) of the European Commission contains an optimal scope, describing AI as "systems that display intelligent behaviour by analysing their environment and taking actions—with some degree of autonomy—to achieve specific goals" (p. 20). Despite the plethora of definitions available, Mikalef and Gupta (2021) note that a common theme in describing AI is a machine seeking to reproduce human cognitive processing of information to interpret, learn, identify, or make inferences from the data.

Additionally, AI is categorized based on its capabilities (IBM, 2023), such as Artificial Narrow Intelligence (ANI), Artificial General Intelligence (AGI), or Artificial Super Intelligence (ASI) (Bahoo et al., 2023; IBM, 2023). Current technology only extends to ANI or weak AI, making AGI and ASI merely theoretical concepts (IBM, 2023). For a detailed flowchart of ANI, AGI, and ASI, refer to Fig. 7.2. Given high school students only have access to ANI, the remainder of our discussion will focus on AI with the capabilities of ANI.

ANI, the first generation of AI tools generally available to society, is characterized by its ability to perform a single task (Kaplan & Haenlein, 2019). Within the realm of ANI, the technology can be further categorized by functionality, such as limited memory AI or reactive machine AI (IBM, 2023). The key distinction between the two is that limited memory AI retains short-term data to improve performance over time, whereas reactive machine AI cannot recall previous data and can only use currently available data (IBM, 2023). Therefore, educational tools powered by AI for students are variations of ANI or weak AI.

7.2.5 *Artificial Intelligence's Relationship with Burnout*

The use of AI in education has been explored from an educator's perspective. Fitria (2021a) and Igbokwe (2023) believe teachers using AI will likely save time and energy by streamlining tasks. For students, innumerable benefits have been proposed when educators integrate AI into the classroom, such as providing personalized learning and virtual mentors (Fitria, 2021a). Moreover, as suggested by Biggs (1999, 2012), students' perceptions of their learning environments, personal skills and abilities, and instructional strategies have an impact on student outcomes. For

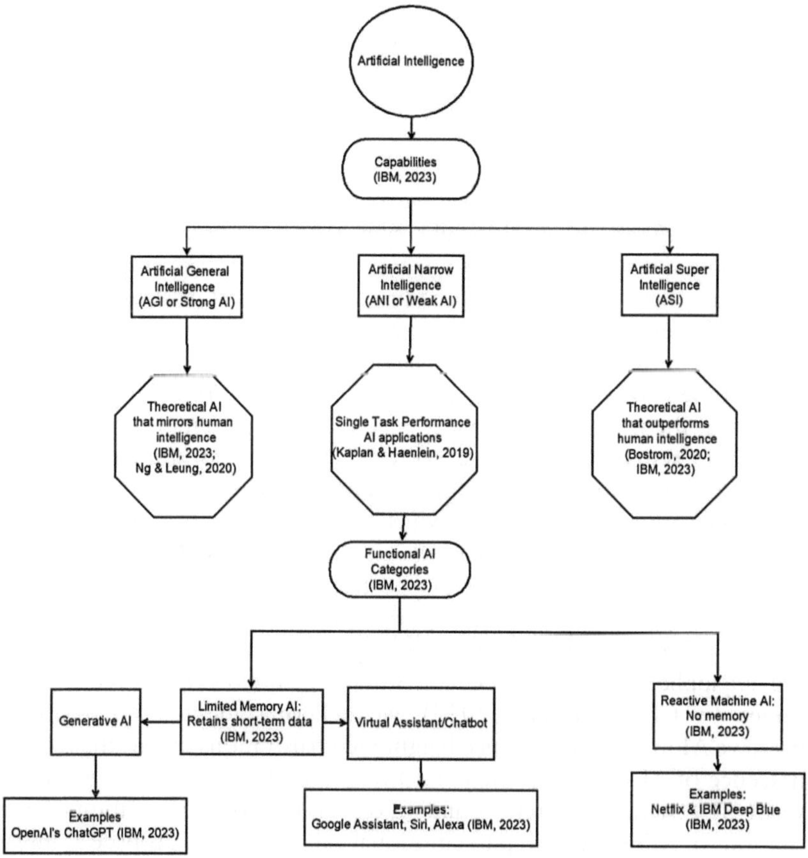

Fig. 7.2 A visual representation of AI

example, in an exploratory study by Chan and Hu (2023) college students in general were receptive to the use of AI and expressed positive perceptions of AI as a tool for brainstorming and research assistance. However, they also found students with positive perceptions of AI use were also aware of the extra time and effort needed to verify the sources and data generated through AI tools (Chan & Hu, 2023). We believe students who leverage AI seeking to minimize resource loss (e.g., time and energy, Fitria, 2021a), will also perceive AI positively. Moreover, they also will be more likely to expend resources to fact-check the information generated through AI tools, ultimately leading to increased emotional exhaustion. As such, we propose the following propositions:

Proposition 1: *Positive perceptions of AI are positively associated with emotional exhaustion.*

Proposition 2: *Negative perceptions of AI are negatively associated with emotional exhaustion.* Additionally, cynicism, the concept of individuals developing indifferent attitudes towards their school tasks or assignments, can occur when students are exposed to chronic academic stress (Walburg, 2014). For example, outside of the college enrollment process, students who leverage tools such as Grammarly can obtain real-time feedback on their homework when feeling cynical about their work. Fitria (2021b) notes that the free tool Grammarly enables students to input their writing into their system and detects, corrects, and teaches students how to improve their writing. Using variations of AI, such as machine learning, the author argues Grammarly produces more accurate recommendations, providing students with quick feedback and less stress. As another example, from a college enrollment perspective, Free Application for Federal Student Aid (FAFSA) launched a virtual assistant named Aidan. This AI tool uses natural language processing to guide students through the financial aid application process (Federal Student Aid, n.d.). However, as mentioned by Chan and Hu (2023), we believe students who positively perceive AI may still experience heightened frustrations due to the relative nascence of AI tools since they need to exert additional due diligence to verify AI-generated content. Building on our previous proposition of perceptions of AI's relationship to emotional exhaustion, we propose a similar relationship for cynicism. Therefore, propositions 3 and 4 are:

Proposition 3: *Positive perceptions of AI are positively associated with cynicism.*

Proposition 4: *Negative perceptions of AI are negatively associated with cynicism.* Despite recent research indicating students utilizing AI are aware of the need to spend time verifying findings and results (Chan & Hu, 2023), we suspect students using AI tools and positively perceiving the benefits of AI will experience a greater sense of academic efficacy. We also believe students will have a greater sense of academic efficacy because they will have engaged more intentionally with the research or academic projects they are using AI for, and therefore feel more confident in their knowledge generating or acquisition skills. Moreover, given the improved accuracy, real-time feedback, and overall support AI can provide students on coursework or in navigating the college admissions process (e.g., Fitria, 2021b; Igbokwe, 2023), we posit positive perceptions of AI will decrease academic inefficacy (increase efficacy). This is consistent with Bandura's (1977) idea that mastery experiences (e.g., completing a task) can increase efficacy. Thus, propositions 5 and 6 are:

Proposition 5: *Positive perceptions of AI are negatively associated with academic inefficacy.*

Proposition 6: *Negative perceptions of AI are positively associated with academic inefficacy.*

7.2.6 Relationship Between Burnout and Intentions to Enroll

Current burnout literature is mostly focused on empirical studies investigating the implications of burnout in people-oriented professional occupations (Maslach & Schaufeli, 2018). Given the original research on burnout was specific to staff burnout in clinics (Freudenberger, 1974) and "people-work" (Maslach & Jackson, 1981, p. 99), this is not surprising. Indeed, scholars later discovered the prevalence of burnout impacts various occupations and industries (Gabriel & Aguinis, 2022), including the restaurant industry (e.g., Han et al., 2016) and accounting professionals (e.g., Chong & Monroe, 2015). Overall, therefore, a substantial portion of the research stream on burnout in various occupations indicates burnout is related to turnover and decreased organizational commitment (e.g., Jung & Kim, 2012).

In the current study, we propose US high school students experiencing burnout are less likely to enroll in college. This notion is based on the idea that the college application and admission process subjects high school students to prolonged stress, which results in burnout and diminished intentions to enroll in college (mirroring the established turnover and absenteeism findings among adult employees). Specifically, academic demands (Walburg, 2014), academic stress, and the pressure to enroll in college (Feld & Shusterman, 2015) can substantially drain resources and lead to burnout symptoms. We propose the consequences of burnout can emerge even before a student has the opportunity to become indifferent to coursework and can instead act as an obstacle to initial enrollment. More specifically, we propose the following propositions:

Proposition 7: *Emotional exhaustion is negatively associated with the likelihood of enrolling in college.*

Proposition 8: *Cynicism is negatively associated with the likelihood of enrolling in college.*

Proposition 9: *Academic inefficacy is negatively associated with the likelihood of enrolling in college.*

7.3 Methods

7.3.1 Participants

Data were collected through Centiment, an online platform that connects researchers who need tasks to be completed (e.g., taking a survey) with participants who are willing to complete the tasks in exchange for modest compensation. The use of Centiment has received support in the literature from a data quality standpoint (Ruiz & Bell, 2022). The initial sample size for this investigation consisted of 196 participants, of whom 23 did not provide informed consent and were thus excluded from the study. Furthermore, in accordance with the work of Howard (2023), we subsequently removed participants from the final sample if either of the following conditions applied: (1) The subject failed multiple attention checks, and (2) the respondent did not complete multiple scales. These exclusions resulted in a final sample size of 160, which exceeded the adequacy threshold of 100 as established by Hair et al. (2019) for statistical analysis of the proposed relationships. The subjects ranged in age from 15 to 19 years, far surpassing the previously established age guidelines for research participation (Borgers et al., 2000). The respondents' descriptive statistics are provided in Table 7.1.

Table 7.1 Descriptive statistics

Variable definition	*Frequency*	*Percent*
Gender		
Agender	2	1.30%
Gender fluid	3	1.90%
Gender queer	2	1.30%
Man	69	43.10%
Nonbinary	2	1.30%
Woman	80	50%
Prefer not to disclose	2	1.30%
Age		
15	1	0.60%
16	60	37.50%
17	63	39.40%
18	35	21.90%
19	1	0.60%
Grade Level		
Freshmen	1	0.60%
Sophomore	30	18.80%
Junior	65	40.60%
Senior	64	40.00%
Location		
Midwest	43	26.90%
Northeast	33	20.60%
South	53	33.10%
West	30	18.80%
Puerto Rico or other US territories	1	0.60%
Parents education		
Neither parent graduated from college	51	31.90%
Both parents graduated from college	57	35.60%
Only father graduated from college	14	8.80%
Only mother graduated from college	25	15.60%
Unknown	13	8.10%
Social Class		
Working class	27	16.90%
Lower middle class	48	30%
Upper middle class	62	38.80%
Upper class	8	5%
Prefer not to disclose	15	9.45%

7.3.2 Measures

In the following sections, we describe our survey measures and discuss common method variance. Further, we present our analytical procedures and measurement model assessment. We conclude with our structural model assessment.

7.3.2.1 Academic Burnout

Following the logic of Emerson et al. (2023), academic burnout was operationalized using the MBI: General Survey—Students (MBI-GS[S]), created by Maslach et al. (2016) and originally designed to be used by adult students at the university level. Within the context of academic burnout, three elements exist: emotional exhaustion, cynicism, and academic inefficacy. To measure these concepts, we first adopted five established scale items to gauge emotional exhaustion, in which respondents were queried about their feelings toward school. Next, cynicism was measured using four scale items pertaining to students' enthusiasm for and interest in their studies. Finally, to measure academic inefficacy, we utilized five scale items associated with students' academic confidence. These items were reverse-scored to permit a more intuitive interpretation of the results (Emerson et al., 2023). In other words, a higher score is associated with lower academic efficacy. In addition, two scale items from the MBI-GS(S) were excluded on the basis of low loadings on their associated constructs in previous studies (e.g., Emerson et al., 2023) and in the interest of scale parsimony. The omitted scale items are as follows: cynicism, "I just want to get my work done and not be bothered; academic inefficacy", "I feel exhilarated when I accomplish something at the university."

7.3.2.2 Artificial Intelligence

To measure the AI constructs, we adapted five items from the General Attitudes Towards Artificial Intelligence Scale (GAAIS) (Schepman & Rodway, 2020) to match the context of this study. The GAAIS assesses individuals' perceptions of AI, including societal and personal benefits, and negative emotions such as discomfort over using AI. As AI is conceptualized as having both positive and negative components, we created two AI constructs in the present research. More specifically, the first AI construct, Positive Perceptions of AI, measures the degree to which respondents' find AI beneficial and their preferences for using AI in school or when searching for a school. The second AI construct, Negative Perceptions of AI, assesses subjects' concerns about AI mining personal data and evaluates whether they would suffer from compulsory AI use.

7.3.2.3 *Intentions to Enroll*

No established measures were available to measure intentions to enroll. As a result, in consultation with several experts in higher education, a scale was developed to measure students' intentions to enroll in college, following established guidelines for scale development (Anaza et al., 2021). First, a representative list of items was generated on the basis of qualitative interviews. Next, the items were subjected to pretesting (Howard, 2019) with a small sample of high school students. The pretest results yielded five scale items focused on the significance and value of pursuing higher education. One scale item was removed because of low loadings. In total, four scale items were included in the Intentions to Enroll in College construct. Composite reliability was calculated to assess the internal consistency reliability of the scale ($\alpha = .788$), which exceeded the guideline of .70 established by Hair et al. (2019).

7.3.2.4 *Common Method Variance*

Common method variance exists "when responses systematically vary because of the use of a common scaling approach on measures derived from a single data source" (Fuller et al., 2016, p. 3192). This phenomenon is problematic, as unrelated constructs may appear to be related (Howard & Henderson, 2023). To minimize the effects of common method variance, the scales used were varied to include different response formats (Podsakoff et al., 2003). In addition, responses to the questions at the end of a survey are likely to be impacted by boredom or fatigue (Lindell & Whitney, 2001). Therefore, to minimize the effects of common method variance, we placed demographic questions, which generally require little cognitive effort, at the end of the questionnaire. As a post hoc measure, we applied Harman's one-factor test (Harman, 1976; Babin et al., 2016) and determined the single-factor solution accounted for less than 50% of the total variance, indicating common method variance is not a meaningful concern.

7.3.2.5 *Analytical Procedures*

Analyses were performed using partial least squares structural equation modeling (PLS-SEM) (Hair et al., 2022) to test the theoretical model depicted in Fig. 7.1. PLS-SEM is the variance-based, prediction-oriented counterpart of covariance-based SEM (Hair et al., 2024) and utilizes total variance to estimate parameters (Hair et al., 2022). PLS-SEM is an appropriate technique for this investigation since it meets the guidelines established by Manley et al. (2021), including the following: (1) research of an exploratory nature that seeks to develop or further extend theory; (2) data consisting of multiple-item latent variables; (3) the study involves a small population, which leads to a smaller sample size; and (4) the ultimate goal of the study is prediction. To assess the quality of the measurement and structural models, the confirmatory composite analysis (CCA) procedure was followed (Hair et al., 2020). Thereafter, we assessed the metrics of the structural model, including out-of-sample prediction for our theoretical structural model using PLSpredict (Hair, 2021).

7.3.2.6 *Measurement Model Assessment*

In our assessment, we began by examining the reliability and convergent validity of the measurement model constructs. The results for both reliability and convergent validity along with the indicator loadings are depicted in Table 7.2. Indicator reliability is determined by squaring each individual indicator loading (Hair et al., 2019) and is a step in the CCA process for reflective measurement models (Hair et al., 2020). We do not explicitly report these results because of space constraints. The Cronbach alpha values ranged from 0.598 for the Positive Perceptions of AI construct to 0.905 for the Emotional Exhaustion construct. The low Cronbach alpha value for Positive Perceptions of AI can be attributed to several factors, including the following: (1) the relatively novel nature of AI in the education domain, (2) the lack of an established scale to measure high school students' perceptions of AI, and (3) the exploratory nature of this study. But most importantly, Cronbach alpha is a poor measure of reliability since individual items are equally weighted in the calculations of the metric (Hair et al., 2022).

In contrast, composite reliability is a superior measure of internal consistency reliability since the items are individually weighted based on the amount of variance they contribute to the calculation (Aibinu & Al-Lawati, 2010; Hair et al., 2020). The composite reliability values ranged from

Table 7.2 Convergent reliability and validity assessment

Construct	*Scale items*	*Loadings*	*CA*[a]	*rho_a*	*CR*[b]	*AVE*[c]
Positive perceptions of AI	There are many beneficial applications of AI.	0.750	0.598	0.626	0.788	0.557
	I would like to use Artificial Intelligence in school.	0.846				
	When trying to find out about a university I am considering attending, I would rather interact with an AI system	0.628				
Negative perceptions of AI	People like me will suffer from using AI.[d]	0.886	0.634	0.651	0.844	0.73
	I am concerned about AI mining my personal data[d]	0.822				
Emotional exhaustion	I feel emotionally drained by my studies.	0.836	0.905	0.922	0.929	0.723
	I feel used up at the end of the day at school.	0.848				
	I feel tired when I get up in the morning and have to face another day at school.	0.823				
	Attending classes every day is really straining for me.	0.878				
	I feel burnt out from my studies.	0.865				
Academic inefficacy	I can effectively solve the problems that arise in my studies.[d]	0.726	0.84	0.868	0.885	0.608
	I feel I am making an effective contribution to my classes.[d]	0.786				
	In my opinion, I am a good student.[d]	0.860				
	I have accomplished many worthwhile things in my life.[d]	0.732				
	While at school, I feel confident I am effective at getting things done.[d]	0.788				
Cynicism	Since entering my junior or senior year in high school, I have become less interested in my studies.	0.757	0.802	0.819	0.867	0.621

(*continued*)

Table 7.2 (continued)

Construct	*Scale items*	*Loadings*	*CA*[a]	*rho_a*	*CR*[b]	*AVE*[c]
	I have become less enthusiastic about my studies.	0.775				
	I have become more cynical about whether my school work contributes to anything meaningful.	0.822				
	I doubt the significance of my studies.	0.797				
Intentions to enroll	I intend to enroll at my top choice school, if I get in.	0.813	0.788	0.795	0.862	0.61
	Enrolling in college will improve my career prospects.	0.803				
	Attending junior college or a university will give me a sense of personal accomplishment.	0.768				
	Enrolling in college is money well spent	0.732				

[a] Cronbach Alpha
[b] Composite Reliability
[c] Average Variance Extracted
[d] Reverse-scored item

0.788 to 0.929, which falls within the acceptable threshold of 0.70 to 0.90, as specified by Hair et al. (2019).

We subsequently evaluated the average variance extracted (AVE) for each construct to assess convergent validity. As depicted in Table 7.2, the AVE values ranged from 0.557 to 0.73, which exceed the minimum benchmark of 0.50 (Fornell & Larcker, 1981). Discriminant validity was examined using the heterotrait-monotrait criterion (HTMT). All the values were below the thresholds of 0.85 (conceptually dissimilar) and 0.90 (conceptually similar) for constructs (Henseler et al., 2015).

As a precautionary step, we conducted an additional post hoc test to assess our model for common method variance. The variance inflation factors (VIFs) were examined for each indicator and did not exceed 2.88. Thus, consistent with the previous assessment of common method variance, all VIFs are below the threshold of 3.0 and therefore CMV is not a meaningful concern in our analysis (Kock, 2015).

7.3.2.7 *Structural Model Assessment*

In line with the work of Emerson et al. (2023) and the CCA procedure (Hair et al., 2020), we followed a multistep structural model assessment after confirming the measurement models. First, we calculated two parameters: (1) R^2 and (2) effect size (f^2). Furthermore, we examined the beta coefficients and their respective *t*-values using a 10,000 resample bootstrapping method recommended by Hair et al. (2022). A closer inspection of the beta coefficients yielded interesting findings. First, we examined the path coefficients between students' perceptions of AI the three dimensions of burnout. Positive perceptions of AI showed a significant positive association with emotional exhaustion. In other words, positive perceptions of AI are a predictor of emotional exhaustion, a notion that supports Proposition 1. Alternatively, negative perceptions of AI had a significant negative association with emotional exhaustion, which supports Proposition 2. Positive perceptions of AI also had a significant positive association with cynicism, thus supporting Proposition 3. On the other hand, negative perceptions of AI had a significant negative association with cynicism, which supports Proposition 4. Positive perceptions of AI did not have any significant association with academic inefficacy, thus failing to support Proposition 5. As was the case with positive perceptions of AI, negative perceptions of AI did not have a significant association with academic inefficacy. Thus, our findings do not support Proposition 6.

Next, we examined the relationship between the three dimensions of burnout and intentions to enroll in college. Emotional exhaustion had a significant positive association with intentions to enroll, a finding that fails to support Proposition 7. On the other hand, cynicism had a significant negative association with intentions to enroll, thus supporting Proposition 8. Finally, academic inefficacy had a significant negative association with intentions to enroll, which supports Proposition 9.

The VIF values are also interpreted to evaluate model multicollinearity. Becker et al. (2015) noted collinearity is not a problem with a VIF score as low as 3.0. Since the VIF values are below this threshold, collinearity is not a concern. The R^2 values indicate positive and negative perceptions of AI explained 13.8%, 10.8%, and 1.5% of the variance in emotional exhaustion, cynicism, and academic inefficacy, respectively. In addition, our theoretical model explained 11.8% of the variance in students' intentions to enroll in college.

The effect size (f^2) indicates how meaningful the relationship is between two variables, with large effect size meaning the relationship has practical

Table 7.3 Structural model tests of propositions

Examined path	*Beta*	*t-value*	*Proposition supported (Y/N)*	*f Square*
Positive perceptions of AI ---> emotional exhaustion	0.22	2.508**	Y	0.053[a]
Negative perceptions of AI ---> emotional exhaustion	−0.35	4.810**	Y	0.136[a]
Positive perceptions of AI ---> cynicism	0.306	3.705**	Y	0.1[a]
Negative perceptions of AI ---> cynicism	−0.203	2.240*	Y	0.044[a]
Positive perceptions of AI ---> academic inefficacy	−0.127	0.874	N	0.016
Negative perceptions of AI ---> academic inefficacy	0.03	0.245	N	0.001
Emotional exhaustion ---> intentions to enroll	0.286	2.923**	N	0.074[a]
Cynicism ---> intentions to enroll	−0.167	1.73*	Y	0.025[a]
Academic inefficacy ---> intentions to enroll	−0.234	1.931*	Y	0.058[a]

[a]Small effect size

*$p < .05$; ** $p < .01$

significance and a small effect size indicating limited practical value (Hair et al., 2019). Effect sizes range from 0.02, which represent a small effect; 0.15 a medium effect; and 0.35 a large effect (Cohen, 1988). Effect sizes are reported in Table 7.3, along with their corresponding path coefficients and *t*-values. Seven of the significant path coefficients had small effect sizes.

Finally, we examined the out-of-sample predictive ability of the model using the PLSpredict procedure (Shmueli et al., 2019), which applies a holdout sample approach to develop predictions for the dependent variable (Manley et al., 2021). All Q^2 values were above zero providing an initial indication of out-of-sample prediction (Hair et al., 2023). To examine the predictive power of our theoretical model, we examine and compare the prediction errors of the PLS-SEM model and the baseline linear model (LM). That is, we compared the root mean squared errors (RMSE) to the LM errors since the error prediction distribution was not highly non-symmetric (Manley et al., 2021).

The $Q^2_{predict}$ statistic for the indicators of the ultimate dependent variable enrollment intentions were examined first. This process involves

Table 7.4 PLS predict results

Indicator	*Prediction error metrics*		
	RMSE[a] PLS-SEM model	*RMSE[a] Linear benchmark model*	*Q^2 predict[b] PLS-SEM model*
Intentions to enroll 1	3.195	3.274	0.006
Intentions to enroll 2	2.884	2.9	0.011
Intentions to enroll 4	2.918	2.902	0.008
Intentions to enroll 5	2.922	2.877	0.022

[a] Smaller values represent higher predictive power
[b] Higher values represent higher predictive power

determining whether the model's predictions outperform the linear model (LM) prediction benchmark (Shmueli et al., 2019). Table 7.4 presents the out-of-sample prediction results. The results indicate the PLS-SEM model had a higher RMSE for indicators 4 and 5 in comparison to the linear benchmark model. Since the same number of the prediction errors of the indicators for the PLS-SEM model were higher than the linear benchmark model, we can conclude the model has medium predictive power (Manley et al., 2021).

7.4 Discussion

Given the looming enrollment cliff for many universities, the present study investigates how, if at all, AI will impact Generation Z's intentions to enroll in college. Thus, the motivation for this study was twofold. First, based on the COR theory and the burnout literature, we sought to explore two research questions related to high school students' perceptions of AI, burnout measured as three sub-constructs (emotional exhaustion, cynicism, and academic inefficacy), and high school students' intentions to enroll in college. Second, we proposed and analyzed nine propositions. Our findings indicate that while high school student's emotional exhaustion did not negatively impact college enrollment intentions, cynicism and academic inefficacy did relate to enrollment. Further, high school students exhibiting positive perceptions of AI were likely to experience emotional exhaustion and cynicism, but not academic inefficacy. In contrast, high school students exhibiting negative perceptions of AI were less likely to experience emotional exhaustion and cynicism, but not academic inefficacy.

7.4.1 *Theoretical and Practical Implications*

Our study poses several theoretical implications. First, it seeks to extend Hobfoll's COR theory (1989) by investigating modern technological advancements' implications for an individual's resource loss or gain. Given limited research conducted on high school students' perceptions of AI, we believe this knowledge is timely and pertinent, and particularly since these students will be the next generation of employees in the workforce. Second, our findings contribute to the burnout literature, as previous empirical studies have primarily focused on people-oriented occupations (Maslach & Schaufeli, 2018) or university level students not living in the U.S. (e.g., Fiorilli et al., 2022). Based on our findings, educational institutions should continue to investigate the impact AI will have on students. Given that cynicism and academic inefficacy had a significant negative impact on college enrollment intentions, we encourage higher education institutions to integrate new tools to streamline the admissions process. If institutions are concerned about enrollment rates, it would be fruitful for high schools to monitor the well-being of their students during the enrollment process. Finally, we recommend the impact of AI on students continue to be explored. Perhaps the implementation of student training on AI and guidelines for usage will impact its relationship with burnout. In addition, as AI becomes more widespread it may have a more robust impact on high school students.

7.4.2 *Limitations*

The current study has several limitations. First, we relied on survey responses to obtain insights into high school students' intentions to enroll in college. While precautionary steps were taken to minimize common method variance and to ensure respondents' engagement, such as employing randomized attention checks during the survey, a longitudinal study might provide a more robust approach to investigating our research questions. The findings of the current study also suggest the current scale utilized to measure perceptions of AI may not be well suited for a generation of individuals who grew up during the post-internet era and may not view the integration of AI in a manner similar to previous generations. Future research should include a re-examination of established scales measuring student's perceptions of the AI to ensure current measures accurately capture potential generational differences.

Furthermore, high school students' perceptions of AI did not have a significant relationship with academic inefficacy. These findings do not eliminate the possibility a relationship may exist in the presence of alternative antecedents or moderators. Given the hypothesized relationships were theoretically grounded, future research could explore various contextual factors in the literature that might explain AI's impact on academic inefficacy. Given the prevalence of burnout and AI, we urge scholars to continue investigating the implications of emotional exhaustion, cynicism, and academic inefficacy.

7.4.3 *Future Research*

7.4.3.1 Social Robots

Our investigation asked respondents about their perceptions of AI. Future research should also investigate Generation Z's attitude toward social robots. Campa (2016) describes social robots as maintaining a humanoid (human-like) or animaloid (animal-like) appearance to interact, on an emotional level, with humans. Through anthropomorphizing, interactions with social robots can be more natural, exchanging social cues or reflexive facial expressions (Breazeal, 2003). Notably, the use of social robots has been studied in the context of healthcare (e.g., Cifuentes et al., 2020), therapy (e.g., Cabibihan et al., 2013), and mental health (e.g., Scoglio et al., 2019). The application of social robots in education, however, has received limited attention.

To aid students through burnout, we suggest future research explore how social robots can be used to support high school students emotionally. In a study conducted by Kirby et al. (2010), generative models were used to elicit emotional responses from social robots during human-robot interactions. Given that efficacy is improved with mastery experiences (Bandura, 1977), social robots can give students positive reinforcement and emotional support, helping them persevere in completing tasks.

From a college enrollment perspective, social robots can potentially be used to develop trusting relationships with prospective students and better support them through their academic transition. Glass (2023) concludes first-generation students need additional support to navigate the college enrollment process, finding that mentoring programs increase the likelihood of high school graduation and college enrollment. In addition, Belpaeme et al. (2018) note that social robots can be used as peer learners or tutors. These thoughts align with the central focus of the work of Luria (2025 [Chap. 4]) that discusses the use of social robots and AI for personalized education. Thus, future research should explore the use of social robots as mentors during the college enrollment process.

7.4.3.2 Higher Education Incorporating AI

The impacts and benefits of AI within Higher Education offer numerous future research opportunities. First, as mentioned in the limitation section, one area of future research is to construct and validate an AI perceptions scale in better alignment with the generational familiarity of and ease with technology adoption. We suspect most members of Generation Z do not recall a time without the Internet and related technology being readily accessible and integrated into their daily lives. Therefore, the current scale measures may not fully capture accurate perceptions of technology and AI use. In tandem with developing and validating a new AI perceptions scale, researchers should explore the use of qualitative research to obtain a more comprehensive and richer understanding of the impacts of AI.

While the purpose of this study was to examine the impact of perceptions of AI on intentions to enroll at an institution of higher education, future research should explore whether perceptions of AI impact first-to second-year retention rates and ultimately graduation rates. Given the upcoming enrollment cliff within higher education, identifying and developing retention tools are of high practical significance for all those involved in the higher education industry. A final area for future research is to explore the relationships between personality and AI perceptions and use.

7.4.3.3 Critical Reflections

Finally, we encourage future scholarship to continue to monitor the ongoing changing political, technological, and economic atmospheres as it relates to higher education and AI use. Many of the chapters included in this text explore aspects of ethical AI use within the current political setting. Specifically, as it relates to the present investigation on the impact of AI usage on burnout and college enrollment intentions, it is critical to account for the present trends for governments and school systems to impose a phone ban across school systems (Selwyn, 2024) and how this restricted access to technology may impact findings in future research.

7.5 Conclusion

In totality, the current study has proposed two research questions that warrant investigation during the rapidly increasing proliferation of AI. From these questions, we developed nine theoretically grounded propositions to explore the perceptions of AI, burnout, and high school students' intentions to enroll in college. Notably, few studies have analyzed Generation Z in the context of burnout and AI. Therefore, our findings provide valuable insights to scholars and practitioners. We urge scholars to continue researching Generation Z and AI, as they are the future of our workforce. Our current findings pose several theoretical implications, such as extending the COR Theory and burnout literature. The present study further proposes practical implications by highlighting the impact of AI on students' intentions to enroll in higher education. Based on our findings, scholars can expand upon our research questions to further develop our understanding of these concepts. Further, we provide recommendations for scholars to explore the use of social robots as a direction for future research. Lastly, the findings of this study highlight the need for higher education institutions and high schools to monitor and investigate high school students' burnout levels and college enrollment intentions.

References

Aibinu, A. A., & Al-Lawati, A. M. (2010). Using PLS-SEM technique to model construction organizations' willingness to participate in e-bidding. *Automation in Construction, 19*(6), 714–724.

Alarcon, G. M. (2011). A meta-analysis of burnout with job demands, resources, and attitudes. *Journal of Vocational Behavior, 79*(2), 549–562.

Anaza, N. A., Saavedra, J. L., Bagherzadeh, R., Hair, J. F., Rawal, M., & Osakwef, N. F. (2021). Customer-brand disidentification: Conceptualization, scale development and validation. *Journal of Business Research, 133*, 116–131.

Arruda, E. P., & Arruda, D. P. (2025). Artificial intelligence in the global educational context: Prospects for pedagogical practice and application in the context of teacher shortages. In S. Papadakis & G. Lampropoulos (Eds.), *Social robots and artificial intelligence in education: Integrating AI in K-12 and higher education* (pp. 37–60). Palgrave Macmillan.

Babin, B., Griffin, M., & Hair, J. F., Jr. (2016). Heresies and sacred cows in scholarly marketing publications. *Journal of Business Research, 69*(8), 3133–3138.
Bahoo, S., Cucculelli, M., & Qamar, D. (2023). Artificial intelligence and corporate innovation: A review and research agenda. *Technological Forecasting and Social Change, 188*, 122264.
Bandura, A. (1977). Self-efficacy: Toward a unifying theory of behavioral change. *Psychological Review, 84*(2), 191–215.
Bask, M., & Salmela-Aro, K. (2013). Burned out to drop out: Exploring the relationship between school burnout and school dropout. *European Journal of Psychology of Education, 28*, 511–528.
Becker, J.-M., Ringle, C. M., Sarstedt, M., & Völckner, F. (2015). How collinearity affects mixture regression results. *Marketing Letters, 26*(4), 643–659.
Belpaeme, T., Kennedy, J., Ramachandran, A., Scassellati, B., & Tanaka, F. (2018). Social robots for education: A review. *Science Robotics, 3*(21), 1–9.
Bhalla, R., Tiwari, P., & Chowdhary, N. (2021). Digital natives leading the world: Paragons and values of Generation Z. In *Generation Z marketing and management in tourism and hospitality: The future of the industry* (pp. 3–23). Springer International Publishing.
Biggs, J. (1999). What the student does: Teaching for enhanced learning. *Higher Education Research & Development, 18*(1), 57–75.
Biggs, J. (2012). What the student does: Teaching for enhanced learning. *Higher Education Research & Development, 31*(1), 39–55.
Borgers, N., de Leeuw, E., & Hox, J. (2000). Children as respondents in survey research: Cognitive development and response quality. *Bulletin of Sociological Methodology, 66*(1), 60–75.
Bostrom, N. (2020). Ethical issues in advanced artificial intelligence. *Machine Ethics and Robot Ethics*, 69–75.
Boyd, N. G., Lewin, J. E., & Sager, J. K. (2009). A model of stress and coping and their influence on individual and organizational outcomes. *Journal of Vocational Behavior, 75*(2), 197–211.
Breazeal, C. (2003). Toward sociable robots. *Robotics and Autonomous Systems, 42*(3–4), 167–175.
Cabibihan, J. J., Javed, H., Ang, M., & Aljunied, S. M. (2013). Why robots? A survey on the roles and benefits of social robots in the therapy of children with autism. *International Journal of Social Robotics, 5*, 593–618.
Campa, R. (2016). The rise of social robots: A review of the recent literature. *Journal of Evolution and Technology, 26*(1), 106–113.
Chan, C. K. Y., & Hu, W. (2023). Students' voices on generative AI: Perceptions, benefits, and challenges in higher education. *International Journal of Educational Technology in Higher Education, 20*(1), 43.
Chong, V. K., & Monroe, G. S. (2015). The impact of the antecedents and consequences of job burnout on junior accountants' turnover intentions: A structural equation modelling approach. *Accounting & Finance, 55*(1), 105–132.

Cifuentes, C. A., Pinto, M. J., Céspedes, N., & Múnera, M. (2020). Social robots in therapy and care. *Current Robotics Reports, 1*, 59–74.

Cohen, J. (1988). *Statistical power analysis for the behavioral sciences.* Lawrence Erlbaum Associates.

Cotton, D. R., Cotton, P. A., & Shipway, J. R. (2024). Chatting and cheating: Ensuring academic integrity in the era of ChatGPT. *Innovations in Education and Teaching International, 61*(2), 228–239.

Cugurullo, F., & Acheampong, R. A. (2024). Fear of AI: An inquiry into the adoption of autonomous cars in spite of fear, and a theoretical framework for the study of artificial intelligence technology acceptance. *AI & Society 39*, (4), 1569–1584.

Davis, F. D. (1989). Perceived usefulness, perceived ease of use, and user acceptance of information technology. *MIS Quarterly, 13*(3), 319–340.

Dempere, J., Modugu, K., Hesham, A., & Ramasamy, L. K. (2023, September). The impact of ChatGPT on higher education. In *Frontiers in Education* (Vol. 8, p. 1206936). Frontiers Media SA.

Duan, Y., Edwards, J. S., & Dwivedi, Y. K. (2019). Artificial intelligence for decision making in the era of Big Data—Evolution, challenges and research agenda. *International Journal of Information Management, 48*, 63–71.

Emerson, D. J., Hair, J. F., Jr., & Smith, K. J. (2023). Psychological distress, burnout, and business student turnover: The role of resilience as a coping mechanism. *Research in Higher Education, 64*(2), 228–259.

Federal Student Aid. (n.d.). *Meet Aidan.* https://studentaid.gov/h/aidan

Feld, L. D., & Shusterman, A. (2015). Into the pressure cooker: Student stress in college preparatory high schools. *Journal of Adolescence, 41*, 31–42.

Fiorilli, C., Farina, E., Buonomo, I., Costa, S., Romano, L., Larcan, R., & Petrides, K. V. (2020). Trait emotional intelligence and school burnout: The mediating role of resilience and academic anxiety in high school. *International Journal of Environmental Research and Public Health, 17*(9), 3058.

Fiorilli, C., Barni, D., Russo, C., Marchetti, V., Angelini, G., & Romano, L. (2022). Students' burnout at university: The role of gender and worker status. *International Journal of Environmental Research and Public Health, 19*(18), 11341.

Fitria, T. N. (2021a). Artificial intelligence (AI) in education: Using AI tools for teaching and learning process. In *Prosiding Seminar Nasional & Call for Paper STIE AAS* (pp. 134–147).

Fitria, T. N. (2021b). Grammarly as AI-powered English writing assistant: Students' alternative for writing English. *Metathesis: Journal of English Language, Literature, and Teaching, 5*(1), 65–78.

Fornell, C., & Larcker, D. F. (1981). Evaluating structural equation models with unobservable variables and measurement error. *Journal of Marketing Research, 18*(1), 39–50.

Fountaine, T., McCarthy, B., & Saleh, T. (2019). Building the AI-powered organization. *Harvard Business Review, 97*(4), 62–73.

Freudenberger, H. J. (1974). Staff burn-out. *Journal of Social Issues, 30*(1), 159–165.

Fuller, C. M., Simmering, M. J., Atinc, G., Atinc, Y., & Babin, B. J. (2016). Common methods variance detection in business research. *Journal of Business Research, 69*(8), 3192–3198.

Gabriel, K. P., & Aguinis, H. (2022). How to prevent and combat employee burnout and create healthier workplaces during crises and beyond. *Business Horizons, 65*(2), 183–192.

Gaur, L., Afaq, A., Singh, G., & Dwivedi, Y. K. (2021). Role of artificial intelligence and robotics to foster the touchless travel during a pandemic: A review and research agenda. *International Journal of Contemporary Hospitality Management, 33*(11), 4079–4098.

Glass, L. E. (2023). Social capital and first-generation college students: Examining the relationship between mentoring and college enrollment. *Education and Urban Society, 55*(2), 143–174.

Hair, J. F. (2021). Next-generation prediction metrics for composite-based PLS-SEM. *Industrial Management & Data Systems, 121*(1), 5–11.

Hair, J. F., Black, W. C., Babin, B. J., & Anderson, R. E. (2019). *Multivariate data analysis* (8th ed.). Cengage.

Hair, J. F., Howard, M. C., & Nitzl, C. (2020). Assessing measurement model quality in PLS-SEM using confirmatory composite analysis. *Journal of Business Research, 109*, 101–110.

Hair, J. F., Hult, G. T. M., Ringle, C. M., & Sarstedt, M. (2022). *A primer on partial least squares structural equation modeling (PLS-SEM)* (3rd ed.).

Hair, J. F., García-Machado, J. J., & Martínez-Avila, M. (2023). The impact of organizational compliance culture and green culture on environmental behavior: The moderating effect of environmental commitment. *Green Finance, 5*(4), 624–657.

Hair, J. F., Sarstedt, M., Ringle, C. M., & Gudergan, S. (2024). *Advanced issues in partial least squares structural equation modeling* (2nd ed.). Sage.

Halbesleben, J. R. (2006). Sources of social support and burnout: A meta-analytic test of the conservation of resources model. *Journal of Applied Psychology, 91*(5), 1134–1145.

Halbesleben, J. R., & Buckley, M. R. (2004). Burnout in organizational life. *Journal of Management, 30*(6), 859–879.

Halbesleben, J. R., Wheeler, A. R., & Paustian-Underdahl, S. C. (2013). The impact of furloughs on emotional exhaustion, self-rated performance, and recovery experiences. *Journal of Applied Psychology, 98*(3), 492–503.

Halbesleben, J. R., Neveu, J. P., Paustian-Underdahl, S. C., & Westman, M. (2014). Getting to the "COR" understanding the role of resources in conservation of resources theory. *Journal of Management, 40*(5), 1334–1364.

Han, S. J., Bonn, M. A., & Cho, M. (2016). The relationship between customer incivility, restaurant frontline service employee burnout and turnover intention. *International Journal of Hospitality Management, 52*, 97–106.

Harman, H. H. (1976). *Modern factor analysis.* University of Chicago Press.

Henseler, J., Ringle, C. M., & Sarstedt, M. (2015). A new criterion for assessing discriminant validity in variance-based structural equation modeling. *Journal of the Academy of Marketing Science, 43*, 115–135.

Hobfoll, S. E. (1989). Conservation of resources: A new attempt at conceptualizing stress. *American Psychologist, 44*(3), 513–524.

Hobfoll, S. E. (2001). The influence of culture, community, and the nested-self in the stress process: Advancing conservation of resources theory. *Applied Psychology, 50*(3), 337–421.

Howard, M. C. (2019). Scale pretesting. *Practical Assessment, Research, and Evaluation, 23*(1), 1–14.

Howard, M. C. (2023). Integrating the person-centered approach with the study of vaccine hesitancy: Applying latent profile analysis to identify vaccine hesitancy subpopulations and assess their relations with correlates and vaccination outcomes. *Vaccine, 41*(33), 4823–4835.

Howard, M. C., & Henderson, J. (2023). A review of exploratory factor analysis in tourism and hospitality research: Identifying current practices and avenues for improvement. *Journal of Business Research, 154*, 113328.

Hoy, M. B. (2018). Alexa, Siri, Cortana, and more: An introduction to voice assistants. *Medical Reference Services Quarterly, 37*(1), 81–88.

IBM. (2023). *Understanding the different types of artificial intelligence.* https://www.ibm.com/blog/understanding-the-different-types-of-artificial-intelligence/

Igbokwe, I. C. (2023). Application of artificial intelligence (AI) in educational management. *International Journal of Scientific and Research Publications, 13*(3), 300–307.

Jung, J., & Kim, Y. (2012). Causes of newspaper firm employee burnout in Korea and its impact on organizational commitment and turnover intention. *The International Journal of Human Resource Management, 23*(17), 3636–3651.

Kaplan, A., & Haenlein, M. (2019). Siri, Siri, in my hand: Who's the fairest in the land? On the interpretations, illustrations, and implications of artificial intelligence. *Business Horizons, 62*(1), 15–25.

Kirby, R., Forlizzi, J., & Simmons, R. (2010). Affective social robots. *Robotics and Autonomous Systems, 58*(3), 322–332.

Kock, N. (2015). Common method bias in PLS-SEM: A full collinearity assessment approach. *International Journal of e-Collaboration, 11*(4), 1–10.

Lampropoulos, G., & Papadakis, S. (2025). Integrating social robots into education. In S. Papadakis & G. Lampropoulos (Eds.), *Social robots and artificial intelligence in education: Integrating AI in K-12 and higher education* (pp. 21–35). Palgrave Macmillan.

Leffer, L. (2023, October 2). *"AI anxiety" is on the rise—Here's how to manage it.* Scientific American. https://www.scientificamerican.com/article/ai-anxiety-is-on-the-rise-heres-how-to-manage-it/
Leiter, M. P., & Maslach, C. (2018). Interventions to prevent and alleviate burnout. In *Current issues in work and organizational psychology* (pp. 32–50). Routledge.
Lindell, M. K., & Whitney, D. J. (2001). Accounting for common method variance in cross-sectional research designs. *Journal of Applied Psychology, 86*(1), 114–121.
Luria, E. (2025). Revolutionizing English language learning through social robots and artificial intelligence: Towards inclusive and personalized education. In S. Papadakis & G. Lampropoulos (Eds.), *Social robots and artificial intelligence in education: Integrating AI in K-12 and higher education* (pp. 61–90). Palgrave Macmillan.
Manley, S. C., Hair, J. F., Williams, R. I., & McDowell, W. C. (2021). Essential new PLS-SEM analysis methods for your entrepreneurship analytical toolbox. *International Entrepreneurship and Management Journal, 17*, 1805–1825.
Maslach, C. (2003). *Burnout: The cost of caring.* ISHK.
Maslach, C., & Jackson, S. E. (1981). The measurement of experienced burnout. *Journal of Occupational Behavior, 2*, 99–113.
Maslach, C., & Leiter, M. P. (2017). Understanding burnout: New models. In C. Cooper & J. Quick (Eds.), *The handbook of stress and health: A guide to research and practice* (pp. 36–56). Wiley.
Maslach, C., & Schaufeli, W. B. (2018). Historical and conceptual development of burnout. In *Professional burnout* (pp. 1–16). CRC Press.
Maslach, C., Jackson, S. E., & Schwab, R. (2016). *MBI-general survey (students) MBI-GS (S).* Mind Garden.
Mikalef, P., & Gupta, M. (2021). Artificial intelligence capability: Conceptualization, measurement calibration, and empirical study on its impact on organizational creativity and firm performance. *Information & Management, 58*(3), 103434.
Miller, D. D., & Brown, E. W. (2018). Artificial intelligence in medical practice: The question to the answer? *The American Journal of Medicine, 131*(2), 129–133.
Ng, G. W., & Leung, W. C. (2020). Strong artificial intelligence and consciousness. *Journal of Artificial Intelligence and Consciousness, 7*(01), 63–72.
Osterman, M. J. K., Hamilton, B. E., Martin, J. A., Driscoll, A. K., & Valenzuela, C. P. (2023). *Births: Final data for 2021.* National Center for Health Statistics.
Podsakoff, P. M., MacKenzie, S. B., Lee, J. Y., & Podsakoff, N. P. (2003). Common method biases in behavioral research: A critical review of the literature and recommended remedies. *Journal of Applied Psychology, 88*(5), 879–903.
Rudolph, J., Tan, S., & Tan, S. (2023). War of the chatbots: Bard, Bing Chat, ChatGPT, Ernie and beyond. The new AI gold rush and its impact on higher education. *Journal of Applied Learning and Teaching, 6*(1), 364–389.

Ruiz, J. B., & Bell, R. A. (2022). Parental COVID-19 vaccine hesitancy in the United States. *Public Health Reports, 137*(6), 1162–1169.

Schepman, A., & Rodway, P. (2020). Initial validation of the general attitudes towards Artificial Intelligence Scale. *Computers in Human Behavior Reports, 1*, 100014.

Scoglio, A. A., Reilly, E. D., Gorman, J. A., & Drebing, C. E. (2019). Use of social robots in mental health and well-being research: Systematic review. *Journal of Medical Internet Research, 21*(7), e13322.

Selwyn, N. (2024, March 24). There is a growing trend for governments declaring nationwide phone 'bans' across school systems, but ... *Critical Studies of Education & Technology*. https://criticaledtech.com/2024/03/24/there-is-a-growing-trend-for-governments-declaring-nationwide-phone-bans-across-school-systems-but/

Selye, H. (1956). *The stress of life* (Revised ed.). McGraw Hill.

Sheikh, H., Prins, C., & Schrijvers, E. (2023). Artificial Intelligence: Definition and background. In *Mission AI: The new system technology* (pp. 15–41). Springer International Publishing.

Shmueli, G., Sarstedt, M., Hair, J. F., Cheah, J. H., Ting, H., Vaithilingam, S., & Ringle, C. M. (2019). Predictive model assessment in PLS-SEM: Guidelines for using PLSpredict. *European Journal of Marketing, 53*(11), 2322–2347.

Szymkowiak, A., Melović, B., Dabić, M., Jeganathan, K., & Kundi, G. S. (2021). Information technology and Gen Z: The role of teachers, the internet, and technology in the education of young people. *Technology in Society, 65*, 101565.

Tambe, P., Cappelli, P., & Yakubovich, V. (2019). Artificial intelligence in human resources management: Challenges and a path forward. *California Management Review, 61*(4), 15–42.

Turing, A. M. (2009). *Computing machinery and intelligence* (pp. 23–65). Springer.

Walburg, V. (2014). Burnout among high school students: A literature review. *Children and Youth Services Review, 42*, 28–33.

Wright, T. A., & Cropanzano, R. (1998). Emotional exhaustion as a predictor of job performance and voluntary turnover. *Journal of Applied Psychology, 83*(3), 486–493.

Yeadon, W., Inyang, O. O., Mizouri, A., Peach, A., & Testrow, C. P. (2023). The death of the short-form physics essay in the coming AI revolution. *Physics Education, 58*(3), 035027.

Yigitcanlar, T., & Cugurullo, F. (2020). The sustainability of artificial intelligence: An urbanistic viewpoint from the lens of smart and sustainable cities. *Sustainability, 12*(20), 8548.

PART II

Artificial Intelligence in Student Learning and Cognitive Development

Part II examines how AI contributes to students' personalized learning and cognitive development through educational technologies. The chapters of this section address how AI-driven technologies can foster agency, creativity, problem-solving, and critical thinking skills in diverse student groups. Collectively, the chapters showcase both the pedagogical promise and the practical challenges of integrating AI into teaching and learning activities. This part includes the following three chapters:

- **Chapter 8: Engaging Young Minds in AI through Co-Designing and Making Their Own AI-Driven Apps in the Classroom** (Juho Kahila, Henriikka Vartiainen, Nicolas Pope, Matti Tedre): Describes a co-design project in Finnish schools where students created AI-driven apps using machine learning tools. It discusses how design-oriented pedagogy enhances students' understanding of AI concepts and fosters ethical awareness.
- **Chapter 9: Exploring Horizons: Study on the Perceptions of Primary Students Regarding the Future of Work in Robotics and STEAM** (Pablo Dúo-Terrón, Jesús López-Belmonte, José-Antonio Marín-Marín, Antonio-José Moreno-Guerrero): Presents research into primary students' perceptions of future careers in robotics and STEAM (Science, Technology, Engineering, Art, and Mathematics), following hands-on experiences with educational robots. It highlights the role of robotics in developing critical and transversal skills.

- **Chapter 10: Elevating EFL Paraphrasing Skills through Generative Artificial Intelligence: An Exploration with ChatGPT** (Anh Nguyen-Nhu Le, Nguyen Hoai Ly Nga): Explores the use of ChatGPT to help students improve their skills in the context of learning English as a Foreign Language (EFL). It evaluates the effectiveness of AI-generated content and discusses how generative AI can be a valuable tool in academic writing.

Although there are some key differences among the chapters of this section, such as physical embodiment versus virtual embodiment, attitudinal changes versus curricular integration, and tool autonomy versus teacher expertise, the following recurrent themes arose:

- Learner agency and maker-centered pedagogy: The chapters focus on promoting students' engagement and agency and comment upon the benefits of adopting a maker-centered pedagogy while capitalizing on AI tools. The initiative to empower students as co-creators rather than passive consumers aligns with design-oriented and constructionist theories, stressing that building artifacts deepens conceptual learning.
- Equity and accessibility: Given the cost, infrastructure, and policy barriers associated with the adoption of AI in education, the chapters comment on the need for a shared commitment to equitable and inclusive access across varied contexts and economic realities.
- Bridging cognitive domains: The ability of AI to serve as a versatile scaffold for cognitive development by stimulating problem-solving, critical reflection, and, metacognitive awareness in content-specific settings is highly regarded.
- Ethical and societal reflections: Whether elicited through co-design discussions on AI bias, future-of-work dilemmas, or educational integrity considerations, all four chapters integrate ethical inquiry as central into AI literacy.

These chapters present a multifaceted landscape in which AI acts as both a tool for enhancing cognition and as a means for pedagogical change. This section invites readers to consider how AI technical affordances can be thoughtfully harnessed to foster student agency, ethical awareness, domain-specific competencies, and cognitive development while also drawing attention to the limitations and challenges associated with their successful integration.

CHAPTER 8

Engaging Young Minds in AI Through Co-Designing and Making Their Own AI-Driven Apps in the Classroom

Juho Kahila, Henriikka Vartiainen, Nicolas Pope, and Matti Tedre

8.1 Introduction

In an era where artificial intelligence (AI) is rapidly reshaping society, the integration of AI education into K-12 curricula has become increasingly important (Rizvi et al., 2023). Today's children are growing up in a world where data-driven services collect data for profiling, tracking, and modeling, raising concerns about online safety, especially for those with limited critical literacy skills (Lupton & Williamson, 2017; Pangrazio & Selwyn,

J. Kahila (✉) • H. Vartiainen
School of Applied Educational Science and Teacher Education, University of Eastern Finland, Joensuu, Finland
e-mail: juho.kahila@uef.fi; henriikka.vartiainen@uef.fi

N. Pope • M. Tedre
School of Computing, University of Eastern Finland, Joensuu, Finland
e-mail: nicolas.pope@uef.fi; matti.tedre@uef.fi

S. Papadakis, G. Lampropoulos (eds.), *Social Robots and Artificial Intelligence in Education*, Digital Education and Learning,
https://doi.org/10.1007/978-3-032-03977-4_8

2019; Tedre, Toivonen, et al., 2021b). As AI continues to influence sectors such as healthcare, finance, education, transportation, and entertainment, it is crucial for the younger generation to understand and navigate this landscape (Keen, 2022; Tedre, Denning, et al., 2021a; Touretzky et al., 2019).

Although the educational landscape is evolving with a growing emphasis on integrating computational thinking into school curricula—thereby preparing students for challenges in an advancing technological landscape, as discussed in Dúo-Terrón et al. (2025 [Chap. 9]), AI is often considered a difficult domain to teach due to its complexity (Vartiainen et al., 2020). There is a lack of empirical research on how to effectively teach AI topics to children (Rizvi et al., 2023). However, recent educational initiatives and research emphasize the importance of introducing these subjects early in educational programs to foster a data-literate society (Long & Magerko, 2020; Shapiro et al., 2018). This is in line with the growing recognition of the need to understand and critically evaluate the digital world we inhabit (Tedre, Toivonen, et al., 2021b). Nevertheless, comprehending the broader and more complex impacts of AI is challenging without an understanding of its basic concepts and principles. To address these challenges, this chapter focuses on an interdisciplinary educational co-design project implemented in Finnish schools. The project aims to introduce key AI concepts, workflows, and societal impacts to young learners through the creation of AI-driven apps. This approach is designed to foster students' understanding of AI and empower them as participants in a digital society shaped by AI technologies (Touretzky et al., 2019; Vartiainen et al., 2021; Wong et al., 2020).

The project combines educational theory with practical application, employs new educational technology, and presents a robust model for introducing AI concepts to young learners. It involved researchers from multiple fields, elementary school teachers from 12 schools, and over 200 fourth and seventh-grade students (Kahila, Vartiainen, Tedre, Arkko, et al., 2024a). A central component of the workshops was the GenAI Teachable Machine, a no-code AI education tool enabling the full app development workflow in the classroom. Developed collaboratively with researchers, software developers, and teachers, this tool provides a beginner-friendly platform that supports the design and implementation of AI-driven apps, thus promoting an engaging learning experience (Pope et al., 2024). However, the tool alone is insufficient for teaching AI; a comprehensive pedagogical framework is necessary. Our approach was grounded in design-oriented pedagogy, in which learners build their conceptual understanding and ways of thinking by creating digital or physical artifacts and projects with the support of technology (Kafai et al., 2020). Nevertheless, this does not imply that children deepen their understanding independently; knowledge creation requires various forms of support (Puntambekar & Kolodner, 2005; Viilo, 2020), and throughout the workshops, various scaffolding strategies were employed (Kahila, Vartiainen, Tedre, Arkko, et al., 2024a).

In the following pages, we will explain how we integrated the GenAI Teachable Machine tool and a robust pedagogical framework into classroom settings through interactive workshops. We provide an example of a children's AI project and discuss how positioning children as designers and creators in workshops serves as a motivating and effective method for teaching basic AI concepts. This discussion also examines how an understanding of AI's core concepts and mechanisms lays the foundation for comprehending the broader impacts of AI and becoming informed and ethical citizens. Through this exploration, the chapter contributes to the discussion on how to integrate AI education into K12 education in an effective and meaningful way for students. Moreover, this chapter aims to serve as a guide for educators, policymakers, and researchers interested in the development of future-centric learning environments.

8.2 The GenAI Teachable Machine

The Google Teachable Machine (GTM) is a widely used and effective tool for introducing machine learning (ML) concepts in K-12 education. It allows children to develop customized image classifiers, creating their own classes, providing samples for training, and testing the model directly in their browsers (Carney et al., 2020; Toivonen et al., 2020). Although GTM is user-friendly, fully utilizing it requires programming skills that both teachers and children need to have (Gresse von Wangenheim et al., 2021). Furthermore, GTM lacks compatibility with many hardware devices commonly found in educational settings and does not fully meet data protection and accessibility standards. To meet our educational goals, we need a tool that supports a comprehensive workflow, facilitates the practical application of students' projects in real-world contexts, and promotes collaborative learning environments without requiring extensive technical expertise.

The GenAI Teachable Machine (Fig. 8.1) was developed to bridge the gap in accessible, collaborative, and comprehensive workflow ML tools. Designed to make learning ML concepts both accessible and engaging,

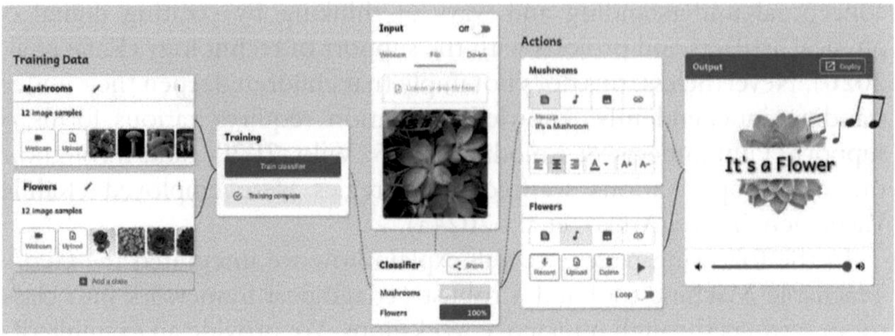

Fig. 8.1 The complete workflow interface of the GenAI Teachable Machine

this tool empowers users to create their own ML-driven apps. It offers an intuitive and integrated approach to teaching the complex and often abstract concepts of ML in classrooms (Pope et al., 2024). Primarily aimed at young learners in K-12 education, it adheres to the “low-floor, high-ceiling, and wide-walls” design philosophy, ensuring accessibility for beginners and support for more complex and diverse projects (Resnick et al., 2005). Specifically designed for children and educational settings, it prioritizes privacy and security, complying with GDPR requirements to ensure that student data is securely handled and that there is no unnecessary data retention or exposure (Pope et al., 2024).

8.3 Design-Oriented Pedagogy

Our pedagogical framework is grounded in design-oriented pedagogy, which builds on sociocultural insights on learning. In this approach, learners develop their conceptual understanding and creative skills through open-ended, creative, and collaborative design processes (Hakkarainen & Seitamaa-Hakkarainen, 2022). According to Hakkarainen et al. (2013), collaborative learning and design can be viewed as an object-oriented process, where social interaction and engagement with materials, tools, and technologies mediate jointly constructed activities. This process involves guiding students to pursue meaningful projects and open-ended design problems that stem from exploration of real-world phenomena and feel personally interesting. Rather than relying on scripted tasks or step-by-step exercises, design-oriented pedagogy entails students building their conceptual understanding and new ways of thinking by inventing and developing solutions to open-ended, real-world problems, which they approach by co-designing and creating inventions and artifacts that address these challenges (Hakkarainen & Seitamaa-Hakkarainen, 2022; Kafai et al., 2020; Vartiainen et al., 2021).

Moreover, design-oriented pedagogy shifts from individual exercises to collaborative learning and knowledge creation, mirroring expert problem-solving processes and disciplinary epistemic practices (Krajcik & Blumenfeld, 2006). Students collaborate to create a shared understanding of the challenges, share and construct ideas, and iteratively develop those ideas through experimentation, prototyping, and testing until a desired outcome is achieved (Hakkarainen & Seitamaa-Hakkarainen, 2022). As Seitamaa-Hakkarainen et al. (2022) note, students develop not only ideas and artifacts but also their collaborative skills, as they make decisions, negotiate roles and responsibilities, and plan, monitor, and reflect on the process's progress. Moreover, transforming abstract ideas into concrete artifacts involves engaging with externalized ideas through writing, drawing, and making prototypes (Seitamaa-Hakkarainen et al., 2010). This externalization of evolving ideas and reasoning processes facilitates joint discussions and evaluations, guiding subsequent actions and the development of potential solutions (Hennessy & Murphy, 1999; Seitamaa-Hakkarainen et al., 2010). To foster social interaction and collaborative knowledge creation, student teams are encouraged to share their ideas, knowledge, and artifacts with the entire class, making co-design and knowledge creation a collective endeavor (Kahila, Vartiainen, Tedre, Arkko, et al., 2024a).

While the students have a large degree of freedom in what they choose to co-design, their evolving activities are scaffolded by the epistemic, material, and social structures that support the learning of basic concepts and disciplinary practices. Scaffolding the co-design process includes multiple affordances, such as curriculum materials, tools, and educational technologies that foster collaboration and enable students to learn disciplinary ways of knowing, doing, and talking (Puntambekar & Kolodner, 2005; Tabak, 2004). In this process, the teacher plays a crucial role in orchestrating the relationships among the tools, materials, and supportive learning technologies, which are interconnected and mediate the evolving activities (Hämäläinen & Vähäsantanen, 2011; Littleton et al., 2012; Viilo, 2020). Building thematic continuity through learning activities, curriculum materials, tools, peer discussions, and teacher facilitation can increase and sustain students' interest and motivation. It also enhances their ability to connect and apply disciplinary knowledge and practices to solve meaningful problems and take agentially central roles in developing their understanding and skills through collaborative activities (Damşa et al., 2010; Hakkarainen & Seitamaa-Hakkarainen, 2022).

8.4 Related Work

The increasing presence of AI technologies in daily life has underscored the need for AI education. Educational initiatives, such as the "Elements of AI" MOOC, have provided a broad and accessible introduction to AI for a general audience (Heintz & Roos, 2021). However, there is growing recognition of the importance of introducing AI concepts to children, as they are interacting with AI technologies more than ever before (Tedre et al., 2021b; Touretzky et al., 2019; Zhou et al., 2020). Research highlights the need to incorporate AI topics into educational programs early on to cultivate a data-literate society (Shapiro et al., 2018), and various guidelines have been developed regarding what AI education in the K-12 context should include. According to Long and Magerko (2020), AI education in K-12 should cover fundamental questions such as the following: "What is AI?" "What can AI do?" "How does AI work?" "How should AI be used?" "How do people perceive AI?" The AI4K12 guidelines (Touretzky et al., 2019) introduce five core concepts that should be included in K-12 AI education: perception (how AI systems sense the world), representation and reasoning (how AI models information and makes decisions), learning (how AI systems improve over time), natural interaction (how AI communicates with humans), and societal impact (the effects of AI on society). These guidelines are designed to ensure that students gain a comprehensive understanding of AI, its mechanisms, and its ethical implications.

In addition to guidelines, several tools and platforms specifically designed for K-12 education have been developed to make learning about AI accessible and engaging for children (Gresse von Wangenheim et al., 2021; Marques et al., 2020). For example, the web-based SmileyCluster learning environment teaches k-means clustering through data visualization, supporting hands-on, collaborative learning (Wan et al., 2020). Another tool, described by Mariescu-Istodor and Jormanainen (2019), is a web-based application that teaches the basics of ML by allowing them to build an object recognition system.

The GTM is one of the most widely used tools in AI education. It provides a user-friendly interface that enables students to experiment with ML concepts with minimal programming knowledge (Toivonen et al., 2020). It has been successfully integrated into educational programs like the "Machine Learning for All" course in Brazil, which introduces middle and high school students to basic ML concepts and guides them in creating image recognition models. The course emphasizes hands-on learning, aligns with AI4K12 guidelines, and also addresses the ethical and societal impacts of AI (Gresse von Wangenheim et al., 2021). The tool has also been used with young children in Nigeria, fostering their data literacy (Sanusi et al., 2024). Similarly, in a Finnish elementary school, sixth-grade students co-designed ML applications related to their interests using the GTM, thereby enhancing their understanding of ML principles through hands-on exploration (Vartiainen et al., 2021).

The crucial role of teachers in AI education has also been highlighted in research. Lin and Van Brummelen (2021) explored integrating AI education into K-12 curricula through co-design workshops with teachers. They suggest that embedding AI topics into core subjects can make AI education more accessible, allowing students to explore AI concepts, reflect on AI ethics, and help teachers confidently introduce AI topics into classroom settings. In a recent study, Jormanainen et al. (2024) emphasize the teacher's role in connecting disciplinary concepts to students' hands-on activities, thereby making complex ML concepts more explicit and relevant for students. Moreover, this study underscores the importance of a distributed scaffolding approach, where the teacher's guidance, technological affordances, and peer interactions work together to make ML concepts accessible and meaningful to learners (Jormanainen et al., 2024).

These studies together underscore the importance of involving educators in the design of AI curricula and the effectiveness of hands-on, interest-driven approaches in teaching AI concepts to young learners. However, our study adds a robust pedagogical framework—design-oriented pedagogy—into the process. Moreover, our work goes further by developing a new educational tool, the GenAI Teachable Machine, specifically designed for classroom use. This tool, co-designed with teachers, is user-friendly, is GDPR-compliant, and covers the entire ML workflow, making it well-suited for the K-12 environment (Pope et al., 2024). By combining teacher involvement, student interest, and a strong pedagogical foundation, our study provides a comprehensive and practical model for integrating AI education into school curricula.

8.5 Implementation of Workshops

This section details workshops that are part of an ongoing series of design-based research (Kahila, Vartiainen, Tedre, & López Pernas, 2024b; Kahila, Vartiainen, Tedre, Arkko, et al., 2024a; Tedre, Toivonen, et al., 2021b; Vartiainen et al., 2020, 2022, 2024) focused on incorporating AI topics into educational settings. Workshops were conducted under a broad research program led by the University of Eastern Finland, designed to enhance children's data agency, defined as individuals' ability and volition to take informed actions that impact their digital worlds (Tedre et al., 2020). This chapter focuses on the first round of AI education held in Eastern Finland in the spring of 2023, attended by over 200 Finnish fourth and seventh grade students. The project followed the guidelines of the Finnish National Board on Research Integrity (TENK, 2019). Before school visits, research permission was secured from the local educational administration and the children's guardians. Guardians received an information sheet describing the voluntary nature of participation, the right to withdraw, and details about data collection, processing, and reporting. The children were also informed about the study and gave verbal consent at the first workshop's start.

The intervention comprised three two-hour workshops, totaling six classroom hours. During these sessions, students were introduced to the basics of ML, data-driven design, and the impacts and ethics of AI. Although researchers and schoolteachers collaboratively taught the student groups, a researcher who is also a classroom teacher led the AI topics, due to the general lack of AI expertise among the schoolteachers. Students' activities were supported by various scaffolding methods, including presentation slides, peer discussions, educational technology, and curriculum materials (for a more detailed description, see Kahila, Vartiainen, Tedre, Arkko, et al., 2024a).

A large part of the project was dedicated to designing and implementing AI applications, which resulted in dozens of apps reflecting students' interests. These included, for example, apps that recognized dance poses, animals, country flags, soccer players, poisonous mushrooms, and other plants. Other apps told jokes, recommended clothing, or assisted color-blind individuals in identifying colors. These applications illustrated how students' interests guided the process and the resulting products, creatively integrating AI to solve problems important to them. In the following sections, we detail the goals and general flow of the three workshops and explore the activities of a group of four fourth-grade students—Mirjam, Leo, Olivia, and Laura—providing examples from their work and discussions, transcribed from video data.

8.5.1 *Workshop*

The first workshop focused on discussing AI and its impact on daily life, as well as introducing the basic concepts of creating an ML classifier. It also familiarized students with the GenAI Teachable Machine, encouraging them to collaboratively explore and experiment with it.

The teacher began the workshop by briefly explaining the types of applications AI has and how they function. He provided practical examples of AI's involvement in everyday activities, products, and services familiar to children, such as recommendation systems and targeted advertising on social media. Following these examples, students completed their first written assignment, analyzing which social media platforms they use and whether these platforms employ AI. The results were then discussed in groups:

Leo: I use Instagram, WhatsApp, TikTok, Facebook
Laura: I use Instagram, YouTube, and WhatsApp
Olivia: I use TikTok, WhatsApp, YouTube, and Pinterest
Leo: What is Pinterest?
Laura: Pinterest is like where you find all sorts of pictures
Olivia: Yeah, I use it a lot to find different things
Laura: I've only searched there twice

(The group members also compared which social media platforms they thought used AI)

Olivia: Pinterest, TikTok, Snapchat, YouTube, and Instagram
Laura: I have Instagram, Snapchat, TikTok

The main goal of this assignment was to demonstrate how AI technologies are already integrated into their daily lives. Similar to most children participating in this study (Kahila, Vartiainen, Tedre, & López Pernas, 2024b), the example group were active social media users. They also recognized that the services they actively use employ artificial intelligence.

After this, the students completed their first group assignment, in which they explored the types of data that social media platforms collect from their users, how this data is utilized, and the potential risks associated with its use. Each group selected one platform for analysis, discussed their thoughts, and documented their findings. The example group chose Instagram. They began by discussing the data collected during user registration, followed by an examination of data collection practices during ongoing use of Instagram:

Laura: Can I say that it collects pictures?
Olivia: Yeah, you can.
Laura: At least it probably saves them.
Olivia: Yeah, right, write that it saves them.

(After that, the group members discussed the use of data and its risks)

Mirjam: What risks do you think are associated with the use of collected data?
Laura: Uh, from the pictures you send … like things can be taken from those pictures for AI

The main aim of the task was to connect the concept of training data to the students' own experiences. Although the members of the example group lacked a clear understanding of how their actions on social media are training data, they already realized that the images they post to social media are somehow used to train AI.

After the assignment, the teacher led a class discussion on how AI systems use various data traces in their operation. The students explored how individuals could be classified into groups like gamers or skateboarders based on their data traces. This discussion introduced the classifier as the second key concept, effectively linking it to the students' daily interactions.

Next, students were introduced to the GenAI Teachable Machine. The teacher demonstrated how the GenAI Teachable Machine works by creating a rock-paper-scissors classifier that recognizes hand positions. He also illustrated the significance of key concepts in practice: how images of different hand signs formed the basis for classification (training data), how accurately the classifier recognized various hand positions (confidence), and how small changes in the environment could impair the classifier's performance (brittleness). Students then worked in small groups to explore these concepts and the operational mechanisms of ML by developing their own classifiers based on their ideas (Fig. 8.2). They used a worksheet that guided the generation and categorization of training data and included questions to help them test and analyze the classifier's performance.

The example group created a classifier that recognized a stuffed toy (Nuffi), a plastic llama, and a glue bottle. They also tested their classifier

Fig. 8.2 Fourth-grade students explore the tool through collaborative work with peers

with other objects and discussed with the teacher why Mirjam's pencil case was so strongly classified as Nuffi:

Olivia: It thought Mirjam's pencil case was Nuffi.
Teacher: What is Nuffi?
Olivia: This (shows Nuffi).
Teacher: Oh, that's Nuffi.
Olivia: And this is Mirjam's pencil case, that's why it thought so.
Teacher: And Mirjam's pencil case is also furry.
Olivia: Yeah...

(After the teacher left, they continued discussing.)

Laura: I wonder what the teacher would look like?
Olivia: Probably like Nuffi since he has the same-colored shirt.

The goal was to familiarize students with the GenAI Teachable Machine and connect their ideas and findings to the basic principles of ML. The example group quickly created a classifier for the objects available. Inspired by this hands-on practice, they began to consider the mechanics of AI, such as which image features influence image recognition. Subsequently, the teacher showed how to add actions to the classifier, building upon the rock-paper-scissors example used earlier. As homework, students were tasked with ideating their own AI applications that could be implemented using the GenAI Teachable Machine.

8.5.2 Workshop

The second workshop was designed to facilitate the ideation, design, and implementation of the children's own classifier-based ML applications. It also aimed to enhance their understanding of the key concepts and workflows of machine learning, further deepening their practical and theoretical knowledge.

Before the second workshop began, the teacher reviewed the students' ideas and their proposed implementations to understand what support the small groups would need. The workshop started with a brief recap of the key concepts and classifier construction steps from the previous session, followed by an outline of the learning goals for this workshop. The workshop then transitioned to small group activities, where students discussed their homework application ideas, aiming to select one idea to further develop as a team. The groups required varying levels of teacher support to choose a feasible idea that met everyone's approval.

The example group had four application ideas to consider. After discussing with the teacher, they found that two of the ideas were not feasible with the GenAI Teachable Machine. The remaining ideas were a dog breed identifier and an animal identifier. Guided by the teacher, they chose the animal identifier since another group was already working on the dog breed identifier.

Once the app idea was decided, each group received a worksheet to guide the design and implementation process. This worksheet provided an overview of the key steps in the development process and included questions designed to focus attention on essential aspects. Teams documented their progress, choices, observations, and conclusions on the worksheet, with the teacher reviewing these questions with the groups:

Teacher: Here it asks what the app does, and what it needs to distinguish between. And if you think about teaching the computer to differentiate between things, does it make sense to have it distinguish between, say, a striped field mouse and a yellow-necked mouse? What do you think?

Laura: No, they are similar.

Teacher: Yes, they look quite similar. So, what kinds of animals should you choose?

Olivia: Different ones.

After the teacher left, the group discussed which animals to train the classifier to recognize. They chose a pig, a squirrel, and an elephant. Throughout the workshop, teachers provided support and guidance as necessary. The example group used images downloaded from the Internet to train the classifier. While documenting their work on the worksheet, they discussed what qualifies as training data for their project"

Olivia: So, we write that we are getting animal pictures from the internet.
Teacher: The animal pictures are the training data used to train it (the classifier).

(Students nod)

Enhancing the children's understanding of key ML concepts—classifier, training data, brittleness, and confidence—was an important goal of the workshop. Teachers used consistent terminology and, whenever possible, connected these concepts to the students' projects and everyday experiences.

Once the groups finalized their plans and received the teacher's approval, they moved into the application implementation phase, using the GenAI Teachable Machine. Teachers and worksheets guided and supported the students' work. The example group's application was designed not just to display text and pictures but also play the sound of the recognized animal (elephant, squirrel, or pig). The students recorded these sounds in a quiet area outside the classroom and then tested and documented the functionality of their application by showing it with pictures of animals:

(Pig sound)

(Everyone laughs)

Laura: It worked well!
Leo: Write that it works … Put the elephant, put the elephant!

(Elephant sound)

(Everyone laughs)

The students were very excited about the AI application they had created. Members of the example group were particularly interested in animals, with three out of their four original ideas related to animals. They were also very skilled at mimicking various animal sounds. Open-ended tasks that involved designing and creating AI applications based on their interests provided an opportunity to create applications that interested them and utilized their special skills, thereby making learning both engaging and meaningful. Furthermore, in addition to collaboration and mutual learning within the group, different small groups also influenced and learned from each other. The example group's animal app captured the attention of others, inspiring some to start designing sounds for their own applications.

8.5.3 Workshop

The third workshop focused on two main objectives: sharing and presenting the groups' apps and projects, and analyzing the benefits, impacts, and risks of AI, including algorithmic bias. This session aimed to enhance students' understanding of AI's practical applications and ethical considerations.

At the start of the workshop, students prepared presentations for their apps using a worksheet that encouraged reflection on their design and learning processes to enhance their understanding of data-driven thinking. The worksheet also prompted teams to consider the ethical implications and broader impacts of their AI applications, highlighting that AI design extends beyond technical skills. The groups documented their thoughts and discoveries throughout the process before presenting their apps to the entire class.

After the presentations, the workshop moved into its second phase, where algorithmic biases were examined with the support of generative AI. The students were introduced to the topic using the familiar rock-paper-scissors example from the first workshop, as well as applications they had developed. These examples were used to illustrate manually classified training data. The teacher then explained how these processes are automated in AI applications that analyze Internet content and learn from vast numbers of images available online. Furthermore, the teacher explained how, in addition to recognizing images (from image to text), this training data can also be used to generate images (from text to image). To exemplify algorithmic bias, the teacher showed images of leaders produced by a generative AI (Midjourney), illustrating how biases are embedded in AI systems:

Teacher: Do you notice anything unusual about these leaders?
Student: They all look pretty much the same.
Teacher: They do all look quite similar, what do they look like?
Student: They all wear black suits. They all have similar mustaches.
Teacher: Yes.

(several suggestions on how all the leaders in the picture look similar: angry, bald, human)

Student: Old.
Teacher: Yes, they are old, what else?
Student: Men.
Teacher: Old men. So, are all leaders old men?
Several Students: Nooo!
Teacher: Exactly, but still, the AI shows that all leaders are old men. That's called bias in training data. So even though not all leaders in real life are old men, most of the images of leaders on the internet show old men. That's why the AI (Midjourney) starts to interpret that leaders look like this.

The teacher also presented additional examples of bias in images generated by artificial intelligence, illustrating how training data can reflect biases, structural injustices, and discrimination, thereby leading to biased computer outcomes. Following this exploration, students worked in small groups using generative AI to create images and analyze algorithmic biases. Using a structured worksheet, they described the biases, discussed their causes, and proposed solutions to correct them. The example group primarily prompted images of various professions such as police officer, soldier, teacher, and firefighter. Since the images were displayed on a shared channel, students could also see the images (and biases) found by others. The example group actively observed and commented on images created by other groups:

Olivia: Here it says chef, hey it says chef and it's only men.
Olivia: Look, teacher, someone typed in 'chef' here, and it only showed men.
Teacher: Yes, that was over there with them (points to another group), they prompted a chef. And according to that, all chefs in the world are men with mustaches.

The students also noticed a stereotypical-looking image of a bully created by another group. Laura remarked: "Look, bully. They don't really look like that. There are no women." For their worksheet assignment, the example group analyzed firefighter images they had generated, noting, "this bias is common. We wrote firefighter and it only showed young men. No women appeared." To address this, they suggested increasing the number of images of female firefighters on the Internet to correct the bias.

This activity aimed to deepen children's understanding of AI's impacts and help apply what they learned to new situations. The children understood the role of training data in developing algorithmic bias and methods to correct it. The students could understand this complex phenomenon because they learned the basic mechanics and concepts of artificial intelligence, such as training data, during the workshops.

8.6 Discussion

This chapter explored the integration of AI education into K-12 curricula through a co-design project in Finnish schools, focusing on the development of AI-driven applications by students. In addition to examining core AI concepts, the project addressed the societal impact of AI technologies. Alongside a general description of the workshops, the study followed the progress of one student group, offering detailed insights into their learning process. The findings provide valuable insights into the effective teaching of AI concepts and their societal implications for young learners.

Contextualizing AI within students' daily lives was crucial for making the learning experience meaningful. By linking AI to familiar services, such as social media, students began to grasp how AI operates in their everyday environments. The example group, active social media users, initially intuited that their social media posts were used as training data. As the workshops progressed, their understanding of training data deepened alongside other key ML concepts and mechanics. For example, the group discussed how their classifier might categorize their teacher, considering factors such as his clothing. As students gained a more nuanced understanding of how AI systems work, they also began to critically evaluate the accuracy and limitations of these technologies.

The study highlights the effectiveness of design-oriented pedagogy as a framework for teaching AI. This approach emphasizes social interaction and hands-on engagement with materials, tools, and technologies to support collaborative learning (Hakkarainen & Seitamaa-Hakkarainen, 2022; Kafai et al., 2020). Instead of guiding students toward predefined solutions, the project encouraged them to pursue open-ended design problems stemming from their own interests. For instance, the example group was interested in animals and chose to create an animal recognition application, incorporating their unique skill of mimicking animal sounds. This autonomy empowered students to build conceptual understanding through projects that held personal meaning for them. Previous research supports the importance of hands-on, interest-driven learning in making AI concepts more accessible (Gresse von Wangenheim et al., 2021; Jormanainen et al., 2024; Vartiainen et al., 2021). The understanding of ML mechanics developed through these projects also facilitated discussions on AI ethics and societal impact, including algorithmic bias. Statistically significant progress was observed in students' ability to explain these biases (Vartiainen et al., 2024).

The project also underscores the crucial role of teachers in facilitating AI education through scaffolding. Teachers connected abstract ML concepts to hands-on activities, making them more explicit and meaningful for students. This aligns with Jormanainen et al. (2024), who emphasize the importance of teacher guidance, tools, and peer interactions in fostering student learning in AI education. Effective scaffolding includes curriculum materials, tools, and technologies that promote collaboration and enable students to apply disciplinary knowledge. By managing these elements, teachers help students connect abstract AI concepts to practical applications, keeping them motivated and engaged in collaborative learning (Hämäläinen & Vähäsantanen, 2011; Puntambekar & Kolodner, 2005). Moreover, teachers played a critical role not only during the workshops but also in co-designing the intervention, ensuring that complex AI concepts were communicated in ways that children could understand. This finding echoes Lin and Van Brummelen (2021), who highlight the importance of co-designing AI curricula with teachers to make AI education more accessible.

While these findings present a largely positive picture of AI education, it is also important to consider broader questions about the role of technology in education. In line with the tradition of critical educational technology (Selwyn, 2010), it is essential to recognize that tools like the

GenAI Teachable Machine are not neutral but socially shaped by wider political, economic, and cultural contexts. Although our workshops aimed to empower students and support ethical reflection, these efforts should also be understood as part of a broader conversation about equity, participation, and the social justice implications of technology in education.

8.6.1 Implications for AI Education in K-12

The findings from this study suggest several implications for AI education in K-12 curricula. First, involving teachers in the design of AI curricula is essential for ensuring that complex AI concepts are communicated effectively. Teachers play a vital role in simplifying difficult ideas and making them engaging for young learners. Co-designing curricula with educators, as recommended by Lin and Van Brummelen (2021), ensures that AI education is both accessible and meaningful.

Second, providing students with autonomy to pursue interest-driven projects enhances their learning experience. Allowing students to work on topics that are personally relevant—such as the example group's animal recognition project—motivates them and makes AI concepts more tangible. This autonomy fosters a deeper understanding of both the technical and societal aspects of AI. Therefore, integrating open-ended, interest-driven projects into AI education should be a priority.

Finally, while tools like the GenAI Teachable Machine are critical, they are most effective when combined with robust pedagogical frameworks like design-oriented pedagogy. To help students become both creators and critical thinkers in the digital world, future educational programs should provide accessible tools along with well-structured pedagogical support to maximize the impact of AI education.

8.7 Conclusions

The integration of AI education within K-12 curricula represents a significant step toward preparing the next generation to navigate and thrive in a data-driven world. This chapter explored an interdisciplinary educational project that introduced Finnish fourth and seventh graders to key AI concepts by engaging them in the design and creation of AI-driven applications based on their own ideas. This hands-on approach, facilitated by the GenAI Teachable Machine, exemplifies the pedagogical potential of blending theoretical knowledge with practical application.

The workshops were designed to connect AI topics to the children's daily experiences, and to teach AI's key concepts and workflows, highlighting AI's practical relevance in their own lives. Through the process of designing and implementing their own AI-driven applications, AI transformed from an abstract concept into a practical tool for addressing real-world issues related to their interests. This hands-on approach made the study of AI concepts and mechanics engaging and meaningful for the children. Moreover, the collaborative projects aimed to teach more than just technical skills, as learning to critically evaluate AI technology is essential for becoming informed and ethical citizens. As children learned the basic concepts and mechanics of AI, they were guided to consider societal and ethical implications of their apps. They deepened their understanding further by generating and analyzing AI-created images to identify algorithmic biases.

The GenAI Teachable Machine played a significant role in the workshops, providing students with a comprehensive workflow to create their own AI-driven apps (Pope et al., 2024). However, the tool alone is insufficient for teaching AI concepts comprehensively. Teachers played a crucial role in managing how teaching materials, peer discussions, and technologies interacted to support student learning. They facilitated students' understanding of AI concepts and their application to real-world scenarios.

While the study yielded valuable insights, it also had some limitations. Conducted in a specific cultural and educational context in Finnish schools, its findings may not be fully generalizable to other settings. Moreover, the short workshop duration may have restricted the depth of students' exploration of AI concepts. Future research could address these limitations by conducting longer interventions and assessing the scalability of this approach in diverse educational settings. Investigating the long-term effects of AI education on students' understanding of ethics and societal implications would also help shape future curricula.

Despite these limitations, this project, which brought together researchers, teachers, and other stakeholders, effectively demonstrates the potential of co-design in educational settings to enhance students' understanding of AI and its societal impacts. By incorporating AI education into K-12 curricula, we can equip young learners not only with essential technical skills but also with the critical thinking necessary to ethically engage with AI technologies. Continued research and development of collaborative pedagogical models like this one are crucial for preparing future generations to navigate and shape our increasingly digital world.

Acknowledgments This study received funding from the Strategic Research Council (SRC) established within the Research Council of Finland, grants #352859 and #352876. The authors thank the January Collective for their continuous support.

References

Carney, M., Webster, B., Alvarado, I., Phillips, K., Howell, N., Griffith, J., Jongejan, J., Pitaru, A., & Chen, A. (2020, April 25). *Teachable machine: Approachable web-based tool for exploring machine learning classification*. Conference on Human Factors in Computing Systems—Proceedings. https://doi.org/10.1145/3334480.3382839

Damşa, C. I., Kirschner, P. A., Andriessen, J. E. B., Erkens, G., & Sins, P. H. M. (2010). Shared epistemic agency: An empirical study of an emergent construct. *Journal of the Learning Sciences, 19*(2), 143–186. https://doi.org/10.1080/10508401003708381

Dúo-Terrón, P., López-Belmonte, J., Marín-Marín, J.-A., & Moreno-Guerrero, A.-J. (2025). Exploring horizons: Study on the perceptions of primary students regarding the future of work in robotics and steam. In S. Papadakis & G. Lampropoulos (Eds.), *Social Robots and Artificial Intelligence in Education: Integrating AI in K-12 and Higher Education* (pp. 213–235). Palgrave Macmillan.

Gresse von Wangenheim, C., Hauck, J. C. R., Pacheco, F. S., & Bertonceli Bueno, M. F. (2021). Visual tools for teaching machine learning in K-12: A ten-year systematic mapping. *Education and Information Technologies, 26*(5), 5733–5778. https://doi.org/10.1007/s10639-021-10570-8

Hakkarainen, K., & Seitamaa-Hakkarainen, P. (2022). Learning by inventing: Theoretical foundations. In *Invention pedagogy—The Finnish approach to maker education* (pp. 15–27). Taylor and Francis. https://doi.org/10.4324/9781003287360-3

Hakkarainen, K., Paavola, S., Kangas, K., & Seitamaa-Hakkarainen, P. (2013). Socio-cultural perspectives on collaborative learning: Towards collaborative knowledge creation. In C. Hmelo-Silver, C. Chinn, C. Chan, & A. O'Donnell (Eds.), *The international handbook of collaborative learning* (pp. 57–73). Routledge.

Hämäläinen, R., & Vähäsantanen, K. (2011). Theoretical and pedagogical perspectives on orchestrating creativity and collaborative learning. In. *Educational Research Review, 6*(3), 169–184. https://doi.org/10.1016/j.edurev.2011.08.001

Heintz, F., & Roos, T. (2021). Elements of AI-teaching the basics of AI to everyone in sweden.iniated (Ed.), *EDULEARN21* (pp. 2568–2572). https://www.classcentral.com/course/independent-elements-of-ai-12469

Hennessy, S., & Murphy, P. (1999). *The Potential for Collaborative Problem Solving in Design and Technology**. https://doi.org/10.1023/A:1008855526312

Jormanainen, I., Vartiainen, H., Kahila, J., & Tedre, M. (2024). Exploring the dynamics of scaffolding in K-12 ML/AI education: Insights from a machine learning workshop. *IEEE International Conference on Advanced Learning Technologies (ICALT), 2024*, 210–212. https://doi.org/10.1109/ICALT61570.2024.00067

Kafai, Y., Proctor, C., & Lui, D. (2020). From theory bias to theory dialogue: Embracing cognitive, situated, and critical framings of computational thinking in K-12 CS education, 11(1). https://doi.org/10.1145/3381887

Kahila, J., Vartiainen, H., Tedre, M., Arkko, E., Lin, A., Pope, N., Jormanainen, I., & Valtonen, T. (2024a). Pedagogical framework for cultivating children's data agency and creative abilities in the age of AI. *Informatics in Education.* https://doi.org/10.15388/infedu.2024.15

Kahila, J., Vartiainen, H., Tedre, M., & López Pernas, S. (2024b). Children's data privacy fears on social media. https://doi.org/10.2139/ssrn.4794092

Keen, C. (2022). Apathy, convenience or irrelevance? Identifying conceptual barriers to safeguarding children's data privacy. *New Media and Society, 24*(1), 50–69. https://doi.org/10.1177/1461444820960068

Krajcik, J. S., & Blumenfeld, P. C. (2006). Project-based learning.

Lin, P., & Brummelen, J. van. (2021). *Engaging teachers to co-design integrated ai curriculum for k-12 classrooms.* Conference on Human Factors in Computing Systems—Proceedings1, 1–12. https://doi.org/10.1145/3411764.3445377

Littleton, K., Scanlon, E., & Sharples, M. (2012). Editorial introduction: Orchestrating inquiry learning: Karen Littleton, Mike Sharples and Eileen Scanlon. In K. Littleton, E. Scanlon, & M. Sharples (Eds.), *Orchestrating inquiry learning* (pp. 1–6). Routledge. https://doi.org/10.4324/9780203136195-6

Long, D., & Magerko, B. (2020). *What is AI literacy? Competencies and design considerations.* Conference on Human Factors in Computing Systems - Proceedings, 1–16. https://doi.org/10.1145/3313831.3376727

Lupton, D., & Williamson, B. (2017). The datafied child: The dataveillance of children and implications for their rights. *New Media & Society, 19*(5), 780–794. https://doi.org/10.1177/1461444816686328

Mariescu-Istodor, R., & Jormanainen, I. (2019). Machine learning for high school students. In *Proceedings of the 19th Koli Calling International Conference on Computing Education Research* (pp. 1–9). https://doi.org/10.1145/3364510.3364520

Marques, L. S., Gresse Von Wangenheim, C., & Hauck, J. C. R. (2020). Teaching machine learning in school: A systematic mapping of the state of the art. *Informatics in Education, 19*(2), 283–321. https://doi.org/10.15388/INFEDU.2020.14

Pangrazio, L., & Selwyn, N. (2019). 'Personal data literacies': A critical literacies approach to enhancing understandings of personal digital data. *New Media and Society, 21*(2), 419–437. https://doi.org/10.1177/1461444818799523

Pope, N., Vartiainen, H., Kahila, J., Laru, J., & Tedre, M. (2024). *A no-code AI education tool for learning AI in K-12 by making machine learning-driven apps.* 2024 IEEE International Conference on Advanced Learning Technologies (ICALT), 105–109. https://doi.org/10.1109/ICALT61570.2024.00037

Puntambekar, S., & Kolodner, J. L. (2005). Toward implementing distributed scaffolding: Helping students learn science from design. *Journal of Research in Science Teaching, 42*(2), 185–217. https://doi.org/10.1002/tea.20048

Resnick, M., Myers, B., NakaKoji, K., Shneiderman, B., Pausch, R., Selker, T., & Eisenberg, M. (2005). Design principles for tools to support creative thinking.

Rizvi, S., Waite, J., & Sentance, S. (2023). Artificial Intelligence teaching and learning in K-12 from 2019 to 2022: A systematic literature review. *Computers and Education: Artificial Intelligence, 4.* https://doi.org/10.1016/j.caeai.2023.100145

Sanusi, I. T., Sunday, K., Oyelere, S. S., Suhonen, J., Vartiainen, H., & Tukiainen, M. (2024). Learning machine learning with young children: Exploring informal settings in an African context. *Computer Science Education, 34*(2), 161–192. https://doi.org/10.1080/08993408.2023.2175559

Seitamaa-Hakkarainen, P., Viilo, M., & Hakkarainen, K. (2010). Learning by collaborative designing: Technology-enhanced knowledge practices. *International Journal of Technology and Design Education, 20*(2), 109–136. https://doi.org/10.1007/s10798-008-9066-4

Seitamaa-Hakkarainen, P., Sormunen, K., Davies, S., Matilainen, J., & Hakkarainen, K. (2022). Collaboration and co-regulation in Invention Projects. www.routledge.com/Routledge-

Selwyn, N. (2010). Looking beyond learning: Notes towards the critical study of educational technology. *Journal of Computer Assisted Learning, 26*(1), 65–73.

Shapiro, R. B., Fiebrink, R., & Norvig, P. (2018). Education: How machine learning impacts the undergraduate computing curriculum. *Communications of the ACM, 61*(11), 27–29. https://doi.org/10.1145/3277567

Tabak, I. (2004). Synergy: A complement to emerging patterns of distributed scaffolding. *Journal of the Learning Sciences, 13*(3), 305–335. https://doi.org/10.1207/s15327809jls1303_3

Tedre, M., Vartiainen, H., Kahila, J., Toivonen, T., Jormanainen, I., & Valtonen, T. (2020). Machine learning introduces new perspectives to data agency in K–12 computing education. *IEEE Frontiers in Education Conference (FIE), 2020*, 1–8. https://doi.org/10.1109/FIE44824.2020.9274138

Tedre, M., Denning, P., & Toivonen, T. (2021a). CT 2.0. *Proceedings of the 21st Koli Calling International Conference on Computing Education Research*, 1–8. https://doi.org/10.1145/3488042.3488053

Tedre, M., Toivonen, T., Vartiainen, H., Jormanainen, I., Valtonen, T., Kahila, J., & Pears, A. (2021b). Pedagogical and technological trajectories for artificial intelligence education. *IEEE Access, 9*, 110558–110572. https://doi.org/10.1109/ACCESS.2021.3097962

TENK (Finnish National Board on Research Integrity). (2019). The ethical principles of research with human participants and ethical review in the human sciences in Finland. https://tenk.fi/sites/default/files/2021-01/Ethical_review_in_human_sciences_2020.pdf

Toivonen, T., Jormanainen, I., Kahila, J., Tedre, M., Valtonen, T., & Vartiainen, H. (2020). *Co-designing machine learning apps in K–12 with primary school children*. 2020 IEEE 20th International Conference on Advanced Learning Technologies (ICALT), 308–310. https://doi.org/10.1109/ICALT49669.2020.00099

Touretzky, D. S., Gardner-Mccune, C., Martin, F., & Seehorn, D. (2019). *K-12 guidelines for artificial intelligence: What students should know*. ISTE Conference.

Vartiainen, H., Tedre, M., & Valtonen, T. (2020). Learning machine learning with very young children: Who is teaching whom? *International Journal of Child-Computer Interaction, 25*. https://doi.org/10.1016/j.ijcci.2020.100182

Vartiainen, H., Toivonen, T., Jormanainen, I., Kahila, J., Tedre, M., & Valtonen, T. (2021). Machine learning for middle schoolers: Learning through data-driven design. *International Journal of Child-Computer Interaction, 29*. https://doi.org/10.1016/j.ijcci.2021.100281

Vartiainen, H., Pellas, L., Kahila, J., Valtonen, T., & Tedre, M. (2022). Pre-service teachers' insights on data agency. *New Media and Society*. https://doi.org/10.1177/14614448221079626

Vartiainen, H., Kahila, J., Tedre, M., López-Pernas, S., & Pope, N. (2024). Enhancing children's understanding of algorithmic biases in and with text-to-image generative AI. *New Media & Society*. https://doi.org/10.1177/14614448241252820

Viilo, M. (2020). *Marjut Viilo Teacher's orchestration in longitudinal collaborative inquiry process* (Issue 77). Helsinki Studies in Education.

Wan, X., Zhou, X., Ye, Z., Mortensen, C. K., & Bai, Z. (2020). SmileyCluster: Supporting accessible machine learning in K-12 scientific discovery. *Proceedings of the Interaction Design and Children Conference, IDC, 2020*, 23–35. https://doi.org/10.1145/3392063.3394440

Wong, G. K. W., Dillenbourg, P., & Huan, J. (2020). Broadening artificial intelligence education in K-12: Where to start? *ACM Inroads, 11*(1), 20–29. https://doi.org/10.1145/3381884

Zhou, X., van Brummelen, J., & Lin, P. (2020). Designing AI learning experiences for K-12: Emerging works, future opportunities and a design framework. ArXiv Preprint. http://arxiv.org/abs/2009.10228

CHAPTER 9

Exploring Horizons: Study on the Perceptions of Primary Students Regarding the Future of Work in Robotics and Steam

Pablo Dúo-Terrón, Jesús López-Belmonte, José-Antonio Marín-Marín, and Antonio-José Moreno-Guerrero

9.1 Introduction

The transformation of educational spaces, along with the use and development of technology, is a current trend and concern of many institutions (OECD, 2018), among which are schools and teacher training centers (Lessard et al., 2015). The learning space at school requires a change in the role of teachers and adaptation to new contexts (Bautista & Borges, 2013). One of the proposals with the most didactic relevance in the

P. Dúo-Terrón • J. López-Belmonte (✉) • J.-A. Marín-Marín •
A.-J. Moreno-Guerrero
University of Granada, Granada, Spain
e-mail: pabloduo@ugr.es; jesuslopez@ugr.es; jmarin@ugr.es; ajmoreno@ugr.es

S. Papadakis, G. Lampropoulos (eds.), *Social Robots and Artificial Intelligence in Education*, Digital Education and Learning,
https://doi.org/10.1007/978-3-032-03977-4_9

educational field is the Classroom of the Future (CF) (González et al., 2022). The changes that classrooms are undergoing during the last decade due to the emergence of digital devices, although they may be beneficial, must be the subject of a pedagogical analysis prior to their introduction that allows us to decide if they really fulfill an educational purpose (Martín-Lucas & Sánchez-Rojo, 2023). In this sense, countries like Spain, through the School Code 4.0 program, are immersed in projects with large financial allocations from Europe to promote active methodologies and advanced digital skills in educational centers, in teachers and students (MEFPD, 2023). But this educational innovation and digitalization requires studies of the spaces of creation and computational thinking (CP) that represent an unprecedented challenge of training citizens who will access the increasingly digitalized and globalized labor market in the future (Dúo-Terrón et al., 2022).

The use and application of tools and techniques in educational environments arises from activities directed by teachers rather than being imposed on learning situations (Hickman, 2016). The transformation of the classroom as a "space" where science, technology, engineering, art and mathematics (STEAM) education takes place contributes to the development of practical skills, encourages creativity in problem-solving and promotes digital innovations (Chiriac, 2019; Kanematsu & Barry, 2016; Marín-Marín et al., 2021).

Although the perceptions of future teachers regarding how to use technology in learning may vary, they mostly agree on the need to use it effectively (Delaney et al., 2014). Learning projects and activities supported by technological tools such as programming, robotics or artificial intelligence have important implications for the future of classroom learning, which is no longer spatially limited or focused exclusively on textbooks (Harris, 2017). In addition, these spaces contribute to the development of computational thinking and digital skills of students (Dúo-Terrón, 2023).

The importance of incorporating CP as part of the curriculum reflects the current demands of society, both for our students and for our educators, and is closely linked to the technological advancement that characterizes all aspects of our daily lives (Del Venegas et al., 2021). Although not all the professions that our students will enter in the near future will require the use of computers or programmable technologies (Segura et al., 2019), the current conception of the CP transcends mere programming or the management of computing devices: it focuses in developing the processes and skills necessary to address and solve complex problems (Wing, 2008).

The current and future labor market demands knowledge in digital and social skills (Silva et al., 2021) that facilitate teamwork and the development of socio-emotional skills (Paleczek, et al., 2021). Therefore, it is crucial that schools prepare students to use new technologies and face contemporary challenges, integrating CP into their study plans from early ages to secondary education, as countries such as Spain already do (Dúo-Terrón et al., 2022; Marín-Marín et al., 2024).

In the educational field, robotics is associated with the development of CP skills (Amaral et al., 2022). Research in this field has been activated due to the possible benefits of robotics and the CP for learning programming through mathematics (Molina-Ayuso et al., 2022). This is because it allows students to explore, reflect, apply and reinforce mathematical concepts, such as negative numbers, coordinate planes and axes (Monteiro et al., 2019), angles, degrees, operations and geometry (Brender et al., 2021), showing them how they are applied in real-world contexts. This methodology reduces the workload of teachers in the teaching-learning process, since students are the main actors in this process (Bender et al., 2023). The project work philosophy based on robotics promotes the "Maker" culture in creation spaces (Makerspaces) (Lu, 2021) or Classrooms of the Future, where creation, development, research, exploration, interaction and presentation (Haduong, 2021) in turn stimulate interest in science, technology, engineering and mathematics (STEM) vocations (Dapozo et al., 2017).

Robotics in CF spaces in primary education has a significant importance in the development of key skills in students, as well as in generating motivation and enthusiasm for learning (Sánchez-Sánchez et al., 2020). On the one hand, robotics in primary education introduces students to areas of STEM in a practical and fun way. Students learn concepts such as programming, mechanics, electronics and problem-solving while working on robotics projects (García-Tudela & Marín-Marín, 2023).

Currently, with the growing influence of the digital society and economy, there is a growing demand for professionals who use robotics (Soto-Solier et al., 2023). This demand arises as machines replace human work, thus generating a need for profiles related to STEAM disciplines. It highlights the motivation and involvement that technology creates in students when designing and creating sustainable cities through robotics for our future (Ferrada Ferrada & Díaz-Levicoy, 2023). However, the lack of people trained in STEM fields has led to the emergence of the STEAM movement, which seeks to integrate the "A" of art and creativity (Conradty

& Bogner, 2020) as an alternative to address this lack and foster collaborative and creative skills in educational spaces.

From an early age, boys are often associated with games and activities related to science and technology, while girls are expected to be more interested in the arts and humanities (Juvera & Hernández López, 2021). The gender gap in STEAM jobs is a phenomenon that reflects a significant disparity in the representation and participation of men and women in these fields (Payeras et al., 2024). Although there have been advances in gender equality in many areas, the presence of women in STEAM careers remains notably lower than that of men (Michuy Gaibor et al., 2023).

In this context, education plays a fundamental role as preparation for the future of students, regardless of gender, equipping young people with relevant skills to face social, economic, political, environmental and cultural challenges (Bernate & Guativa, 2020). The importance of art in the development of skills is recognized, both in mathematics and in social and civic skills, providing an effective method for knowledge retention (Diego-Mantecón et al., 2021). Therefore, the educational approach must be adapted to promote comprehensive training that combines elements of STEAM, thus preparing students for success in their professional and personal lives in an increasingly digitalized world (Amaya et al., 2023).

In addition, it promotes critical thinking and problem-solving, where students face challenges in which they must apply critical thinking and develop creative solutions (Pineda-Caro, 2022). This helps them develop problem-solving skills that are useful in many areas of life. Also, develop collaboration and teamwork skills to design, build and program robots. This fosters collaboration, communication, and collaborative work skills among students (Kanobel & Carranza, 2023), as they must learn to work together to achieve a common goal.

Finally, it stimulates creativity and motivation, because robotics gives students the opportunity to create and design their own robots (García Fuentes et al., 2023), which allows them to express their ideas in a tangible way. The practical and playful nature of robotics captures the interest of students and motivates them to actively participate in the learning process. Seeing their creations come to life and perform specific tasks gives them a sense of accomplishment and satisfaction, which motivates them to continue learning and exploring.

Robotics involves skills such as problem-solving, creativity, collaboration and critical thinking, all of which are essential in the twenty-first century. In the CF, where the development of these skills is prioritized, the integration of robotics can help students acquire and develop these skills effectively.

For this reason and because there is little evidence in the scientific literature with student experiences in the CF, this study aims to do the following:

- Compare the concerns about job access trends of 12 primary education students prior to and after an experience with robots.
- Find out if students like creation spaces or CF to do CP projects more than being fixed in the classroom.

This chapter develops the methodology used at both the pedagogical and the investigative levels. Details are provided regarding the design used, the sample of students, as well as the procedure, development of the experiences and data analysis. The chapter concludes with a discussion of the findings and a final conclusion that reveals the scope and prospects of this educational technology and active methodology.

9.2 Methodology

This investigation has several phases. At first, authorization is requested to carry out a robotics project in a primary education center in the City of Ceuta that has the Classroom of the Future seal recognized by the Ministry of Education, Vocational Training and Sports (MEFPyD) of Spain, which it can currently be consulted on the web.

Subsequently, permission is requested from the educational center and the students' families to carry out the research and address ethical considerations based on the Declaration of Helsinki (Orozco & Lamberto, 2022).

Next, the questions are asked before the experimental phase and at the end of it, the same questions are asked again, always accompanied by the researcher and supported by a teacher specialized in computational thinking and ambassador of the Classroom of the Future in Spain.

9.2.1 Sample

The educational center where the sample was selected consists of two lines at the sixth-grade level of primary education with 47 students. Of the total, 6 students from each class were randomly selected by lottery, with the condition that there was parity and reducing research biases. The participants were finally 12 students from the sixth year of primary education from a public educational center in the city of Ceuta. Of the total, 6 were boys and 6 girls to address gender equality. Additionally, 2 of them are students with specific educational support needs.

9.2.2 *Instrument*

Students are asked via a Google Form to write down three to five jobs they would like to do in the future before the experience. This process is also carried out after the experience.

Furthermore, to assess the impact of carrying out projects in the CF, students are asked before and after the experience the following question: Do I like changing creation spaces and classrooms to do projects more than being fixed in the classroom? Students have the option to respond with Yes, No, same or similar.

9.2.3 *Experimentation Phase*

The students' first contact with a robot should be to know what the sensors and actuators are, as well as to present the robot, the platform and its basic visual programming blocks.

Taking the CF spaces as a work reference and in a staggered manner, during seven learning activities and with different challenges, the students linked mathematical concepts such as turns, degrees, geometric figures, divisions and repetitions with programming blocks such as loops, patterns and color sensors, and distances to end up carrying out and presenting STEAM projects with the Matatalab VinciBot robot and LEGO pieces that they will end up presenting to their classmates. The objectives of this experience were as follows:

- Know the basic concepts of programming and robotics.
- Simplify repetitive steps, such as loops.
- Incorporate mathematical competencies and skills into programming.
- Collaboratively produce an innovative and creative product using computational thinking.
- Develop creative thinking.
- Present a design.

The resources used were interactive digital screen and laptops, traditional whiteboard, five robot games and LEGO pieces, grid panel to slide the robot, Matatalab website, A3 sheets for designing line-following circuits, markers and glasses.

The way the students were grouped varied in pairs or trios. In addition, the teacher's role was as a mentor in the block programming phase in the first phases of the activities and, later, as a guide in the preparation of assembly, design and robot programming.

9.2.3.1 Activity 1: My First Steps with the Robot (Interact Zone)

On the interactive whiteboard, the teacher interacts with the sensors and actuators of a robot, simulating the sensory organs and the human locomotor system (Fig. 9.1).

Subsequently, the VinciBot Robot and the Matatalab programming platform are presented.

9.2.3.2 Activity 2: My First Programs with the Robot (Investigates Zone)

The first steps in programming are aimed at using forward, backward and rotating movements. Using the floor grid panel provided with VinciBot, students will program the following instructions:

- Go from point A to point B without turning.
- Go from point A to point B with turns.
- Go from point A to point B with turns and returns.

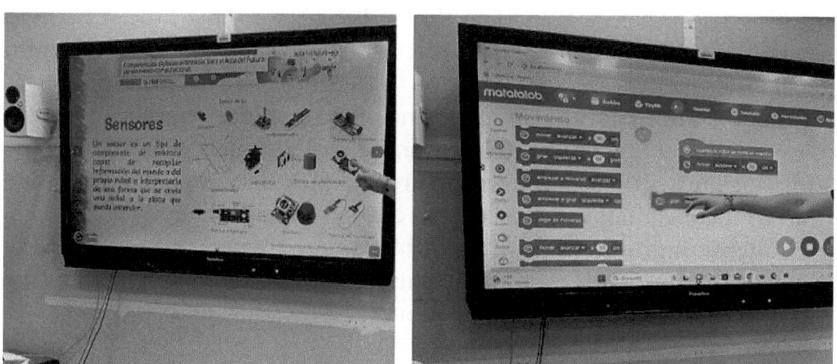

Fig. 9.1 Presentation of sensors and platform

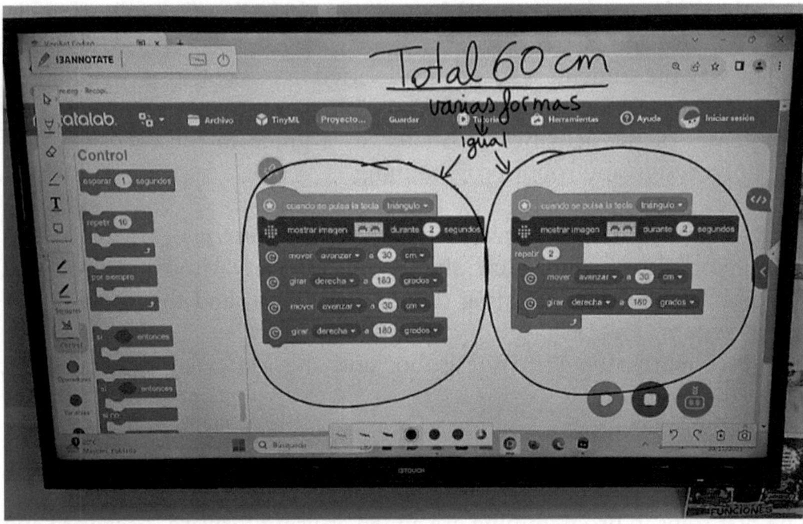

Fig. 9.2 Math activity with programming blocks

In this session, we coordinate with the mathematics teacher to explain the combined operations in parentheses, for example going from point A to point B with a distance of 20 cm and the return is equal to (20+20) or 2×20 (Fig. 9.2).

9.2.3.3 Activity 3: My First Project (Explora Zone)

In this session, before launching the challenge, we show how the sound programming blocks work. The teacher issues a challenge on the grid panel (Fig. 9.3).

The VinciBot robot is an ambulance that is in the "purple" grid but receives an emergency call from a house in the "blue grid" where the VinciBot must stop and sound the siren to pick up the patient and then head to the hospital, which is located on the "green" grid.

Students in groups of three to four begin programming and then present it to their peers. If a group finishes first, it is extended by asking the robot to return to the starting position.

Fig. 9.3 Math activity and floor panel

9.2.3.4 Activity 4: Create Mandalas (Create Zone)
In the first part of the session, we explain to the students how to program and draw geometric figures with the Vincibot, the total route corresponds to the perimeter.

Later we explain how to save programming blocks to make a square, triangle, pentagon. With repeated loops and rotating the corresponding degrees to add 360 degrees. For example, a four-sided square is to be repeated four times (move X steps and rotate 90°). In this part of the session, we draw the figures on A3 size paper and pose a challenge to the groups: How do you make a circle?

Finally, we remember the previous steps but this time we loop geometric figures to make mandalas that are presented to other groups (Fig. 9.4).

9.2.3.5 Activity 5: Sumo Wrestling (Developed Zone)
This activity is divided into three sessions of one hour:

A. First, students are shown the color sensor and test with their hands how the Matatalab detector shows the distance between the robot and the hand and how it varies. Additionally, they are challenged to make the VinciBot's lights change based on the color panel on the robot.

Fig. 9.4 Exhibition of mandalas by groups

They are challenged that depending on the color detected, the name of the color will appear on the main screen.

B. In this second part, students learn to make a robot turn 180°, that is, turn around when it detects black on the ground, making back-and-forth movements. Later they perform the same task but in a closed circle, but this time the robot will rotate randomly between 135° and 270°, so it will not perform back-and-forth movements.

C. In the third and last part, with previous knowledge, students must program a robot without leaving the circle to eject plastic cups from the circle. How? (Fig. 9.5).
    - It will always be rotating on its axis.
    - If it detects an obstacle (plastic cup) less than 30 cm away, it will advance until it detects a black line on the ground.
    - It will move back 10 cm and turn again until it detects a new obstacle.

Fig. 9.5 Students programming the robot to eject glasses

9.2.3.6 Activity 6: Creation of STEAM Projects (Create Zone)

These three hours task consist of three parts:

(a) Explanation and demonstration of the LEGO Kit with the robot.
(b) Selection by vote of the robot project that the students want to assemble.
(c) Program the programming blocks with Matatalab.

Below we show images of the team creation phase of a dancer from one of the groups (Fig. 9.6).

9.2.3.7 Activity 7: Presentation of Projects and Evaluation (Zone Presents)

The different groups present to their classmates the project carried out with their robot. In this case, in Figs. 9.7 and 9.8 we can see a robot programmed so that, when pressing the button on the remote control, it automatically shoots a basket into the plastic cup. The group called this project "the basketball scorer."

When working on projects in CF spaces, it does not make sense to carry out a written evaluation "exam on paper"; therefore, we use a competency evaluation rubric (Fig. 9.9) that allows us to evaluate the team members throughout the experimentation phase.

Fig. 9.6 STEAM project with LEGO pieces and robot

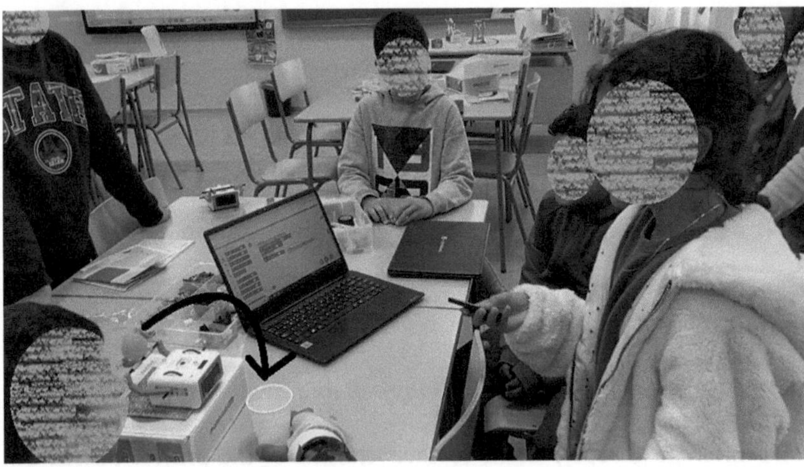

Fig. 9.7 Presentation of the basketball scoring robot

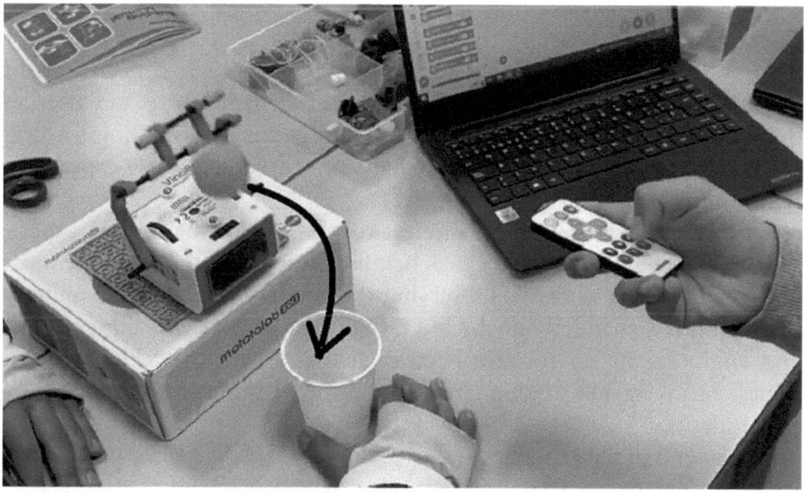

Fig. 9.8 Robot dunking in the glass

aula del futuro

RÚBRICA PARA EVALUAR LAS COMPETENCIAS DE NUESTRO ALUMNADO VINCULADA A LA SITUACIÓN DE APRENDIZAJE__________ NOMBRE DEL ALUMNO/A: __________ CURSO: 6ºC

ESPACIOS ADF	0 PUNTOS	1 PUNTO	2 PUNTOS	3 PUNTOS
INTERACTÚA	**No interactúa en actividades grupo-clase ni en pequeño grupo.** Es incapaz de controlar sus emociones y la de los demás.	**Interactúa cuando el docente propone retos o plantea situaciones,** con su equipo difícilmente controla sus emociones y no interactúa, y en ocasiones, pide un cambio de grupo.	Interactúa cuando el docente lo propone, pero **no propone retos ni ideas con su equipo** esperando que el resto tome la iniciativa, aunque controla sus emociones.	Interactúa cuando el docente propone retos y con su equipo, controlando sus emociones y la del resto de compañeros/as.
EXPLORA/ INVESTIGA	**No sabe utilizar o buscar información** (libros, diccionarios, internet…) adoptando una actitud pasiva.	**Suele preguntar al docente o compañeros para que le ayuden,** pero no investiga por sí solo o en grupo en ningún momento.	El docente o **compañeros son los que guían al estudiante** el método para investigar un contenido. Posteriormente, es capaz de investigar.	Posee habilidades y conocimientos para investigar sobre un contenido. **Sabe qué método utilizar (libros, diccionarios, internet…)**
DESARROLLA/ CREA	**No desarrolla ni crea** productos, tanto individualmente como en grupo, además, suele molestar..	Solo desarrolla y crea cuando trabaja **de manera individual,** pero **no en grupo,** en ocasiones, suele molestar.	**Desarrolla y crea productos de manera individual.** De manera **colectiva interviene muy poco.**	**Desarrolla y crea** cuando trabaja de manera **individual y en grupo.** Suele aportar ideas.
PRESENTA	**Se niega a presentar o exponer** un producto final de proyectos de diseño en diferentes formatos.	**Presenta** o expone proyectos en diferentes formatos **con poca motivación** por un corto periodo.	**Presenta** y/o expone proyectos en diferentes formatos realizados en el aula. Está **motivado** para exponer pero **no tiene fluidez oral comunicativa.**	**Presenta** el producto final de los proyectos de diseño en diferentes formatos **de manera motivada,** explicando los pasos seguidos con un **vocabulario fluido y una buena entonación.**

CALIFICACIÓN TOTAL:

intef INSTITUTO NACIONAL DE TECNOLOGÍAS EDUCATIVAS Y DE FORMACIÓN DEL PROFESORADO

Rúbrica diseñada por Pablo Dúo, derivada del Aula del Futuro INTEF

Fig. 9.9 Evaluation rubric by CF spaces. (Source: https://procomun.intef.es/ode/view/es_2023091712_9102600)

9.3 Results

The results corresponding to the first objective "Compare concerns about job access trends of 12 Primary Education students prior to and after an experience with robots" of the study "Compare concerns about job access trends of 12 Primary Education students prior to and after an experience with robots" can be seen in Table 9.1.

In the result of the jobs of interest to the students before the experience, it can be seen that police, teacher, firefighter and cook were the most in demand by the students. Observing the gender difference, the military profession is demanded only by boys. Only two professions, airplane pilot and engineer, appear to have been carried out with the STEAM disciplines, curiously only among girls, among the 21 professions proposed by the students.

At the end of the experience, the results vary. The first three professions most in demand by students are police, which has low interest, soccer player, and, in third place, computer science among three students. Previously, this job was not in the minds of students. If we look at the rest of the professions, it stands out that related to STEAM disciplines we find doctor, programmer, engineer, mathematician and scientist. The profession of computer scientist is demanded by two girls and a boy, that of doctor by two girls, that of programmer by a boy and a girl, and mathematician and scientist, both by a boy.

To know the results of the second objective, "Find out if the students like creation spaces or CF to do projects more than being fixed in the classroom." Table 9.2 shows that before the experience for 8 students out of the 12, changing spaces for them was equal to similar, even 1 student stated that he preferred the fixed classroom. At the end of the experience, 10 of the 12 students answered that they prefer to change to the creation spaces to carry out projects, only 2 students affirm that they find these spaces the same or similar to those in the fixed classroom.

9.4 Discussion

Currently and due to the influence of society and the digital economy, there is a growing need for specialized professionals in the field of robotics, in response to the progressive replacement of human work by machines. This generates a growing demand for profiles such as computer scientists, programmers, systems engineers and project leaders. It highlights how

Table 9.1 Jobs of interest to students

Pret-test trades	*n*	*M*	*F*	*Post-test trades*	*n*	*M*	*F*
Police	9	3	6	Police	5	2	3
Teacher	9	4	5	Soccer player	4	4	0
Firefighter	6	3	3	Computer scientist	3	1	2
Cook	4	2	2	Cook	2	1	1
Military	4	4	0	Doctor	2	0	2
Doctor	3	1	2	Military	2	1	1
Vehicle mechanic	2	2	0	Programmer	2	1	1
Vet	2	0	2	Teacher	2	0	2
Soccer player	2	2	0	Architect	1	0	1
Airplane pilot	2	0	2	Gardener	1	1	0
Interior designer	1	0	1	Engineer	1	0	1
Artist	1	0	1	Airplane pilot	1	0	1
Steward	1	1	0	Influencer	1	1	0
Fisherwoman	1	0	1	Vet	1	0	1
ATM	1	1	0	Singer	1	0	1
Fashion designer	1	0	1	Babysitter	1	0	1
Nurse	1	0	1	Vehicle mechanic	1	1	0
Engineer	1	0	1	Interior designer	1	0	1
Lawyer	1	0	1	Pharmaceutical	1	0	1
Ballet teacher	1	0	1	Mathematical	1	1	0
Waitress	1	0	1	Scientist	1	1	0

*n= Total responses; M=Male gender and F=Female gender.

Table 9.2 Question about projects in the CF

Do I like changing creation spaces and classrooms to do projects more than being fixed in the classroom?	*Pre-test (n=12)*	*Post-test (n=12)*
Yes	3	10
No	1	0
Same or similar	8	2

technology motivates and involves students by allowing them to design and create sustainable cities through robotics and CF spaces, with a view to the future. However, the lack of qualified people in STEM fields has led to the emergence of the STEAM movement in education from an early age, which seeks to integrate art and creativity as a solution to this lack, thus promoting collaborative and creative skills in educational environments, which we proceed to discuss.

In relation to the first objective of the study, the results of this study show that students have little knowledge in trades where STEAM disciplines play an important role. After experience, the authors of this research agree with the studies by Chiriac (2019), Kanematsu and Barry (2016), and Marín-Marín et al. (2021), which point out that learning situations where computational thinking and teamwork intervene encourage the development of practical skills, encourage creativity in problem-solving and promote digital innovations that translate into a broader vision toward professions unknown to students to access the labor market in the future.

Furthermore, this study supports the idea that STEAM disciplines through robotics support in the early educational levels provides an essential foundation as preparation for the future of students, regardless of gender, equipping young people with relevant skills to face access challenges to STEAM fields, as Bernate and Guativa (2020) point out. The results of this study after the experience indicate that working with robotics allows students to discover and comprehensively train success in their professional and personal lives in an increasingly digitalized world, as stated by Amaya et al. (2023).

In relation to the second objective of the study, creation spaces motivate students to carry out teaching-learning activities with technology, as pointed out by Sánchez-Sánchez et al. (2020). Although the traditional classroom is what is best known and used by students, as can be seen in the assessment before the experience. The authors consider that these results before the experience are derived from teacher training. Therefore, if teachers do not have skill and training in the pedagogical use of technological tools such as programming, robotics or artificial intelligence, this will have an impact on the implications for the future of students in access to STEAM disciplines, as argued by Harris (2017). Likewise, as Dúo-Terrón (2023) points out, the use of creation spaces such as the Classroom of the Future or makerspaces contribute to the development of computational thinking and digital skills of students working cooperatively and collaboratively (Dúo-Terrón, 2023).

9.5 Conclusion

The conclusions of this study are that carrying out robotics projects transversally in areas such as mathematics in creation spaces such as the Classroom of the Future allows students to open up concerns and interests in accessing trades that are unknown to them such as computing, programmers, mathematicians, scientists and engineers. In addition, there is an increase in interest that allows the gender gap to be reduced in STEAM fields from an early age such as primary education. Furthermore, for the majority of students at the end of the experience, working in the traditional classroom is not the same as in spaces like the Classroom of the Future that works transversally and with collaborative projects with technology.

As limitations of the study, it is considered that the sample used of 12 students with 6 students of each sex may be small, although this type of project must be worked with a small ratio or with human resources or support within the classroom from other professionals who are experts in the field subject.

Therefore, educational administrations should consider teacher training in computational thinking and creation spaces because these methodologies not only provide fundamental skills and competencies that are increasingly important in the digital society but also prepare students for jobs in STEAM fields to access the labor market and reduce the gender gap.

The findings of this study align with various discussions presented in related chapters within this part. Specifically, Kahila et al. (2025 [Chap. 8]) explore the engagement of young learners in artificial intelligence (AI) through co-designing AI-driven applications and share common ground with our study in terms of fostering digital literacy and computational thinking from an early stage. The Finnish schools' initiative described in Kahila et al. (2025 [Chap. 8]), where students created their own AI applications using the GenAI Teachable Machine, echoes the objectives of our research in demonstrating the importance of equipping students with hands-on experiences that enhance their understanding of emerging technologies. Similar to the findings in our study, this chapter highlights the role of educators in guiding students through the ethical and societal implications of AI, reinforcing the necessity of teacher training in digital education, a key consideration discussed in our conclusion.

Additionally, our study supports the notion that STEAM disciplines should be integrated across educational levels to better prepare students for future careers in technology-driven fields. This is reflected in the work of Le and Nga (2025 [Chap. 10]), which explores the use of Generative AI, specifically ChatGPT, in enhancing EFL students' paraphrasing skills. Although focused on language learning, Chap. 10 underscores the broader theme of leveraging AI technologies as educational tools. The study presented in this chapter identifies both the potential and the limitations of AI-based learning, paralleling our findings regarding the use of robotics and computational thinking in STEAM education. Just as Le and Nga (2025 [Chap. 10]) emphasize the need for a balanced integration of AI with traditional teaching methods to optimize learning outcomes, our study suggests that teacher training in technological tools is critical to ensuring effective pedagogical implementation.

Both related chapters support the central argument of our study: that the incorporation of AI, robotics and computational thinking into educational settings provides students with essential skills for the future job market. The discussions in Kahila et al. (2025 [Chap. 8]) and Le and Nga (2025 [Chap. 10]) reinforce our conclusions that learning environments enriched with technology, such as the Classroom of the Future and makerspaces, enhance student motivation, creativity and digital proficiency. Furthermore, the research presented in these chapters emphasizes the role of educators in facilitating meaningful learning experiences, aligning with our call for increased teacher training to address the challenges of integrating STEAM disciplines effectively.

The alignment between our study and these chapters underscores the growing necessity of integrating AI and digital tools into educational frameworks. By doing so, students gain practical and cognitive skills that prepare them for an increasingly digitalized society, reducing the gender gap and broadening career prospects in STEAM fields.

References

Amaral, C., Yonezawa, W., & Barros, D. (2022). Pensamiento computacional y formación docente: Desafíos y posibilidades didácticas utilizando la herramienta Scratch. *Universidad Nove Julho, 40*, 17.

Amaya, K. L. A., Rivadeneira, R. O. A., Espino, A. M. E., Chávez, Z. R. M., Cabrera, F. O., & de la Torre, D. Q. (2023). *Tecnología educativa para desarrollar la metodología STEAM.* Editorial Mar Caribe. https://doi.org/10.17613/yan1-wg80

Bautista, G., & Borges, F. (2013). Smart classrooms: Innovation in formal learning spaces to transform learning experiences. *Bulletin of the IEEE Technical Committee on Learning Technology, 15*(3), 18–21. https://bit.ly/2Wf0ruT

Bender, J., Zhao, B., Dziena, A., & Kaiser, G. (2023). Integrating Parsons puzzles within Scratch enables efficient computational thinking learning. *Research and Practice in Technology Enhanced Learning, 18*, 022. https://doi.org/10.58459/rptel.2023.18022

Bernate, J., & Guativa, J. A. V. (2020). Desafíos y tendencias del siglo XXI en la educación superior. *Revista de Ciencias Sociales, 26*(2), 141–154. https://doi.org/10.31876/rcs.v26i0.34119

Brender, J., El-Hamamsy, L., Bruno, B., Chessel-Lazzarotto, F., Zufferey, J. D., & Mondada, F. (2021). Investigating the role of Educational Robotics in formal mathematics education: The case of geometry for 15-year-old students. In *Technology-enhanced learning for a free, safe, and sustainable world* (pp. 67–81). Springer International Publishing. https://doi.org/10.1007/978-3-030-86436-1_6

Chiriac, T. (2019). Promoting robotics in education: Towards a systemic integration of digital innovations in the school curriculum (Moldova initiatives). *14° Conference reader. E-Learning: Unlocking the gate to education around the globe*. Prague, Czech Republic, Jun 20–21, 2019, 179–191.

Conradty, C., & Bogner, F. X. (2020). STEAM teaching professional development works: Effects on students' creativity and motivation. *Smart Learning Environments, 7*(1), 1–20. https://doi.org/10.1186/s40561-020-00132-9

Dapozo, G. D., Greiner, C. L., Petris, R. H., Espíndola, M. C., & Empresa, A. M. (2017). Promoción del Pensamiento Computacional para Favorecer la Formación de Recursos Humanos en Disciplinas STEM. *Red de Universidades con Carreras en Informática: Buenos Aires, Argentina, 2017*, 737–742. http://sedici.unlp.edu.ar/handle/10915/62342

Del Venegas, J. J., Díaz, L. A., Tosina, R. Y., & Ramos, V. L. (2021). La dimensión educativa de la robótica: Del desarrollo del pensamiento al pensamiento computacional en el aula. *Campo Abierto, Revista de Edcuación, 40*(2), 221–233. https://doi.org/10.17398/0213-9529.40.2.221

Delaney, S., Trapani, F., Chandler, P., & Redman, C. (2014). *Contemporary practices of technology and its affordances: Perceptions of pre-service teachers on the utilization of technology in teaching and learning practice* (Version 1). Deakin University. https://hdl.handle.net/10536/DRO/DU:30108933

Diego-Mantecón, J., Blanco, T., Ortiz-Laso, Z., & Lavicza, Z. (2021). STEAM projects with KIKS format for developing key competences. *Comunicar, 29*(66), 33–43. https://doi.org/10.3916/C66-2021-03

Dúo-Terrón, P. (2023). Analysis of scratch software in scientific production for 20 years: Programming in education to develop computational thinking and STEAM disciplines. *Education Sciences, 13*(4), 404. https://doi.org/10.3390/educsci13040404

Dúo-Terrón, P., Hinojo-Lucena, F. J., Moreno-Guerrero, A. J., & López-Belmonte, J. (2022). Impact of the pandemic on STEAM disciplines in the sixth grade of primary education. *European Journal of Investigation in Health, Psychology and Education, 12*(8), 989–1005. https://doi.org/10.3390/ejihpe12080071

Ferrada Ferrada, C., & Díaz-Levicoy, D. A. (2023). Robótica, programación y una aproximación a la educación ambiental. *Transformación, 19*(1), 30–52.

García Fuentes, O., Raposo, R. M., & Martínez Figueira, M. E. (2023). El enfoque educativo STEAM: una revisión de la literatura. *Revista Complutense de Educación, 34*(1), 191–202. https://doi.org/10.5209/rced.77261

García-Tudela, P. A., & Marín-Marín, J.-A. (2023). Use of Arduino in primary education: A systematic review. *Education Sciences, 13*(2), 134. https://doi.org/10.3390/educsci13020134

González, A., Cerezo, I., Llamas, F., & Revuelta, F.I. (2022). Las Aulas del Futuro como espacios favorecedores del cambio educativo en la Educación Superior. *REIDOCREA, 11*(59), 675–683. https://doi.org/10.30827/Digibug.77653

Haduong, P. (2021). Learning together: Collaboration and community in PK–12 computing education. *Proceedings of the 17th ACM Conference on International Computing Education Research.* 401–402. https://doi.org/10.1145/3446871.3469771

Harris, U. S. (2017). Virtual partnerships: Engaging students in E-service learning using computer-mediated communication. *Asia Pacific Media Educator, 27*(1), 103–117. https://doi.org/10.1177/1326365X17701792

Hickman, L. (2016). Educational occupations and classroom technology. *European Journal of Pragmatism and American Philosophy, VIII* (1). https://doi.org/10.4000/ejpap.446

Juvera, J., & Hernández López, S. (2021). Steam in childhood and the gender gap: A proposal for non-formal education. *EDU REVIEW: International Education and Learning Review, 9*(1), 9–25. https://doi.org/10.37467/gka-revedu.v9.2712

Kahila, J., Vartiainen, H., Pope, N., & Tedre, M. (2025). Engaging young minds in AI through co-designing and making their own ai-driven apps in the classroom. In S. Papadakis & G. Lampropoulos (Eds.), *Social robots and artificial intelligence in education: integrating AI in K-12 and higher education* (pp. 64–79). Palgrave Macmillan.

Kanematsu, H., & Barry, D. M. (2016). For the future. In *STEM and ICT education in intelligent environments. Intelligent systems reference library* (Vol. 91). Springer. https://doi.org/10.1007/978-3-319-19234-5_26

Kanobel, M. C., & Carranza, P. (2023). Construyendo un horno solar: perceC-Piones docentes que participaron de una actividad de educación STEAM. *Revista electrónica de divulgación de metodologías emergentes en el desarrollo de las STEM, 5*(1), 165–184. https://www.revistas.unp.edu.ar/index.php/rediunp/article/view/927

Le, A. N.-N., & Nga, N. H. L. (2025). Elevating EFL paraphrasing skills through generative artificial intelligence: An exploration with ChatGPT. In S. Papadakis & G. Lampropoulos (Eds.), *Social robots and artificial intelligence in education: Integrating AI in K-12 and higher education* (pp. 121–136). Palgrave Macmillan.

Lessard, S., Schaefer, L., Huber, J., Murphy, M. S., & Clandinin, D. J. (2015). *Composing a life as a teacher educator. International teacher education: Promising pedagogies* (Part C) (Advances in research on teaching, Vol. 22C), (pp. 235–252). Emerald Group Publishing Limited, Bingley. https://doi.org/10.1108/S1479-368720150000022016

Lu, Y. (2021). Scratch teaching mode of a course for college students. *International Journal of Emerging Technologies in Learning (iJET), 16*(05), 186–200. https://doi.org/10.3991/ijet.v16i05.21269

Marín-Marín, J. A., Moreno-Guerrero, A. J., Dúo-Terrón, P., & López-Belmonte, J. (2021). STEAM in education: A bibliometric analysis of performance and co-words in Web of Science. *IJ STEM, 8*(41). https://doi.org/10.1186/s40594-021-00296-x

Marín-Marín, J.-A., García-Tudela, P. A., & Duo-Terrón, P. (2024). Computational thinking and programming with Arduino in education: A systematic review for secondary education. *Heliyon, 10*(8), e29177. https://doi.org/10.1016/j.heliyon.2024.e29177

Martín-Lucas, J., & Sánchez-Rojo, A. (2023). Technology and teaching space: A pedagogical approach to the classroom of the future. In F. J. García-Peñalvo & A. García-Holgado (Eds.), *Proceedings TEEM 2022: Tenth International Conference on Technological Ecosystems for Enhancing Multiculturality. TEEM 2022* (Lecture notes in educational technology). Springer. https://doi.org/10.1007/978-981-99-0942-1_51

MEFPD. (2023). Ministerio de Educación, Formación Profesional y Deportes. Resolución de 19 de diciembre de 2023, de la Secretaría de Estado de Educación. *Boletín Oficial del Estado, 308*, 171715–171723, December 26. https://www.boe.es/boe/dias/2023/12/26/pdfs/BOE-A-2023-26314.pdf

Michuy Gaibor, M. F., Molina Verdugo, A. L., Peña, P. M., & Vistin Vistin, J. M. (2023). El enfoque STEAM para la mejora de conocimientos y motivación desde un aprendizaje interdisciplinario en estudiantes de primaria. *Prometeo Conocimiento Científico, 3*(2), e51. https://doi.org/10.55204/CPc.v3i2.e51

Molina-Ayuso, Á., Adamuz-Povedano, N., Bracho-López, R., & Torralbo-Rodríguez, M. (2022). Introducción al pensamiento computacional con Scratch para la formación del profesorado español de Matemáticas de Educación Primaria. *Education Sciences, 12*(12), 899. https://doi.org/10.3390/educsci12120899

Monteiro, C., Catarino, P., Soares, A.A., & Fonseca, B. (2019). Vector equation of a line: A scratch application. EDULEARN19 Proceedings, pp. 3911–3914. https://doi.org/10.21125/edulearn.2019.1003

OECD. (2018). *The future of education and skills. Education 2030.* The future we want. (Consultado el 4 de marzo de 2024). https://www.oecd.org/education/2030/E2030%20Position%20Paper%20(05.04.2018).pdf

Orozco, H., & Lamberto, J. (2022). La ética en la investigación científica: consideraciones desde el área educativa. *Perspectivas, 10*(19), 11–21. https://perspectivas.unermb.web.ve/index.php/Perspectivas/article/view/355

Paleczek, L., Maitz, K., Danielowitz, C., & Husny, M. (2021). Scratch options! Using programming to approach so-cial-emotionally challenging situations in Grade 4 class-rooms. In *Proceedings of the 20th European Conference on e-Learning. A Virtual Conference Supported by University of Applied Sciences HTW*, Berlin, Germany, 28–30 October 2021; pp. 658–661.

Payeras, M., Jacob, M., Florido, C., & Domínguez, A. M. (2024). La orientación preuniversitaria ¿Causa de la brecha de género en los estudios universitarios del ámbito STEM? En Libro de resúmenes de trabajos a IRED'23. III Conferencia internacional de investigación y V Jornadas de investigación e innovación educativa, pp. 195. http://hdl.handle.net/11201/163727

Pineda-Caro, D. Y. (2022). Enfoque STEAM: Retos y oportunidades para los docentes. *Revista internacional de pedagogía e innovación educativa, 3*(1), 229–244. https://doi.org/10.51660/ripie.v3i1.115

Sánchez-Sánchez, T., Serrano Sánchez, J. L., & Rojo Acosta, F. (2020). Influence of educational robotics on motivation and cooperative learning in primary education: A case study. *Innoeduca. International Journal of Technology and Educational Innovation, 6*(2), 141–152.

Segura, J. A., Nebot, M. Á. L., Mon, F. E., & Novella, M. G. V. (2019). El debate sobre el pensamiento computacional en educación. *RIED-Revista Iberoamericana de Educación a Distancia, 22*, 171. https://doi.org/10.5944/ried.22.1.22303

Silva, D. S., Melo, S. L., & Basto Diniz, J. R. (2021). Developing a didactic sequence for introducing computational thinking in the early years of elementary school. In *Proceedings of the 2021 XVI Latin American Conference on Learning Technologies (LACLO)*, Arequipa, Peru, 19–21 October 2021; pp. 526–529. https://doi.org/10.1109/LACLO54177.2021.00095

Soto-Solier, P. M., Villena-Soto, V., & Molina Muñoz, D. (2023). PerceCPiones de los futuros docentes sobre la integración de la robótica creativa en Educación Primaria. *Pixel-Bit. Revista De Medios Y Educación, 67*, 283–318. https://doi.org/10.12795/pixelbit.96781

Wing, J. M. (2008). Computational thinking and thinking about computing. *Philosophical Transactions of the Royal Society A: Mathematical, Physical and Engineering Sciences, 366*(1881), 3717–3725. https://doi.org/10.1098/rsta.2008.0118

CHAPTER 10

Elevating EFL Paraphrasing Skills Through Generative Artificial Intelligence: An Exploration with ChatGPT

Anh Nguyen-Nhu Le and Nguyen Hoai Ly Nga

10.1 Introduction

In the domain of English as a Foreign Language (EFL) education, the development of effective writing skills remains a fundamental goal. Among these skills, paraphrasing stands out as particularly challenging for university-level students. Paraphrasing not only requires a deep understanding of the original text but also demands advanced linguistic abilities to restate content in a new form while preserving its original meaning. This skill is crucial for academic success and avoiding plagiarism (Hyland, 2001; Keck, 2006). Recent advancements in Information and Communication Technology (ICT) have brought about significant

A. N.-N. Le (✉)
Ho Chi Minh City University of Education, Ho Chi Minh, Vietnam
e-mail: anhlnn@hcmue.edu.vn

N. H. L. Nga
University of Economics Ho Chi Minh City, Ho Chi Minh, Vietnam
e-mail: nganguyenly@ueh.edu.vn

S. Papadakis, G. Lampropoulos (eds.), *Social Robots and Artificial Intelligence in Education*, Digital Education and Learning,
https://doi.org/10.1007/978-3-032-03977-4_10

transformations in educational methodologies, integrating tools that provide innovative approaches to traditional teaching and learning processes. One such advancement is the application of GenAI in language education.

GenAI is an advanced technology that uses machine learning to create new content based on patterns in existing data, such as text, images, audio, and video (Khosravi et al., 2023; Le et al., 2024). GenAI have shown promise in various educational settings, offering personalized learning, automated content generation, and interactive learning environments (Fitria, 2021; Fu et al., 2020; Pokrivcakova, 2019). Existing research has demonstrated the benefits of AI-powered tools in supporting language learning (Law, 2024) and fostering self-regulated learning practices (Gašević et al., 2023). ChatGPT, powered by Generative Pretrained Transformers (GPTs), leverages large-scale language models to understand and generate humanlike text based on extensive training data (Pavlik, 2023). Its capabilities in natural language processing enable it to provide real-time feedback and generate paraphrased content that aligns with specific instructional guidelines (Brown et al., 2020). However, the specific application of AI-powered tools like ChatGPT in paraphrasing instruction remains underexplored. While studies have investigated the effectiveness of various paraphrasing strategies and classification frameworks (Burstein et al., 2012; Chen et al., 2015; Keck, 2006), there is a notable gap in understanding how AI can be utilized to improve paraphrasing skills in an EFL context.

This study aims to fill this gap by evaluating ChatGPT's proficiency in generating high-quality paraphrases and its impact on EFL students' paraphrasing abilities. Utilizing a qualitative study design, the research employs a five-level paraphrasing assessment framework based on the rubric suggested by Chen et al. (2015). This framework assesses paraphrasing quality from minimal revisions to comprehensive restatements involving phrase replacements and sentence structure rearrangements. By integrating AI with conventional teaching methods, this research seeks to propose a blended approach that leverages the strengths of both human instruction and AI capabilities. Through a detailed analysis of ChatGPT's output, the study aims to provide empirical evidence on the effectiveness of GenAI in supporting paraphrasing skill development. The findings will offer valuable insights for educators and curriculum developers, highlighting the potential of AI tools in enhancing language learning outcomes and suggesting strategies for their effective integration into EFL education.

This chapter is structured as follows. Section 10.2 explores the concept of paraphrasing in the EFL context, highlighting its importance for language learners. Section 10.3 introduces GenAI and discusses the role of ChatGPT in education, with a particular focus on its application in EFL instruction. Section 10.4 delves into prompt engineering and its relevance to EFL teaching methodologies. Section 10.5 addresses the issue of thematic progression in English writing, a common challenge for EFL learners. Section 10.6 outlines the research methodology, followed by Section 10.7, which presents the key findings. Section 10.8 offers a discussion of the results, and finally, Section 10.9 concludes the chapter, summarizing the key insights and implications for future research and practice.

10.2 Paraphrasing in the EFL Context

Paraphrasing involves rephrasing and rewording a passage originally written or spoken by someone else, to incorporate their ideas into your own writing. In the context of English language teaching and learning, English as a Second Language (ESL) learners have to use both cognitive and linguistic skills to paraphrase. First, ESL learners need to understand what the text is saying, which puts their reading skill to work (Wette, 2010). The next step is to extract the main parts of the reading source text and turn them into their own words. To do this, ESL learners often use synonyms (e.g. "students" to replace "learners"), change the word form (e.g. "in the mountains" to "mountainous regions"), change the word order (e.g. "the collapse of a city" to "a city fell apart"), or a combination of all of the above (Bailey, 2018). In addition, ESL students should change sentence structures, such as changing from active tense to passive tense, from direct to indirect speech, or using other sentence structures correctly (Rogers, 2007).

Paraphrasing is an essential skill that ESL students need to develop as it helps them avoid plagiarism (Hyland, 2001; Keck, 2006). Furthermore, the ability to paraphrase effectively is a crucial indicator of advanced language proficiency, requiring significant language processing capabilities (Keck, 2006; Shea, 2000; Wette, 2010; Zhang, 2020). A strong paraphrasing ability demonstrates a writer's understanding of the original text and their capacity to produce syntactically and semantically accurate sentences. As a result, ESL teachers often use paraphrasing as a tool to assess students' reading comprehension and academic writing skills (Li &

Casanave, 2012; Pecorari, 2010). Poor paraphrasing practices, which are commonly referred to as "patchwriting," involve making minimal changes to the original text while still closely mimicking the original author's style, such as replacing each word with a synonym (Jamieson, 2015).

10.2.1 Paraphrasing Classification and Assessment

Three major studies have explored paraphrasing from a product-oriented perspective, aiming to classify the various types of paraphrased work (Burstein et al., 2012; Chen et al., 2015; Keck, 2006).

Keck (2006) conducted a study on paraphrasing techniques in summary writing tasks among 79 native English speakers and 74 ESL writers. The findings revealed distinct patterns between the two groups and established four categories of paraphrasing: Near Copy, Minimal Revision, Moderate Revision, and Substantial Revision. Keck's classification system defined Near Copy, the lowest level of paraphrasing, as containing over 50% of words from the original text in long copied strings. "Minimal Revision" included 20–49% of words from the source, while "Moderate Revision" contained 1–19%. The "Substantial Revision," which was considered the highest level of paraphrasing, contained no unique links to the original text. Keck found that most native English speakers understudy avoided "Near Copy" paraphrases and frequently used "Moderate and Substantial Revisions." In contrast, the majority of ESL writers included at least one Near Copy paraphrase, with half using two or more and did not employ Moderate or Substantial Revisions.

Burstein et al. (2012) conducted an analysis on paraphrasing in the responses of Test of English as a Foreign Language (TOEFL) writing exam participants, examining the differences in paraphrasing strategies between native English speakers and ESL speakers. Their classification of responses included four types: "syntactic paraphrasing," "lexical paraphrasing," "conceptual paraphrasing," and "global paraphrasing" (p. 7). Among these classifications, the lower level of paraphrasing skill relates to "lexical paraphrasing," in which readers can identify the lexical substitutions of the original text, and "syntactic paraphrasing," in which readers can identify syntactic reordering. By contrast, "conceptual paraphrasing" and "global paraphrasing" are regarded as higher level, which the scholars described as: "Paraphrase that cannot be easily characterized by any syntactic or word-based classification" (p. 8), and "Paraphrase of the gist of the reading and the lecture that could not be isolated to specific text segments." Furthermore, the researchers discovered that native English speakers more frequently employed "conceptual paraphrase."

Chen et al. (2015) developed a scoring rubric, which was modified from TOEFL's Integrated Writing Rubrics to assess the performance of student participants in their study (Table 1 in the Appendix). According to this scoring rubric, paraphrasing is assessed at three main levels: single-word replacements, phrase replacements, and sentence restructuring, with a maximum of nine points available for restating entire paragraphs. At the highest level (Level 5), the authors highlighted that the overall meaning and quality of the paraphrasing is the crucial factor, and paraphrasing quality is not guaranteed by simply changing more phrases or replacing them with more synonyms. Based on the examples of patchwriting and good paraphrased texts (Burstein et al., 2012; Chen et al., 2015; Jamieson, 2015; Pecorari, 2010), it is essential to reduce word count while preserving key terms, remove unnecessary words and phrases, and replace lengthy noun phrases with shorter equivalents. Additionally, altering word order and sentence structure helps to avoid mere word substitutions (i.e. avoid "patchwriting"), ensuring that the original meaning of the text is maintained.

Generally, the study of paraphrasing in English writing has largely taken a product-oriented approach, concentrating on various methods to classify different types of paraphrased content (Burstein et al., 2012; Chen et al., 2015; Keck, 2006). However, there is a lack of research on the application of these classification frameworks in linguistics, particularly in evaluating paraphrased text generated not by ESL learners, but by GenAI technology.

10.2.2 Automated Paraphrasing Tools

The earlier version of the paraphrasing function of GenAI was commonly referred to as automated paraphrasing tools (APTs) in research articles. These web-based tools use AI algorithms to transform one text into another. The technology behind paraphrasing tools started with "article spinning," a technique that converts old content into new or original content by replacing words and phrases with synonymous expressions (Bailey, 2018). Zhang et al. (2014) noted that some online paraphrasing tools were initially developed to facilitate text spinning as a search engine optimization (SEO) technique, enabling the generation of multiple versions of a single article to be posted on various websites with backlinks to the promoted site. However, Prentice and Kinden (2018) pointed out that the simple technique of replacing each word with a synonym, the output of some APTs cannot recognize the key terminology in various fields, resulting in a paraphrased text with completely different meaning.

10.3 GenAI in Education

GenAI is a sophisticated technology that has revolutionized digital content creation (Le et al., 2024). GenAI leverages machine learning frameworks to produce new content by analyzing and learning from existing data, using techniques rooted in statistics and probabilities (Khosravi et al., 2023). This innovative technology can create diverse types of content, such as text, images, audio, and video, by recognizing patterns within the training examples provided (Abukmeil et al., 2022; Gui et al., 2023).

GenAI's application in education has opened new avenues for enhancing learning experiences (Pokrivcakova, 2019). It offers various benefits, such as personalized learning, automated content generation, and interactive learning environments, which can significantly aid educators and students (Fitria, 2021; Fu et al., 2020). However, the integration of AI in educational settings also presents challenges, including the need for substantial data, ethical concerns, and the requirement for educators to adapt to new technological tools (Yang & Kyun, 2022). Examples of GenAI in education include automated essay scoring systems, intelligent tutoring systems, and tools for generating customized learning materials.

Extant literature has identified two major GenAI techniques: Generative Adversarial Networks (GANs) and GPTs (Abukmeil et al., 2022; Gui et al., 2023). GANs, a commonly used GAI technique, involve two neural networks—a generator and a discriminator—that work together to create synthetic data, such as images, that are indistinguishable from real data (Gui et al., 2023). This technique is predominantly utilized for tasks like voice generation, graphics, and video creation (Abukmeil et al., 2022).

In contrast, GPT models focus on natural language processing (NLP) and are designed to generate humanlike text by learning from vast amounts of publicly available digital content (Aydin & Karaarslan, 2023). GPT models, such as GPT-3, use extensive training data to produce text that can range from paragraphs to full research articles, demonstrating creativity and coherence in multiple languages (Brown et al., 2020). These models are capable of engaging in humanlike conversations, making them useful for applications like customer service chatbots and fictional characters in video games (Aydin & Karaarslan, 2023; Korngiebel & Mooney, 2021; Pavlik, 2023).

The recent development of GPT-3, a more sophisticated version of earlier models, has enhanced the capabilities of Generative AI. With 175 billion parameters, GPT-3 significantly surpasses previous models in terms of scale and performance, making it a powerful tool for a wide range of applications, including education, engineering, journalism, and healthcare (Brown et al., 2020). GPT-3's ability to generate coherent and contextually appropriate text has positioned it as a fundamental NLP engine behind tools like ChatGPT, which have garnered attention for their potential in various fields (AlShaikh & Hewahi, 2021; Pavlik, 2023; Qadir, 2023; Tate et al., 2024; Williamson et al., 2023).

10.3.1 ChatGPT in Education

ChatGPT is an AI chatbot developed by OpenAI, released in November 2022. It combines the functionalities of a chatbot with the capabilities of a GPT, specifically utilizing LLMs designed to understand, interpret, and generate humanlike text based on extensive input data (Cao et al., 2023). By training on billions of sentences from diverse sources—such as websites, customer data, and past reports—LLMs acquire a comprehensive understanding of data analysis and context, excelling in natural language processing tasks (Cooper, 2023). Through extensive pretraining on a substantial collection of text data from books, articles, and websites, ChatGPT gains a deep understanding of natural language connections and structures, enabling it to produce logical and authentic responses during conversations. Built upon OpenAI's GPT models, particularly GPT-3.5 and GPT-4, ChatGPT has been fine-tuned for conversational applications, making it a robust tool for various interactive tasks (BaïDoo-Anu & Owusu Ansah, 2023; Cooper, 2023; Le et al., 2024).

10.3.2 ChatGPT in EFL instruction

The integration of large language models and AI-powered tools like ChatGPT has opened up new possibilities in EFL instruction. Research has shown that using natural language prompts, large language models can help students develop their ability to ask thought-provoking questions, enhancing their critical thinking skills (Gašević et al., 2023). AI chatbots, including ChatGPT, have emerged as promising tools for enhancing linguistic output and reducing anxieties in language learning (Bao & Li, 2023).

ChatGPT's advanced capabilities have made it a valuable tool in various educational contexts, including EFL instruction. For instance, in creative writing, AI-powered tools like ChatGPT can provide inspiration and support during the writing process, helping students to generate ideas and improve their writing skills (Kangasharju et al., 2022). Additionally, ChatGPT can enhance the assessment process by generating quizzes and providing real-time feedback, thus complementing traditional assessment methods and improving the overall learning experience (Kasneci et al., 2023).

ChatGPT can deliver personalized feedback and support to EFL learners by providing real-time advice on vocabulary, grammar, and syntax (Le et al., 2024). This individualized feedback helps students enhance their writing skills in a manner tailored to their specific needs (Fu et al., 2020). Traditional methods of providing personalized feedback can be time-intensive and demanding for educators. ChatGPT alleviates this challenge by offering efficient and effective feedback, significantly aiding students' writing development (Le et al., 2024; Yang & Kyun, 2022). By generating humanlike text customized to each student's needs, ChatGPT fosters their progress and helps improve their language skills over time (Fitria, 2021).

One significant advantage of using ChatGPT in EFL instruction is its ability to engage and motivate learners through its natural language capabilities (Castelli & Manzoni, 2022). ChatGPT creates a more immersive and interactive writing experience, making practice sessions feel more genuine and realistic (Le et al., 2024). This helps maintain students' interest and motivation in EFL writing. Traditional EFL writing activities often struggle to keep students engaged, but ChatGPT's ability to generate humanlike text can provide a more engaging and motivating learning environment (Fu et al., 2020).

ChatGPT supports the development of students' language skills by providing ample opportunities for practice and feedback. Activities designed to leverage ChatGPT, such as writing responses to prompts or completing sentences, help students practice their writing skills and receive valuable feedback (Kangasharju et al., 2022; Le et al., 2024). This continuous cycle of practice and feedback is crucial for language development (Bondarenko, 2017; Le et al., 2023; Vu et al., 2023). Additionally, offering guidance on interpreting and using ChatGPT's feedback can further enhance students' learning experiences and outcomes (Barrot, 2023).

Using ChatGPT in EFL instruction offers several benefits, including personalized feedback, engagement through natural language capabilities, and support for language skills development through continuous practice and feedback (Kasneci et al., 2023; Wu et al., 2023). Nonetheless, ethical considerations, such as data privacy and potential biases in AI and NLP technologies, must be carefully managed to ensure the ethical and effective use of ChatGPT in language education (Klimova et al., 2023).

10.4 Prompt Engineering and EFL Instruction

Prompt engineering is an element of AI literacy education, involving the creation of specific instructions that optimize a chatbot's performance. As defined by Liu et al. (2023), prompt engineering is the process of crafting instructions that unlock the full capabilities of models like ChatGPT. This skill is essential because appropriate prompts can enhance ChatGPT's basic and advanced natural language processing abilities (Ouyang et al., 2024; White et al., 2024). However, factors such as prompt length, key words, word order, and emotional tone can significantly impact the effectiveness of the chatbot's responses (Wu et al., 2023).

The construction of effective prompts is often a trial and error process, especially for nontechnical users, highlighting the need for expertise and practice in this area (Dang et al., 2022; Zamfirescu-Pereira et al., 2023). The most effective prompts are typically handcrafted by humans, and ignorance of prompt engineering can limit the benefits of ChatGPT (Chang et al., 2024). Therefore, developing effective AI literacy and prompt engineering education at both classroom and institutional levels is crucial for realizing the full potential of ChatGPT in educational settings (Marvin et al., 2024).

Prompts generally consist of two main components: context and natural language (Marvin et al., 2024). Context provides a knowledge base that grounds the chatbot's output, often using input data or output examples to facilitate learning from minimal examples (Parnami & Lee, 2022). Natural language involves explicit instructions written in the user's native language, allowing the chatbot to understand and perform tasks based on abstract descriptions and human concepts (Reynolds & McDonell, 2021). Direct task specification, or natural language instruction, uses commands like "generate," "list," or "write" to communicate desired outputs, making interactions with the chatbot more intuitive and humanlike (Wang et al., 2023).

Effective prompt engineering can also unlock advanced capabilities like chain-of-thought (COT) reasoning. Users can enable chatbots to break down problems into logical steps, enhancing their problem-solving abilities (Kojima et al., 2022). For ChatGPT, strategies such as adopting specific perspectives or roles can further optimize its responses. Ultimately, the key to effective prompting lies in careful word choice and the strategic layering of context and instructions (Atlas, 2023).

Effective prompt design in EFL contexts involves creating rich, contextual scenarios that provide sufficient background information, allowing the chatbot to generate accurate and relevant responses. For instance, setting up a context for a conversation topic before asking ChatGPT to simulate a dialogue can greatly enhance the learning experience. Natural language instructions tailored to specific language learning objectives can significantly improve EFL instruction. Commands like "explain the grammar rules for past tense" or "generate a dialogue between two friends discussing their weekend plans" can help students practice and understand various aspects of the English language. Clear and specific instructions ensure that ChatGPT generates responses that are both accurate and educationally valuable.

One-shot and few-shot learning techniques are particularly effective in EFL instruction. By presenting a few examples of correct language usage, students can better understand and internalize language patterns. For example, providing sentences that use new vocabulary words and asking ChatGPT to generate additional sentences with those words can reinforce learning. Chain-of-thought reasoning can also be beneficial for EFL learners. Prompts like "Explain the steps to form a complex sentence in English" encourage ChatGPT to break down the language construction process into manageable steps, aiding students in grasping more complex structures. This step-by-step guidance is particularly useful for students struggling with advanced grammar and syntax.

Additionally, assigning specific roles or perspectives to ChatGPT can create more engaging and realistic language practice scenarios. For example, prompting ChatGPT to act as a travel agent helping a student plan a trip can provide practical language application and improve conversational skills. Carefully crafted prompts that integrate both context and natural language instructions can significantly enhance the effectiveness of ChatGPT as a tool for EFL instruction, supporting students in developing their language skills more effectively. However, in combination with crafting prompts, it is also important to evaluate the AI's text output. Typically, when evaluating an English text, four factors can be considered to assess its quality. These include relevance and content (Flowerdew, 2013), grammar and syntax range (Azar & Hagen, 2009; Quirk, 2012), word choice (Schmitt & Schmitt, 2020), and thematic progression (Halliday & Hasan, 2013).

Notably, a recent study by Chui (2023) suggests that ChatGPT can provide examples and instant feedback on student paraphrases, with no spelling or grammar errors in its responses. Chui cautioned that students must be aware of certain limitations, including the potential for improved paraphrases to still contain patchwriting and the occasional provision of inaccurate explanations. To delve deeper into these issues, our study will focus on examining the word choice and thematic progression in the paraphrased output.

10.5 Problem of Thematic Progression in English Writing

When it comes to the analysis of English writing, thematic progression analysis is the common method to evaluate the coherence of an English text. A "theme" serves as the main point; it is that which the clause is concerned, or it is what the clause is about. The remainder of the message, the part in which the theme is developed, is called as rheme (Gerot & Wignell, 1994). Thematic progression, as described by Eggins (1994), refers to how themes and rhemes develop throughout a text. The problem of "brand new theme" occurs when writers place new information in the theme position without connecting it to previous themes or rhemes. Below is an example:

Exams are the results of last-minute studying (1). Most of them are tests of memories rather than skills (2). Schools usually offer a lot of courses (3). Teachers need to teach a lot of skills (4).

From the example, it can be seen that the sentence (1) and sentence (2) share the same themes, which are about exams. However, in sentence (3), the appearance of "Schools," and, in sentence (4), the appearance of "Teachers" are brand new themes. These new themes did not explain why "Most of them are test of memories rather than skill," nor give any further connections to this message.

Notably, several studies conducted to analyze the quality of academic text written by ESL students (Fitriati & Gayatri, 2021; Kuswoyo & Susardi, 2017; Thang & Tam, 2020; L. Wang, 2007; Yunita, 2018) revealed that this error in having a total new theme that are not related to the previous ones was a common error among ESL students.

Generally, the study of paraphrasing in English writing has largely taken a product-oriented approach, concentrating on various methods to classify different types of paraphrased content (Burstein et al., 2012; Chen et al., 2015; Keck, 2006). In terms of analyzing text, aside from lexical errors and grammatical errors, the analysis of theme rheme progress in English text written by ESL students is also getting more attention. However, aside from one study conducted by Chui (2023), there is a lack of research on the application of these classification frameworks, particularly in evaluating paraphrased text written not by ESL learners, but by GenAI technology. Therefore, this study aims to analyze and evaluate the English text written by ChatGPT. Since the responses of ChatGPT is flawless in terms of spelling or grammatical errors, the researchers will analyze its output in terms of how it remains the terminology, the overall main idea, and its coherence and cohesion following the theme-rheme progression.

10.6 Methodology

10.6.1 Introduction

This methodology section outlines the research design employed to evaluate the effectiveness of different paraphrasing prompts when used with ChatGPT 4.0. The primary aim of this study is to assess how well the AI can produce high-quality paraphrases that adhere to specific guidelines, including reducing word count, retaining key terms, eliminating superfluous words, combining sentences, and maintaining the original meaning. By adopting a qualitative approach, the study provides an in-depth analysis of the AI's performance through iterative refinement of prompts and systematic evaluation of generated outputs. This approach enables a comprehensive examination of the AI's paraphrasing capabilities and highlights areas for potential improvement. The following sections detail the research design, data collection methods, procedural steps, and analytical frameworks employed in this study.

10.6.2 Research Design

A qualitative research design (Creswell & Creswell, 2022) has been utilized in this study to explore the effectiveness of different paraphrasing prompts when used with ChatGPT 4.0. A qualitative approach was chosen due to its suitability for examining the nuanced performance of GenAI in producing paraphrased texts and identifying the specific linguistic and structural changes made by the AI (Vianello et al., 2023). The research was conducted in three iterative stages, each involving the development and refinement of prompts based on previous literature and the analysis of AI-generated paraphrases. The iterative nature of the design allowed for continuous improvement of the prompts and a thorough examination of the AI's performance under varying instructions. Figure 10.1 summarizes the three stages of the research design.

Stage 1—Initial Prompt Development and Testing: In the first stage, a prompt was developed based on synthesized guidelines from previous literature on effective paraphrasing. This prompt, referred to as Prompt 1.0, was designed to instruct the AI to reduce word count, retain key terms, remove unnecessary words, and avoid single-word replacements. The AI's output was then analyzed to identify areas of strength and weakness.

Stage 2—Prompt Refinement and Further Testing: Based on the findings from Stage 1, the prompt was modified to better address identified weaknesses, such as the need for more complex sentence structures and the avoidance of repetitive grammatical patterns. This resulted in the creation of Prompt 1.1 and Prompt 1.2, each progressively refining the

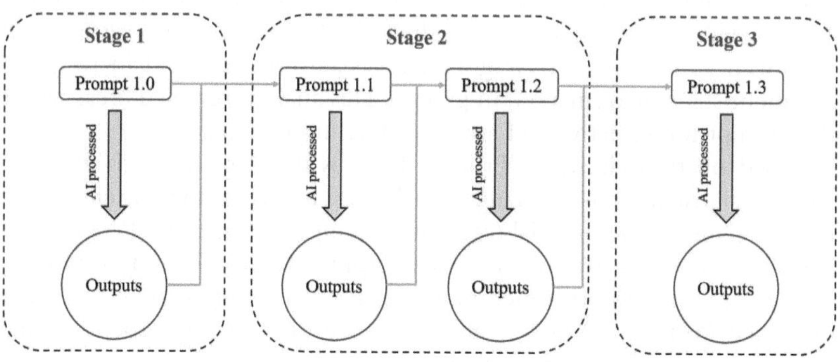

Fig. 10.1 Summary of the three iterative stages of the research design

instructions to the AI. The outputs generated from these prompts were analyzed for improvements in adherence to the guidelines.

Stage 3—Final Prompt Development and Validation: In the final stage, insights from the previous iterations were combined to develop Prompt 1.3, which integrated all effective elements from the earlier prompts. This prompt aimed to produce high-quality paraphrases by instructing the AI to use fewer words, retain key terms, remove unnecessary phrases, combine sentences with relative clauses, change word forms, and use varied sentence structures. The AI's performance with this prompt was validated using three additional texts to ensure consistency and effectiveness across different content.

10.6.3 Data Collection

The data for this study comprised four original texts selected from the textbook in an English course for ESL undergraduate students at a university in Ho Chi Minh City, Vietnam, for their diverse content and varying levels of complexity. These texts covered different topics—tourism and hospitality, digital banking, time management, and financial accounting—to test the versatility of ChatGPT 4.0 in paraphrasing different types of material. The criteria for selecting these texts included their relevance to real-world contexts, the presence of varied sentence structures, and the inclusion of both simple and complex sentences.

The initial stage began with the creation of Prompt 1.0, which was designed based on synthesized guidelines from previous literature on effective paraphrasing techniques. Following the generation of Output 1.0, a detailed analysis was conducted. A color-coding system was employed to identify similar grammatical patterns (marked in red) and divergent structures (marked in green). This analysis revealed areas where the prompt was effective and areas needing improvement.

Based on these insights, Prompt 1.1 was developed, which included additional instructions to change word forms and combine sentences with relative clauses. The original text was then input again using Prompt 1.1, and the new output was recorded as Output 1.1. This second output was analyzed to assess improvements and identify remaining weaknesses, based on which Prompt 1.2 was created, focusing on reducing repetition and enhancing sentence variety. Using Prompt 1.2, the original text was input once more, and the resulting paraphrase was recorded as Output 1.2. Finally, Prompt 1.3 was developed by integrating the successful elements from all previous prompts and was then used to paraphrase three additional texts, covering different topics to ensure consistency and effectiveness across diverse content. The outputs from these texts were recorded as Output 2, Output 3, and Output 4.

Each output underwent a thorough analysis to verify adherence to the instructions and to identify any persistent issues. The iterative process of refining the prompts based on detailed analyses ensured that the final prompt was capable of generating high-quality paraphrases. Throughout this process, the AI's paraphrased outputs were systematically collected and analyzed to ensure adherence to the given instructions and to identify areas of improvement. This iterative approach allowed for the continuous refinement of prompts and ensured the reliability and validity of the data collected.

10.6.4 Data Analysis

The data analysis involved multiple levels of evaluation to assess the quality and effectiveness of the paraphrased outputs generated by ChatGPT 4.0. This section outlines the analytical approaches used to examine grammatical patterns, sentence structures, theme-rheme flow, and overall adherence to the paraphrasing guidelines.

Grammar analysis: Each paraphrased output was first subjected to a grammar analysis using a color-coding system. Red was used to highlight grammatical patterns that were remarkably similar to those in the original text, while green indicated divergent grammar structures. This visual coding allowed for a clear comparison between the original and paraphrased texts, making it easier to identify where the AI had successfully altered grammatical constructions and where it had not. This step was crucial for evaluating the AI's ability to rephrase content while maintaining grammatical correctness.

Sentence structure analysis: The next step involved a detailed analysis of sentence structures in both the original and paraphrased texts. Each sentence was broken down to identify its grammatical components, such as subjects, predicates, clauses, and phrases. The original text often contained a mix of simple and complex sentences, and the analysis focused on whether the AI could replicate this variety in its paraphrases. Particular attention was paid to the AI's ability to combine sentences using relative clauses and to change word forms, such as converting adjectives to nouns and vice versa. Successful examples of complex sentence formation were noted, alongside instances where the AI failed to sufficiently alter the original structure.

Theme

rheme analysis: To assess the flow of information within and across sentencesa theme-rheme analysis was conducted. This analysis focused on the logical progression of ideasensuring that each sentence smoothly transitioned from the known information (theme) to the new information (rheme). Theme-rheme errorswhich disrupt this flowwere identified and analyzed. For instancesentences that awkwardly repeated structures or disrupted the logical progression of ideas were marked for further review. An example of such an error was found in Output 2where the flow between sentences was less coherent due to repetitive structures

Assessment of paraphrasing quality: The overall quality of the paraphrased outputs was assessed using established paraphrasing assessment frameworks from linguistic scholars, such as Keck (2006), Burstein et al. (2012), and Chen et al. (2014). These frameworks provided criteria for evaluating the preservation of the main ideas, accuracy, and grammatical correctness of the paraphrases. Each output was compared against these criteria to ensure that the AI maintained the original meaning of the text while introducing sufficient lexical and structural changes.

10.7 Findings

In our first attempt to paraphrase Text 1, we applied our synthesis of guidelines from previous literature to create the first prompt:

Prompt 1.0

"Here is the instruction for rewriting: 1/ Use fewer words than the original text, 2/ Keep the key terms, 3/ Remove unnecessary words and phrases, 4/ Replace long noun phrases with shorter phrases, 5/ Change the word order and sentence structure to avoid single-word replacements, 6/ Keep the original meaning of the text. Act as a writing assistant, use the instruction to rewrite the following text:"

Next, we recorded the first paraphrased text from ChatGPT (4.0 version) and named it as Output 1.0. For grammar analysis, we employed a color-coding system: red denotes remarkably similar grammatical patterns, while green indicates divergent grammar structures.

Text 1	*Output 1.0*
Social media has revolutionized the tourism and hospitality industry. Platforms such as Instagram and Facebook enable travelers to instantly share their experiences with a global audience, often swaying the travel decisions of their friends and followers. Many hotels and restaurants now actively participate in social media to boost their visibility and reputation. Travelers frequently rely on online reviews and ratings to choose accommodations and dining options. Social media campaigns can draw a large number of tourists to specific destinations. Often, destinations employ influencers to promote their attractions and services, reaching a broader audience swiftly. Additionally, social media serves as a platform for customers to express their complaints and feedback. Managing an effective online presence can greatly influence a business's success in the tourism and hospitality sector. In summary, social media has become a crucial tool for growth and customer engagement in this industry.	Social media has transformed tourism and hospitality. Platforms like Instagram and Facebook let travelers quickly share their experiences, influencing others' travel choices. Hotels and restaurants use social media to enhance visibility and reputation. Travelers depend on online reviews for accommodation and dining decisions. Social media campaigns attract tourists to specific spots, with destinations hiring influencers to promote attractions, reaching more people fast, which is also a place for customer feedback and complaints. An effective online presence is vital for success in this sector. Overall, social media is essential for growth and customer engagement in tourism and hospitality.

According to Output 1.0, ChatGPT fulfilled most prompt requirements: reducing words, retaining key terms (e.g. "tourism and hospitality," "social media campaigns," "visibility," "reputation," "customer engagement"), and eliminating superfluous words (e.g. "many," "actively," "frequently"). It also fully addressed Point 4 by replacing "their friends and followers" with "others," replacing the long phrase "a platform for customers to express their complaints and feedback" to a shorter phrase: "for customer feedback and complaints."

Regarding Point 5, all words are composed differently and there is no single-word replacement. However, the requirement on changing word order and sentence structure was not adhered to fully. For instance, the first four sentences still have exactly the same word order and grammar structure with the original text. Moreover, the original conclusion sentence, "In summary, social media has become a crucial tool for growth and customer engagement in this industry" also remained unchanged in both word order and structure in Output 1.0.

By contrast, regarding the next two sentences, we found that they are good examples of grammatical structure reconstruction. Below is our analysis of the sentence structure of each one:

Original text:

1. "Social media campaigns can draw a large number of tourists to specific destinations."

Structure: Simple sentence. This sentence has a subject ("Social media campaigns") and a predicate ("can draw a large number of tourists to specific destinations").

2. "Often, destinations employ influencers to promote their attractions and services, reaching a broader audience swiftly."

Structure: Complex sentence. This sentence has a main clause ("destinations employ influencers to promote their attractions and services") and a participial phrase ("reaching a broader audience swiftly") that adds additional information.

3. "Additionally, social media serves as a platform for customers to express their complaints and feedback."

Structure: Simple sentence. This sentence has a subject ("social media") and a predicate ("serves as a platform for customers to express their complaints and feedback").

Good paraphrased text:

Social media campaigns attract tourists to specific spots, with destinations hiring influencers to promote attractions, reaching more people fast, which is also a place for customer feedback and complaints.

Structure: Complex sentence. This sentence has a main clause ("Social media campaigns attract tourists to specific spots"), two participial phrases, and a relative clause.

Accordingly, the original text contains two simple sentences and one complex sentence. The AI successfully combined these into one complex sentence with a relative clause.

Regarding the sixth instruction to keep the original meaning intact, ChatGPT demonstrated excellent performance. Therefore, we decided to remove the instructions "avoid single-word replacement" of Point 5 and "keep the original meaning" of Point 6 for our next prompt (Prompt 1.1), and elaborated mostly on Point 5. Based on the good paragraphed text, we aim to guide ChatGPT to produce more sentence combinations, having more complex structures such as relative clauses, and more word form changes.

In the second attempt, we modified the first prompt and created Prompt 1.1 as follows: "Rewrite, combine some sentences with relative clauses, change the word forms such as from adjectives to nouns and vice versa"; and we analyzed Output 1.1:

Text 1	*Output 1.1*
Social media has revolutionized the tourism and hospitality industry. Platforms such as Instagram and Facebook enable travelers to instantly share their experiences with a global audience, often swaying the travel decisions of their friends and followers. Many hotels and restaurants now actively participate in social media to boost their visibility and reputation. Travelers frequently rely on online reviews and ratings to choose accommodations and dining options. Social media campaigns can draw a large number of tourists to specific destinations. Often, destinations employ influencers to promote their attractions and services, reaching a broader audience swiftly. Additionally, social media serves as a platform for customers to express their complaints and feedback. Managing an effective online presence can greatly influence a business's success in the tourism and hospitality sector. In summary, social media has become a crucial tool for growth and customer engagement in this industry.	Social media, which has transformed the tourism and hospitality industry, enables travelers to quickly share their experiences on platforms like Instagram and Facebook, influencing their friends' and followers' travel decisions. Many hotels and restaurants, aiming to boost visibility and reputation, actively engage on social media. Online reviews and ratings, which travelers rely on to select accommodations and dining options, have become crucial. Social media campaigns, capable of attracting large numbers of tourists, often employ influencers who promote attractions and services, reaching a broad audience swiftly. Additionally, social media serves as a platform for customers' complaints and feedback. An effective online presence, essential for success in this sector, greatly influences a business. In essence, social media has become a vital tool for growth and customer engagement in tourism and hospitality.

In this Output 1.1, ChatGPT successfully combined sentences using relative clauses, creating more complex structures without grammatical errors. However, we found that the frequent repetition of these structures in the output resulted in some awkwardness. Notably, the two original sentences "Additionally, social media serves as a platform for customers' complaints and feedback" and "In summary, social media has become a crucial tool for growth and customer engagement in this industry" remained unchanged in word order and structure in both attempts. It appears that in its current version (4.0), ChatGPT identifies sentences containing the conjunctive adverbs "Additionally" and "In summary" as main ideas and does not combine or modify their structure.

In our third attempt, we modified the prompt to reduce the repetition of sentence structure as follows: "Rewrite, combine some sentences with relative clauses, change the word forms such as from adjectives to nouns and vice versa, use different structures for each sentence" (Prompt 1.2).

Text 1	*Output 1.2*
Social media has revolutionized the tourism and hospitality industry. Platforms such as Instagram and Facebook enable travelers to instantly share their experiences with a global audience, often swaying the travel decisions of their friends and followers. Many hotels and restaurants now actively participate in social media to boost their visibility and reputation. Travelers frequently rely on online reviews and ratings to choose accommodations and dining options. Social media campaigns can draw a large number of tourists to specific destinations. Often, destinations employ influencers to promote their attractions and services, reaching a broader audience swiftly. Additionally, social media serves as a platform for customers to express their complaints and feedback. Managing an effective online presence can greatly influence a business's success in the tourism and hospitality sector. In summary, social media has become a crucial tool for growth and customer engagement in this industry.	Social media's revolution in the tourism and hospitality industry allows travelers to share their experiences on platforms like Instagram and Facebook, influencing their friends' and followers' travel decisions. Many hotels and restaurants, seeking to boost visibility and reputation, actively engage on social media. The importance of online reviews and ratings has increased as travelers rely on them to choose accommodations and dining options. Social media campaigns, which often attract large numbers of tourists, frequently use influencers to promote attractions and services, thereby reaching a wider audience swiftly. Additionally, social media is a platform where customers express complaints and feedback. The effective management of an online presence is crucial for business success in this sector. Thus, social media has become an essential tool for growth and customer engagement in tourism and hospitality.

Accordingly, Output 1.2 revealed that ChatGPT did very well on replacing word forms such as from the verb "has revolutionized" to "revolution," from the verb "can greatly influence" to an adjective "is crucial for," from the gerund "Managing an effective online presence" to a noun phrase "The effective management of an online presence." ChatGPT also demonstrated extensive adherence to combining sentences with relative clauses to compose more complex sentence structures: "Social media's revolution in the tourism and hospitality industry allows travelers to share their experiences on platforms like Instagram and Facebook, influencing their friends' and followers' travel decisions" and "Social media campaigns, which often attract large numbers of tourists, frequently use influencers to promote attractions and services, thereby reaching a wider audience swiftly."

Nevertheless, the two sentences: "Additionally, social media is a platform where customers express complaints and feedback" and "Thus, social media has become an essential tool for growth and customer engagement in tourism and hospitality." still have the same sentence structure and word order with the original ones, even though this is the third attempt.

Subsequently, we combined all prompts in three attempts (from Prompt 1.0 to Prompt 1.2) to create the finalized six-step instruction:

Prompt 1.3

Here is the instruction for rewriting: 1/ Use fewer words than the original text, 2/ Keep the key terms, 3/ Remove unnecessary words and phrases, 4/ Replace long noun phrases with shorter phrases, 5/ Combine some sentences with relative clauses and change the word forms such as from adjectives to nouns and vice versa, 6/ Use different structures for each sentence. Act as a writing assistant, use the instruction to rewrite the following text:

We then used Prompt 1.3 for the remaining three original texts and recorded their results, which is presented below:

Text 2	*Output 2*
Digital banking has transformed the way individuals handle their finances. With the advent of online banking, customers have the convenience of accessing their accounts anytime, anywhere. Mobile banking apps empower users to check their balance, transfer funds, and pay bills with just a few taps. The need for physical branches has diminished, saving time and resources. Advanced security features like biometric authentication and encryption safeguard users' information. Many banks provide personalized financial advice through their apps. Online banking also facilitates quicker loan approvals and simplifies investment options. Customers value the convenience and efficiency of digital banking services. Banks benefit from reduced operational costs and the ability to reach a wider customer base. In conclusion, digital banking has made financial management more accessible and efficient for everyone.	Digital banking has revolutionized financial management. Online banking allows customers to access their accounts anytime, anywhere. Mobile banking apps let users check balances, transfer funds, and pay bills easily. Physical branches are less needed, saving time and resources. Biometric authentication and encryption protect user information. Many apps offer personalized financial advice. Online banking speeds up loan approvals and simplifies investments. Customers appreciate the convenience and efficiency, while banks benefit from lower costs and a broader reach. In summary, digital banking makes financial management more accessible and efficient for all.
Text 3	Output 3
Time management is an essential skill for achieving workplace success. Efficient time management enables employees to prioritize tasks and meet deadlines effectively. It helps individuals focus on critical tasks without feeling overwhelmed. Utilizing tools such as calendars and to-do lists can enhance time management. Proper planning reduces stress and boosts productivity. Good time management contributes to a better work-life balance. Employees who manage their time well tend to be more satisfied with their jobs. Managers highly value time management skills and often promote employees who exhibit them. Training programs can assist employees in developing these skills. In summary, time management is vital for attaining both personal and professional objectives.	Effective time management is crucial for workplace success. It allows employees to prioritize and meet deadlines efficiently, focusing on important tasks without being overwhelmed. Using tools like calendars and to-do lists improves time management, reduces stress, and increases productivity. It also leads to a better work-life balance and greater job satisfaction. Managers value these skills and often promote those who demonstrate them. Training programs can help employees develop effective time management. In short, time management is essential for achieving personal and professional goals.

Text 4	Output 4
Financial accounting is vital for any business to monitor its financial health. It encompasses recording, summarizing, and reporting financial transactions. The main documents utilized in financial accounting are the balance sheet, income statement, and cash flow statement. The balance sheet displays a company's assets, liabilities, and equity at a specific point in time. The income statement outlines the company's revenues and expenses over a period, revealing profit or loss. The cash flow statement offers insights into cash inflows and outflows. Accountants rely on these documents to ensure precise financial reporting. Proper financial accounting aids businesses in making informed decisions and ensures compliance with legal and regulatory standards. In essence, financial accounting is essential for transparency and the smooth operation of any business.	Financial accounting, crucial for monitoring a business's financial health, involves recording, summarizing, and reporting transactions. Key documents include the balance sheet, income statement, and cash flow statement. The balance sheet shows assets, liabilities, and equity at a specific time, while the income statement details revenues and expenses, revealing profit or loss. The cash flow statement illustrates cash movements. These documents ensure accurate financial reporting, which aids informed decision-making and legal compliance. In conclusion, financial accounting is essential for transparency and business operations.

The three-output data revealed that when following these six steps, GenAI effectively paraphrased the remaining content. Specifically, it excelled in removing superfluous words, restructuring and combining sentences to show a variety of grammatical changes, as highlighted in green. Most importantly, it preserved the main idea of each source text. However, we noticed that there are several theme-rheme errors in the output. Theme-rheme errors occur when the flow of information within sentences or across sentences is disrupted, making it difficult for readers to follow the logical progression of ideas. Below is an example of theme-rheme error from the paraphrased text:

> Digital banking has revolutionized financial management. Online banking allows customers to access their accounts anytime, anywhere. Mobile banking apps let users check balances, transfer funds, and pay bills easily. (Output 2)

Suggested correction:

Digital banking has revolutionized financial management. In order to manage their account anytime and anywhere, customers can use online banking apps installed on their mobile phones. These apps also let users check balances, transfer funds, and pay bills easily.

Apparently, the correction of theme-rheme errors requires a more sophisticated level of writing and editing proficiency from AI users, as these errors transcend mere lexical or grammatical inaccuracies. Another limitation that we observed from all output is that the concluding sentences from the four original texts kept their grammar structure and word order in the corresponding outputs. This finding reveals that the current version of this AI (ChatGPT 4.0) cannot change or combine sentences with a conjunctive adverb, such as "In conclusion," "In essence," and "In summary."

10.8 Discussion

With the suggested prompt (Prompt 1.3), the responses of ChatGPT indicated that this GenAI tool effectively rewrites the text, preserves core ideas, maintains field-specific terminology, and does not generate any lexical and grammatical errors. Therefore, the findings in this study did not align with Prentice and Kinden (2018), who noted that some APTs merely replace words with synonyms and often misinterpret key terminology, resulting in distorted meanings, as ChatGPT demonstrates a more advanced algorithm to paraphrasing. Conversely, data analysis from the paraphrased output indicates that ChatGPT's paraphrasing capabilities surpass those of earlier automated paraphrasing tools (APTs). Our analysis shows that its paraphrased output aligns with paraphrasing assessment frameworks from previous linguistic scholars (Burstein et al., 2012; Chen et al., 2015; Keck, 2006). In particular, the paraphrased output from this AI understudy can be sorted into the "conceptual paraphrasing" term proposed by Burstein et al. (2012), matching the category of "Minimal Revision" (including 20–49% of words from the source), and matching the highest level (Level 5) of the paraphrasing framework proposed by Chen et al. (2015).

In terms of limitations, the GenAI understudy showcased main errors in maintaining proper theme-rheme progression, particularly in having continuous errors of "brand new theme." This error in having a brand new unrelated theme aligns with previous studies that analyzed the academic texts written by ESL/EFL learners (Fitriati & Gayatri, 2021; Kuswoyo & Susardi, 2017; Thang & Tam, 2020; Wang, 2007; Yunita, 2018). Moreover, the AI cannot reach the level of rewriting a text that matches the description of "Substantial Revision" in the study of Keck (2006), in which "over 85% of a typical Substantial Revision was made up of words that did not occur in the original excerpt" (p. 270).

Beyond these pedagogical insights, it is also important to reflect on the broader implications of employing GenAI tools like ChatGPT in education. As Selwyn (2024) argues, the increasing integration of GenAI into educational contexts raises critical concerns regarding the political economy and social justice dimensions of technology use. While this study utilized ChatGPT for its pedagogical affordances, it exists within a larger ecosystem dominated by commercial technology providers. The development and deployment of such tools are often shaped by corporate priorities, which may influence how educators adopt them, sometimes without full consideration of ethical implications or long-term consequences.

Furthermore, although ChatGPT was found to be supportive in helping EFL learners improve paraphrasing, it is crucial to recognize that AI systems can perpetuate algorithmic biases that disproportionately affect learners from marginalized linguistic or cultural backgrounds. As Mitchell (2019) notes, these systems are often trained on data that reflect dominant norms, potentially leading to unequal learning outcomes or misrepresentations of diverse learner voices. This highlights the importance of employing AI tools with critical awareness and ensuring that their integration into classrooms does not inadvertently reinforce educational inequalities.

In addition to the findings of this study, related chapters in this volume further illustrate the growing emphasis on integrating AI meaningfully in educational contexts. Kahila et al. (2025 [Chap. 8]), for example, demonstrate how involving young learners in co-designing AI-driven applications fosters not only technical understanding but also ethical awareness and data agency. While our study focuses on tertiary-level language learners rather than school-age students, the emphasis on learner empowerment and critical engagement with AI aligns with our goal of promoting paraphrasing skills through active interaction with GenAI. Similarly, Dúo-Terrón et al. (2025 [Chap. 9]) highlight how experiential learning with robotics can shape primary students' understanding of future work and technological change. This resonates with our approach of encouraging

EFL learners to interact directly with AI tools like ChatGPT to build future-relevant skills such as digital literacy, adaptability, and reflective learning. Together, these studies demonstrate the value of pedagogical designs that position learners as active participants in an AI-mediated world, regardless of age group or discipline. In acknowledging these dimensions, this study affirms the need for more ethically grounded and socially responsive approaches to GenAI in language education—ones that prioritize inclusive design, pedagogical transparency, and educator agency.

10.9 Conclusion

In conclusion, we propose Prompt 1.3 as the optimal prompt for generating high-quality paraphrases with the 4.0 version of ChatGPT, which was based on our sentence structure analysis and our revised guidelines over three attempts. Additionally, the response quality to this prompt was evaluated with three other texts, demonstrating its consistency and effectiveness across different sources. Our analysis shows that its paraphrased output aligns well with paraphrasing assessment frameworks from other linguistic scholars (Burstein et al., 2012; Chen et al., 2015; Keck, 2006), particularly in preserving the main idea of each source text without lexical or grammatical errors. Therefore, this study reinforces that effective AI prompt engineering can unlock advanced capabilities, allowing users to deconstruct problems into logical steps, thereby improving their problem-solving skills (Atlas, 2023; Chui, 2023; Kojima et al., 2022). The findings of the study also support existing literature on the impact of GenAI in providing writing feedback, particularly with paraphrasing (Fitria, 2021; Fu et al., 2020; Kangasharju et al., 2022; Le et al., 2024; Yang & Kyun, 2022).

The findings of our study yield multifaceted implications. When a well-developed prompt is given, the GenAI tool understudy demonstrated its capability in rewriting accurate and varied grammatical structures, making it a useful tool for EFL/ESL students to learn and practice paraphrasing. However, the AI exhibited limitations in certain areas, notably in rephrasing sentences that begin with conjunctive adverbs, as well as in maintaining proper theme-rheme flow. These limitations highlight aspects where the algorithms of GenAI need more refinement. Meanwhile, these limitations can offer EFL/ESL teachers' and educators' insights in identifying AI-generated paraphrases, enabling them to give feedback and guidance to their students when necessary.

Due to the small scale of the research, one limitation is the restricted number of English text inputs and prompt trials. This constraint in data may affect the generalizability of the findings across diverse genres and lengths. Additionally, the assessment process did not incorporate multiple human raters to establish a specific inter-rater agreement rate for the identified errors in ChatGPT's paraphrased output. Therefore, to improve the robustness of research findings, future studies should consider these following aspects. Applying the suggested prompt (Prompt 1.3) to a larger and more diverse dataset across various text genres and academic fields would provide a more comprehensive result. Additionally, employing more experienced raters would enhance the reliability of error identification and establish inter-rater agreement. Afterward, a longitudinal analysis is also suggested, as it could track improvements in paraphrasing capabilities of ChatGPT's different models, due to the fact that these models are regularly updated. Finally, a comparative analysis of the suggested prompt should benchmark ChatGPT's performance against other GenAI models regarding its paraphrasing function for a more thorough evaluation.

References

Abukmeil, M., Ferrari, S., Genovese, A., Piuri, V., & Scotti, F. (2022). A survey of unsupervised generative models for exploratory data analysis and representation learning. *ACM Computing Surveys, 54*(5), 1–40. https://doi.org/10.1145/3450963

AlShaikh, F., & Hewahi, N. (2021). *AI and machine learning techniques in the development of Intelligent Tutoring System: A review*. 403–410.

Atlas, S. (2023). *ChatGPT for higher education and professional development: A guide to conversational AI*. DigitalCommons@URI. https://digitalcommons.uri.edu/cba_facpubs/548/?utm_source=digitalcommons.uri.edu%2Fcba_facpubs%2F548&utm_medium=PDF&utm_campaign=PDFCoverPages

Aydin, Ö., & Karaarslan, E. (2023). Is ChatGPT leading generative AI? What is beyond expectations? *Academic Platform Journal of Engineering and Smart Systems, 11*(3), 118–134. https://doi.org/10.21541/apjess.1293702

Azar, B. S., & Hagen, S. A. (2009). *Understanding and using English grammar* (4th ed.). Pearson Education.

BaiDoo-Anu, D., & Owusu Ansah, L. (2023). Education in the era of generative artificial intelligence (AI): Understanding the potential benefits of ChatGPT in promoting teaching and learning. *Journal of AI, 7*(1), 52–62. https://doi.org/10.61969/jai.1337500

Bailey, S. (2018). *Academic writing—A handbook for international students* (5th ed.). Routledge.

Bao, Y., & Li, B. (2023). A preliminary study on graduate student instructors' exploration, perception, and use of ChatGPT. *International Journal of Computer-Assisted Language Learning and Teaching, 13*(1), 1–23. https://doi.org/10.4018/IJCALLT.332873

Barrot, J. S. (2023). Using ChatGPT for second language writing: Pitfalls and potentials. *Assessing Writing, 57*, 100745. https://doi.org/10.1016/j.asw.2023.100745

Bondarenko, I. (2017). The role of positive emotions and type of feedback in self-regulation of learning goals achievement: experimental research. *Procedia—Social and Behavioral Sciences, 237*, 405–411. https://doi.org/10.1016/j.sbspro.2017.02.080

Brown, T. B., Mann, B., Ryder, N., Subbiah, M., Kaplan, J., Dhariwal, P., Neelakantan, A., Shyam, P., Sastry, G., Askell, A., Agarwal, S., Herbert-Voss, A., Krueger, G., & Henighan, T. (2020). Language models are few-shot learners. In H. Larochelle, M. Ranzato, R. Hadsell, M. F. Balcan, & Lin (Eds.), *Advances in neural information processing systems 33 (NeurIPS 2020)*. NeurIPS. https://proceedings.neurips.cc/paper_files/paper/2020/file/1457c0d6bfcb4967418bfb8ac142f64a-Paper.pdf

Burstein, J., Flor, M., Tetreault, J., Madnani, N., & Holtzman, S. (2012). Examining linguistic characteristics of paraphrase in test-taker summaries. *ETS Research Report Series, 2012*(2). https://doi.org/10.1002/j.2333-8504.2012.tb02300.x

Cao, Y., Li, S., Liu, Y., Yan, Z., Dai, Y., Yu, P. S., & Sun, L. (2023). *A comprehensive survey of AI-Generated Content (AIGC): A history of generative AI from GAN to ChatGPT* (arXiv:2303.04226). arXiv. http://arxiv.org/abs/2303.04226

Castelli, M., & Manzoni, L. (2022). Special issue: Generative models in artificial intelligence and their applications. *Applied Sciences, 12*(9), 4127. https://doi.org/10.3390/app12094127

Chang, K., Xu, S., Wang, C., Luo, Y., Xiao, T., & Zhu, J. (2024). *Efficient prompting methods for large language models: A survey* (arXiv:2404.01077). arXiv. http://arxiv.org/abs/2404.01077

Chen, M.-H., Huang, S.-T., Chang, J. S., & Liou, H.-C. (2015). Developing a corpus-based paraphrase tool to improve EFL learners' writing skills. *Computer Assisted Language Learning, 28*(1), 22–40. https://doi.org/10.1080/09588221.2013.783873

Chui, H. C. (2023). *ChatGPT as a tool for developing paraphrasing skills among ESL learners. 11*(2). https://doi.org/10.24191/cplt.v11i2.21723

Cooper, G. (2023). Examining science education in ChatGPT: An exploratory study of generative artificial intelligence. *Journal of Science Education and Technology, 32*(3), 444–452. https://doi.org/10.1007/s10956-023-10039-y

Creswell, J. W., & Creswell, J. D. (2022). *Research design: Qualitative, quantitative, and mixed methods approaches* (6th ed.). Sage Publications.

Dang, H., Mecke, L., Lehmann, F., Goller, S., & Buschek, D. (2022). *How to prompt? Opportunities and challenges of zero- and few-shot learning for human-AI interaction in creative applications of generative models* (Version 1). arXiv. https://doi.org/10.48550/ARXIV.2209.01390

Dúo-Terrón, P., López-Belmonte, J., Marín-Marín, J.-A., & Moreno-Guerrero, A.-J. (2025). Exploring horizons: Study on the perceptions of primary students regarding the future of work in robotics and steam. In S. Papadakis & G. Lampropoulos (Eds.), *Social robots and artificial intelligence in education: Integrating AI in K-12 and higher education* (pp. 213–235). Palgrave Macmillan.

Eggins, S. (1994). *An introduction to systemic functional linguistics.* Pinter Publishers; Distributed in the United States and Canada by St. Martin's Press.

Fitria, T. N. (2021). The use technology based on artificial intelligence in English teaching and learning. *ELT Echo: The Journal of English Language Teaching in Foreign Language Context, 6*(2). https://doi.org/10.24235/eltecho.v6i2.9299

Fitriati, S. W., & Gayatri, N. M. (2021). Thematic progression in EFL learners' writing: A literature review. *Lingua Cultura, 15*(2), 257–262. https://doi.org/10.21512/lc.v15i2.7645

Flowerdew, J. (2013). *Discourse in English language education* (1. publ). Routledge.

Fu, S., Gu, H., & Yang, B. (2020). The affordances of AI-enabled automatic scoring applications on learners' continuous learning intention: An empirical study in China. *British Journal of Educational Technology, 51*(5), 1674–1692. https://doi.org/10.1111/bjet.12995

Gašević, D., Siemens, G., & Sadiq, S. (2023). Empowering learners for the age of artificial intelligence. *Computers and Education: Artificial Intelligence, 4*, 100130. https://doi.org/10.1016/j.caeai.2023.100130

Gerot, L., & Wignell, P. (1994). *Making sense of functional grammar.* Gerd Stabler.

Gui, J., Sun, Z., Wen, Y., Tao, D., & Ye, J. (2023). A review on generative adversarial networks: Algorithms, theory, and applications. *IEEE Transactions on Knowledge and Data Engineering, 35*(4), 3313–3332. https://doi.org/10.1109/TKDE.2021.3130191

Halliday, M. A. K., & Hasan, R. (2013). *Cohesion in English.* Routledge, Taylor & Francis Group.

Hyland, F. (2001). Dealing with plagiarism when giving feedback. *ELT Journal, 55*(4), 375–381. https://doi.org/10.1093/elt/55.4.375

Jamieson, S. (2015). Is it plagiarism or patchwriting? Toward a nuanced definition. In T. A. Bretag (Ed.), *Handbook of academic integrity* (pp. 1–13). Springer Singapore. https://doi.org/10.1007/978-981-287-079-7_68-1

Kahila, J., Vartiainen, H., Pope, N., & Tedre, M. (2025). Engaging young minds in AI through co-designing and making their own AI-driven apps in the classroom. In S. Papadakis & G. Lampropoulos (Eds.), *Social robots and artificial intelligence in education: Integrating AI in K-12 and higher education* (pp. 189–212). Palgrave Macmillan.

Kangasharju, A., Ilomäki, L., Lakkala, M., & Toom, A. (2022). Lower secondary students' poetry writing with the AI-based poetry machine. *Computers and Education: Artificial Intelligence, 3*, 100048. https://doi.org/10.1016/j.caeai.2022.100048

Kasneci, E., Sessler, K., Küchemann, S., Bannert, M., Dementieva, D., Fischer, F., Gasser, U., Groh, G., Günnemann, S., Hüllermeier, E., Krusche, S., Kutyniok, G., Michaeli, T., Nerdel, C., Pfeffer, J., Poquet, O., Sailer, M., Schmidt, A., Seidel, T., et al. (2023). ChatGPT for good? On opportunities and challenges of large language models for education. *Learning and Individual Differences, 103*, 102274. https://doi.org/10.1016/j.lindif.2023.102274

Keck, C. (2006). The use of paraphrase in summary writing: A comparison of L1 and L2 writers. *Journal of Second Language Writing, 15*(4), 261–278. https://doi.org/10.1016/j.jslw.2006.09.006

Khosravi, H., Viberg, O., Kovanovic, V., & Ferguson, R. (2023). Generative AI and learning analytics. *Journal of Learning Analytics, 10*(3), 1–6.

Klimova, B., Pikhart, M., & Kacetl, J. (2023). Ethical issues of the use of AI-driven mobile apps for education. *Frontiers in Public Health, 10*, 1118116. https://doi.org/10.3389/fpubh.2022.1118116

Kojima, T., Gu, S. S., Reid, M., Matsuo, Y., & Iwasawa, Y. (2022). Large language models are zero-shot reasoners. In S. Koyejo, S. Mohamed, A. Agarwal, D. Belgrave, K. Cho, & Oh (Eds.), *Advances in neural information processing systems 35 (NeurIPS 2022)*. NeurIPS. https://proceedings.neurips.cc/paper_files/paper/2022/hash/8bb0d291acd4acf06ef112099c16f326-Abstract-Conference.html

Korngiebel, D. M., & Mooney, S. D. (2021). Considering the possibilities and pitfalls of Generative Pre-trained Transformer 3 (GPT-3) in healthcare delivery. *Npj Digital Medicine, 4*(1), 93. https://doi.org/10.1038/s41746-021-00464-x

Kuswoyo, H., & Susardi, S. (2017). Problems on SFG thematic progression in ESL academic writing. *Leksema: Jurnal Bahasa Dan Sastra, 2*(1), 1–13. https://doi.org/10.22515/ljbs.v2i1.655

Law, L. (2024). Application of generative artificial intelligence (GenAI) in language teaching and learning: A scoping literature review. *Computers and Education Open, 6*, 100174. https://doi.org/10.1016/j.caeo.2024.100174

Le, A. N. N., Bo, L. K., & Nguyen, N. M. T. (2023). Canva-based E-portfolio in L2 writing instructions: Investigating the effects and students' attitudes. *CALL-EJ, 24*(1), 41–62. https://www.scopus.com/inward/record.uri?eid=2-s2.0-85148765778&partnerID=40&md5=4ef977c62a391ab5b6286ff77c6164ed

Le, A. N. N., Nguyen, V. N., Nguyen, M. T.-X., & Bo, L. K. (2024). Exploring the use of ChatGPT as a tool for developing eportfolios in ESL classrooms. In S. Papadakis (Ed.), *IoT, AI, and ICT for educational applications* (pp. 51–76). Springer Nature. https://doi.org/10.1007/978-3-031-50139-5_3

Li, Y., & Casanave, C. P. (2012). Two first-year students' strategies for writing from sources: Patchwriting or plagiarism? *Journal of Second Language Writing, 21*(2), 165–180. https://doi.org/10.1016/j.jslw.2012.03.002

Liu, Y., Du, H., Niyato, D., Kang, J., Cui, S., Shen, X., & Zhang, P. (2023). Optimizing mobile-edge AI-generated everything (AIGX) services by prompt engineering: fundamental, framework, and case study. *IEEE Network*, 1–1. https://doi.org/10.1109/MNET.2023.3335255

Marvin, G., Hellen, N., Jjingo, D., & Nakatumba-Nabende, J. (2024). Prompt engineering in large language models. In I. J. Jacob, S. Piramuthu, & P. Falkowski-Gilski (Eds.), *Data intelligence and cognitive informatics* (pp. 387–402). Springer Nature. https://doi.org/10.1007/978-981-99-7962-2_30

Mitchell, M. (2019). *Artificial intelligence: A guide for thinking humans* (1st ed.). Farrar, Straus and Giroux.

Ouyang, T., MaungMaung, A., Konishi, K., Seo, Y., & Echizen, I. (2024). *Stability analysis of ChatGPT-based sentiment analysis in AI quality assurance* (Version 1). arXiv. https://doi.org/10.48550/ARXIV.2401.07441

Parnami, A., & Lee, M. (2022). Few-shot keyword spotting with prototypical networks. *2022 7th International Conference on Machine Learning Technologies (ICMLT)*, 277–283. https://doi.org/10.1145/3529399.3529443

Pavlik, J. V. (2023). Collaborating with ChatGPT: Considering the implications of generative artificial intelligence for journalism and media education. *Journalism & Mass Communication Educator, 78*(1), 84–93. https://doi.org/10.1177/10776958221149577

Pecorari, D. (2010). *Academic writing and plagiarism*. Bloomsbury Publishing.

Pokrivcakova, S. (2019). Preparing teachers for the application of AI-powered technologies in foreign language education. *Journal of Language and Cultural Education, 7*(3), 135–153. https://doi.org/10.2478/jolace-2019-0025

Prentice, F. M., & Kinden, C. E. (2018). Paraphrasing tools, language translation tools and plagiarism: An exploratory study. *International Journal for Educational Integrity, 14*(1), 11. https://doi.org/10.1007/s40979-018-0036-7

Qadir, J. (2023). Engineering education in the era of ChatGPT: Promise and pitfalls of generative AI for education. *IEEE Global Engineering Education Conference (EDUCON), 2023*, 1–9. https://doi.org/10.1109/EDUCON54358.2023.10125121

Quirk, R. (Ed.). (2012). *A comprehensive grammar of the English language (24. impression)*. Longman, Pearson Education.

Reynolds, L., & McDonell, K. (2021). Prompt programming for large language models: Beyond the few-shot paradigm. *Extended Abstracts of the 2021 CHI Conference on Human Factors in Computing Systems*, 1–7. https://doi.org/10.1145/3411763.3451760

Rogers, B. (2007). *Complete guide To TOEFL test-writing iBT*. Cengage Learning.

Schmitt, N., & Schmitt, D. (2020). *Vocabulary in language teaching* (2nd ed.). Cambridge University Press.

Selwyn, N. (2024). On the limits of artificial intelligence (AI) in education. *Nordisk Tidsskrift for Pedagogikk Og Kritikk, 10*(1). https://doi.org/10.23865/ntpk.v10.6062

Shea, P. (2000). Leveling the playing field: A study of captioned interactive video for second language learning. *Journal of Educational Computing Research, 22*(3), 243–263. https://doi.org/10.2190/3NEY-BNT0-FB28-VBWY

Tate, E., Phillips, W., & Keough, S. (2024). Don't panic! ChatGPT doesn't have all the answers. *Journal of the North American Management Society, 13*(1) https://thekeep.eiu.edu/jnams/vol13/iss1/2

Thang, N. V., & Tam, N. T. M. (2020). The realization of thematic progression in online IELTS writing task 2 samples In An English-learning website: A case study. *VNU Journal of Foreign Studies, 36*(6). https://doi.org/10.25073/2525-2445/vnufs.4634

Vianello, A., Laine, S., & Tuomi, E. (2023). Improving trustworthiness of AI solutions: A qualitative approach to support ethically-grounded AI design. *International Journal of Human–Computer Interaction, 39*(7), 1405–1422. https://doi.org/10.1080/10447318.2022.2095478

Vu, N. N., Tam, D. T., Anh, L. N. N., & Lien, N. T. H. (2023). The impact of 4English Mobile App of EFL students' reading performance in a secondary education context. In H. P. Bui & R. Kumar (Eds.), *Advances in wireless technologies and telecommunication* (pp. 67–87). IGI Global. https://doi.org/10.4018/978-1-6684-7034-3.ch005

Wang, L. (2007). Theme and Rheme in the Thematic Organization of Text: Implications for Teaching Academic Writing. *The Asian EFL Journal, 9*(1), 164–176. https://www.asian-efl-journal.com/main-editions-new/theme-and-rheme-in-the-thematic-organization-of-text-implications-for-teaching-academic-writing/index.htm

Wang, Ming. Chuan., Lin, Yi. Tang., Huang, Cai. Yan., Huang, Chih. Yi., & Yeh, Pu. K. (2023). Empirical assessment of user satisfaction with American and Chinese AI Chatbots. *International Journal of Membrane Science and Technology, 10*(3), 3199–3211. https://doi.org/10.15379/ijmst.v10i3.3170

Wette, R. (2010). Evaluating student learning in a university-level EAP unit on writing using sources. *Journal of Second Language Writing, 19*(3), 158–177. https://doi.org/10.1016/j.jslw.2010.06.002

White, L., Balart, T., Amani, S., Shryock, K. J., & Watson, K. L. (2024). *A preliminary exploration of the disruption of a generative AI systems: Faculty/staff and student perceptions of ChatGPT and its capability of completing undergraduate engineering coursework* (Version 1). arXiv. https://doi.org/10.48550/ARXIV.2403.01538

Williamson, B., Macgilchrist, F., & Potter, J. (2023). Re-examining AI, automation and datafication in education. *Learning, Media and Technology, 48*(1), 1–5. https://doi.org/10.1080/17439884.2023.2167830

Wu, T., He, S., Liu, J., Sun, S., Liu, K., Han, Q.-L., & Tang, Y. (2023). A brief overview of ChatGPT: The history, status quo and potential future development. *IEEE/CAA Journal of Automatica Sinica, 10*(5), 1122–1136. https://doi.org/10.1109/JAS.2023.123618

Yang, H., & Kyun, S. (2022). The current research trend of artificial intelligence in language learning: A systematic empirical literature review from an activity theory perspective. *Australasian Journal of Educational Technology*, 180–210. https://doi.org/10.14742/ajet.7492

Yunita, S. (2018). Theme and thematic progression in students' recount texts. *Indonesian Journal of Applied Linguistics, 7*(3), 524. https://doi.org/10.17509/ijal.v7i3.9797

Zamfirescu-Pereira, J. D., Wong, R. Y., Hartmann, B., & Yang, Q. (2023). Why Johnny can't prompt: How non-AI experts try (and fail) to design LLM prompts. *Proceedings of the 2023 CHI Conference on Human Factors in Computing Systems*, 1–21. https://doi.org/10.1145/3544548.3581388

Zhang, E. D. (2020). An investigation of novice ESL writers' cognitive processes and strategy use of paraphrasing. *Language Testing in Asia, 10*(1), 9. https://doi.org/10.1186/s40468-020-00104-3

Zhang, Q., Wang, D. Y., & Voelker, G. M. (2014). DSpin: Detecting automatically spun content on the Web. *Proceedings 2014 Network and Distributed System Security Symposium*. Network and Distributed System Security Symposium, San Diego, CA. https://doi.org/10.14722/ndss.2014.23004

PART III

Artificial Intelligence Tools, Implications, and Future Trends in Education

Part III explores AI tools in education and their policy implications, addressing future trends and discussing emerging technologies, such as chatbots and generative AI. The chapters focus on both the merits and the challenges of integrating AI across K-12 and higher education contexts. This part includes the following four chapters:

- **Chapter 11: An Open-Source and Open-Design Robot for STEM Education in K-12** (Iraklis Varlamis, Christos Chronis, Chryssa Sofianopoulou, Eleftheria Papageorgiou): Introduces FOSSBot, an open-source educational robots developed for K-12 STEM education. It discusses how FOSSBot promotes AI literacy and enhances learning through hands-on programming and data visualization tasks.
- **Chapter 12: Chatbots in Schools: Opportunities and Considerations** (Damian Maher): Examines how chatbots can support teaching and learning in schools, providing immediate feedback, assisting with lesson planning, and personalizing learning experiences. It also considers ethical issues and the well-being of students in the context of using AI.
- **Chapter 13: Use of Artificial Intelligence in Foreign Language Education in Higher Education Settings** (Blanka Klimova, Marcel Pikhart): Explores the transformative role of AI in foreign language education in higher education. It focuses on AI-driven tools and discusses how they can enhance language skills such as writing, speaking, and listening.

- **Chapter 14: Integrating Generative AI in Education: Affordances, Issues, and Directions** (Jiaqi Wang, Steven Watson): Discusses the integration of generative AI in educational contexts, addresses key challenges and opportunities, and offers directions for future research. It emphasizes the importance of generative AI literacy and the application of AI in classrooms.

Although there are some key differences among the chapters of this section, such as authoritative role versus supportive role, local fit versus global adoption, innovation versus pedagogical suitability, and human assessment versus automatic assessment, the following recurrent themes arose:

- Personalized and adaptive learning: The chapters of this section comment upon the variety of AI tools that can be used in education to support personalized and adaptive learning. Additionally, they discuss suitable frameworks for effectively introducing and integrating AI tools in educational settings.
- Ethical and societal implications: Concerns about bias in training data, student privacy, academic integrity, accessibility, and equity appear in all chapters, thus highlighting the societal and ethical implications associated with the use of AI in education.
- Evolving educator roles: The shift toward more student-centered learning environments in which educators have a more supportive role is highlighted. Within these AI-driven initiatives, educators play a vital role which requires their acquiring new digital and AI literacy skills, adopting new pedagogical strategies, as well as modifying and revising their practices.
- New frameworks and theories: Throughout the chapters, the need to develop new pedagogical approaches, frameworks, and theories that account for AI and AI-driven tools is highlighted. Additionally, it is pointed out that to effectively adopt and integrate AI into educational settings, its use should be based on established theories and frameworks.
- Education stakeholders' involvement: The chapters highlight the need for education stakeholders to communicate and collaborate when developing AI-driven solutions, designing AI-mediated interventions, introducing AI tools in teaching and learning activities, and applying them in classrooms.

These chapters highlight an educational environment in which AI does not focus on simple gadgets or tools but on theoretically grounded systems. Readers are encouraged to evaluate the potentials of AI, its technical capabilities, ethical imperatives, and human centered values when considering adopting or creating cohesive AI strategies, frameworks, and tools that support educational activities.

CHAPTER 11

An Open-Source and Open-Design Robot for STEM Education in K-12

Iraklis Varlamis, Christos Chronis, Chryssa Sofianopoulou, and Eleftheria Papageorgiou

11.1 Introduction

Educational robots represent a transformative force in STEM education within K-12 settings, reshaping traditional learning paradigms and triggering students' interest in problem-solving, exploration, and discovery. These robots, equipped with advanced sensors, actuators, and programming capabilities, go beyond conventional teaching methods, offering immersive and hands-on experiences that bridge the gap between theoretical concepts and practical applications (Benitti, 2012; Chatzichristofis, 2023). By leveraging the power of robotics, educators can cultivate essential skills such as critical thinking, problem-solving, and collaboration, preparing students for the challenges of the future. Moreover, in today's rapidly advancing technological landscape, teaching AI concepts to students in K-12 education is critical. As artificial intelligence becomes

I. Varlamis (✉) • C. Chronis • C. Sofianopoulou • E. Papageorgiou
Department of Informatics and Telematics, Harokopio University of Athens, Athens, Greece
e-mail: varlamis@hua.gr; chronis@hua.gr; csofian@hua.gr; csi23310@hua.gr

S. Papadakis, G. Lampropoulos (eds.), *Social Robots and Artificial Intelligence in Education*, Digital Education and Learning,
https://doi.org/10.1007/978-3-032-03977-4_11

increasingly integrated into various aspects of society, from healthcare to transportation, equipping students with a fundamental understanding of AI principles is essential for their future success. By introducing AI concepts at an early age, students not only develop critical thinking and problem-solving skills but also gain insight into the ethical and societal implications of AI technologies. Moreover, fostering AI literacy in K-12 education lays the groundwork for a more inclusive and equitable future, where all individuals are equipped to engage with and contribute to the development of AI-driven innovations. Therefore, integrating AI education into K-12 curricula not only prepares students for the jobs of tomorrow but also empowers them to become informed and responsible global citizens in an AI-driven world.

Regarding the theoretical background of the study of the use of robots in STEM education, there are fundamental theories and educational frameworks supporting the integration of robotics into the academic curriculum that merit discussion. Educators can rely on them to create more interesting, engaging learning environments for their students. Constructivism learning theory was developed by Jean Piaget and is based on the idea that knowledge is built upon other existing knowledge through active processes involving sensory input and hands-on experiences (Fernando & Marikar, 2017). Students, in this case, can engage with tangible objects like robots in order to construct ideas and knowledge to build their understanding (Khine, 2017) instead of gaining facts to construct knowledge in their minds. Constructionism learning theory builds on constructivism and was developed by Papert (1991). Students can construct new knowledge based on previous experiences in a practical way, for example, through their interaction with STEM tools like robots through hands-on learning experiences (Chronis & Varlamis, 2022). Experiential learning theory was introduced by John Dewey and later expanded by David Kolb. Active experiences and experimentation (Kolb, 1983), in combination with their data and observations, can be converted into effective learning. The theory aligns with the use of robotics in STEM because students, through the execution of different experiments in science classes, can gain knowledge. Vygotsky's theory of the Zone of Proximal Development (ZPD) emphasizes the difference between what a student can achieve with or without guidance (Fani & Ghaemi, 2011) and therefore how robots could help in the learning process. Robots could be used to provide challenges, understand the students' skills, and then offer help according to the students' abilities. Interdisciplinary learning combines knowledge from different disciplines in order to think of ways to solve problems (Stentoft, 2017). An example would be the construction, programming, or use of a robot, which requires knowledge from different disciplines like mathematics, physics, and engineering. An important advantage of repeated exposure to interdisciplinary thought is that learners develop enhanced critical thinking and metacognitive skills as well as an understanding of the relations among perspectives derived from different disciplines (Ivanitskaya et al., 2002). Albert Bandura's social learning theory suggests that observation, imitation, and modeling play a primary role in the learning process (Bandura, 1977). Students can learn from each other's successes or failures when needing to collaborate with each other to solve common problems or achieve common goals. Self-determination theory (SDT) was developed by Ryan and Deci (2000) and states that students' engagement is directly connected with the sense of three specific psychological needs: autonomy, competency, and relatedness (Moore et al., 2020). Robotics contribute to these elements, as students make their own decisions throughout the learning process, develop skills through active experiences, and work together in teams. Last but not least, the Technological Pedagogical Content Knowledge (TPACK) framework is a framework by which technology can be effectively integrated into teaching and learning processes (Rahman et al., 2018). The integration of educational robotics in classes can help students visualize and understand abstract content knowledge in a tangible and concrete manner, offer kinesthetic learning experiences, and promote active learning (Rahman et al., 2017). Based on the theories and frameworks mentioned above, educational robots can be used for interactive teaching and hands-on experiences through which students improve their problem-solving (Mamatnabiyev et al., 2024), critical thinking, metacognitive skills, and practical expertise, and therefore gain knowledge. As a result, it seems that educational robots are a tool for improving the students' learning process if technology is skillfully used by the teachers and aligned with the students' educational needs (Benitti, 2012).

In this chapter, we discuss the potential of educational robots, focusing on an open-source and open-design educational robot, called FOSSBot (Chronis & Varlamis, 2022), as a prime example, and explore how it can revolutionize STEM education, empowering learners to become creators, innovators, and leaders in an increasingly technology-driven world. In addition to its versatility in STEM education, FOSSBot serves as an ideal platform for introducing AI concepts to students in K-12 settings. By incorporating AI-driven functionalities and programming interfaces, FOSSBot provides students with hands-on experience in AI concepts and principles, preparing them to navigate and contribute to the evolving landscape of artificial intelligence.

11.2 FOSSBot Presentation

FOSSBot emerges as a new tool for STEM education in K-12, building on the advantages of open-source innovation, such as flexibility, expandability, and accessibility (Alasbali & Benatallah, 2015). It was launched as an open-source project (https://github.com/eellak/fossbot) by the Greek Free Open Source Society (https://gfoss.eu/) during the Google Summer of Code initiative and has since been further extended, mainly by Harokopio University of Athens, Greece, to cater to a wide range of educational needs. Its open software and hardware architecture allows seamless integration into diverse educational contexts, offering educators the flexibility to tailor learning experiences to suit their students' needs. It is a low-cost, do-it-yourself (DIY) robot that combines commercially available electronic materials and 3D-printable plastic parts, making its assembly and disassembly easy and its repair affordable. FOSSBot not only serves as a versatile educational tool but also promotes a culture of discovery-based learning, empowering students to become active participants in their educational journey and fostering the skills they need to thrive as future innovators and citizens.

FOSSBot is suitable for kindergarten and primary school settings and facilitates the development of educational scenarios across all levels of education. Its open-source code enables easy adaptation to support multiple teaching scenarios, enhancing its versatility. Equipped with a diverse array of sensors, including distance, line, accelerometer, gyroscope, IR, light sensor, two odometers, and a battery sensor, FOSSBot offers a comprehensive platform for interactive learning experiences. Additionally, it has a microphone and a speaker, a front-mounted case for a marker or a pencil, and compatibility with LEGO bricks, which expands its utility in the classroom. The top view of the robot is presented in Fig. 11.1, where some of the sensors, the wheels, its plastic parts, and the LEGO-compatible case are visible. This versatility empowers educators to craft engaging lessons and activities spanning various scientific disciplines. With a total cost of approximately €150 for printing and sourcing materials, coupled with the freely available code, FOSSBot presents a cost-effective alternative to

Fig. 11.1 Top view of the FOSSBot educational robot

commercial solutions, making innovative STEM education accessible to a broader audience.

FOSSBot's unique features make it a powerful educational tool that engages students across various age groups and academic levels. Almost all its plastic components, except for the wheels, are printable on a 3D printer. Symbols are printed on the outer surface to aid in sensor identification and simplify setup for educators; printed meshes contribute to efficient cooling of electrical parts, while strategically placed features like the charging port, on/off switch, and a towing loop enhance usability and durability. FOSSBot offers four distinct modes of operation, catering to diverse learning needs from preschool to high school levels. Whether through its user-friendly interface for preschoolers, block-based graphical programming for primary students, notebook coding for high schoolers, or direct Python coding for advanced users, FOSSBot provides a flexible approach to programming education. Built on a modular software stack, FOSSBot leverages cutting-edge technologies such as Google Blockly, Python Jupyter, and Docker for continuous software development and updates, ensuring its relevance in the ever-evolving landscape of educational robotics. This combination of innovative hardware design, versatile software interfaces, and continuous development underscores FOSSBot's potential to revolutionize STEM education and empower students to become active participants in their learning journey.

Another key feature of FOSSBot is its digital twin, which has been created on the open-source simulator Coppelia. The simulator can also be downloaded for free and provides the same programming interface and capabilities as the physical robot since it employs the same software stack. Using the simulator, a teacher or a student can first get used to programming the robot's digital twin, and later print and assemble the robot to test the same code on the physical version. The main contributions of this study can be summarized as follows:

- The introduction of FOSSBot in STEM education within K-12 settings as an educational tool to enhance the integration of AI concepts across various academic disciplines at an early age.
- Providing students with hands-on experience in AI concepts and principles, using FOSSBot, preparing them to navigate and contribute to the evolving landscape of artificial intelligence.
- Students' interaction with STEM tools through hands-on learning experiences, constructing new knowledge in a practical way (Chronis & Varlamis, 2022), rooted in the constructionism learning theory developed by Papert (1991).
- The adaptability of FOSSBot's design, its low cost, and its modular components make it suitable for each school grade, which allows educators to integrate AI concepts in different subjects and students to interact with AI and understand its principles.

In conclusion, this chapter aims to illuminate the transformative potential of FOSSBot in STEM education, underscoring its adaptability and innovation as a catalyst for AI literacy and interdisciplinary learning. By delving into its unique features, modes of operation, and continuous software development, we aim to showcase how FOSSBot transcends traditional educational boundaries, empowering both educators and students to embrace the evolving landscape of robotics and artificial intelligence. Through the exploration of FOSSBot's capabilities, we seek to inspire educators to harness its potential as a versatile educational tool, fostering a generation of critical thinkers and innovators prepared to tackle the challenges of the future.

11.3 The Importance of Open-Source Robots in Education

Open hardware and electronics and open-source robots have emerged as a cornerstone of modern education, revolutionizing the way students engage with technology and learn fundamental concepts in STEM fields (Betancur-Vásquez et al. 2021). These open solutions are characterized by flexibility, accessibility, and collaborative development, and democratize education by empowering educators to tailor learning experiences to meet the diverse needs of their students. FOSSBot is a pioneering example of how open design can transform STEM learning since it adopts a flexible software and hardware stack that can be adapted to the needs of any educational scenario, and offers three programming interfaces of varying complexity. In the following, we delve into the significance of open-source robots in education, with a particular focus on FOSSBot as an educational tool for K-12 students.

In terms of modern education, the significance of open-source and low-cost robots cannot be overstated. Robots that are used in STEM education must seamlessly integrate into various computer science fields, be adaptable for teachers across disciplines like mathematics and physics, foster an open environment for student contributions, and afford each student the opportunity for personal exploration. While several existing commercial platforms, like LEGO Mindstorms (Afari & Khine, 2017), Thymio (Riedo et al., 2013), Edison, Ozobot, BeeBot, Sphero, and TETRIX, offer robust support, their high price tags pose a barrier to equitable access. Thus, open platforms like Otto DIY (https://www.ottodiy.com), Poppy (Lapeyer et al, 2013), esieabot (Heiss et al., 2023), and Hydra (Tsalmpouris et al., 2021) have gained popularity to cover the need for affordable and accessible robotics solutions in education. By embracing open-source and low-cost robots, educators can effectively bridge theory and practice, fostering creativity, collaboration, and lifelong learning among students across various academic disciplines (Kalaitzidou & Pachidis, 2023).

Robot simulations can play a pivotal role in modern education, offering a virtual platform for students to explore and experiment with robotics concepts in a safe, cost-effective, and accessible manner. Simulation environments like RoSoS or RoboSim (Gucwa & Cheng, 2018) have been developed for robotic competitions and offer an easy and fun tool to engage students in robotics. These simulations allow students to design, program, and test robots without the constraints of physical hardware, enabling iterative learning and experimentation. Moreover, simulations can replicate real-world scenarios with high fidelity, allowing students to tackle complex challenges and scenarios that may be impractical or impossible to replicate in a traditional classroom setting.

Open-source simulator platforms like Gazebo, V-REP, or ARGoS (Pitonakova, et al., 2018) provide even more possibilities, allowing users to simulate any robot in different environments and setups, thus expanding the limits for educators and students. By engaging with simulations, students develop critical thinking, problem-solving, and programming skills while gaining a deeper understanding of robotics principles. Additionally, simulations democratize access to robotics education, breaking down geographical and financial barriers and allowing students from diverse backgrounds to participate in hands-on learning experiences. Overall, the importance of robot simulations in education lies in their ability to enhance learning outcomes, foster innovation, and prepare students for the challenges of the future in an increasingly technology-driven world (Tselegkaridis & Sapounidis, 2021).

Combining a physical robot with simulation technology offers a powerful educational approach that maximizes learning opportunities for both students and teachers. By integrating both modalities, students gain the benefits of hands-on experimentation with a physical robot, allowing them to develop practical skills in robot assembly, programming, and problem-solving. This tangible interaction fosters a deeper understanding of robotics concepts and enhances student engagement and motivation (Zhong et al., 2023). Simultaneously, simulation technology complements the physical experience by providing a flexible and scalable environment for experimentation and exploration. Students can simulate various scenarios and test their programming algorithms in a virtual space, enabling them to iterate and refine their designs without the constraints of physical hardware. For teachers, this combination offers the advantage of versatility and scalability in lesson planning. They can seamlessly transition between physical and virtual environments based on classroom needs, ensuring that every student has access to meaningful learning experiences. FOSSBot uniquely offers this option, providing educators and students with a comprehensive platform that integrates both physical robotics and simulation technology, thereby maximizing learning outcomes in STEM education.

11.4 Working with FOSSBot

11.4.1 Robot Assembly

FOSSBot belongs to the DIY category, with the logic that it can easily be built by other people besides its creators, and its disassembly and reassembly process can be part of an educational process in the classroom. This is possible since FOSSBot is made of electronic parts that are low cost and widely available while the plastic parts are printable. Therefore, they can be replaced easily in case a component is defective or wears out. The following list contains the electronic parts and the printing material that will be needed for the robot assembly and are presented in Fig. 11.2:

- 3D printing thread: different types of plastic can be used for 3D printing the plastic parts. PETG is proposed.
- Rubber wheels: two rubber wheels (66 × 26mm) are needed.
- DC motors: Two DC motors are needed, each one for each rubber wheel.
- Sensors: IR obstacle sensor, light sensor, ultrasonic sensor, noise sensor, Gyro, accelerometer, photoresistor, odometers, battery sensor
- Storage memory: microSD memory card
- Two coded disc encoders
- Three rechargeable lithium batteries and a battery holder
- DC steppers

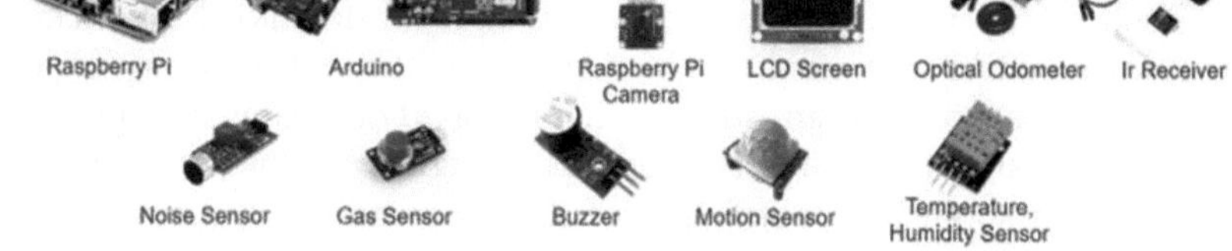

Fig. 11.2 Electronics of FOSSBot

- Cables: female to female, female to male
- Buzzer
- Speaker
- RGB LED
- Power supply
- On/off switch

In order to assemble the robot, one has to follow specific steps that are listed below:

1. Connect the motors and the odometers. Each odometer must be connected with one motor and with one rubber wheel.
2. Create the H-bridge and the step-down.
3. Connect the battery holder and the BMS.
4. Add the wheels to the caster. The two rubber wheels are placed on the left and right side of the robot, and the third metal wheel is on the bottom.
5. Connect the on/off switch, the speaker, and the charging port.
6. Add the IR sensors, the ultrasonic sensors, the photoresistor, and the RGB LED.

More detailed information on the electronics assembly, and instructions are available on the GitHub page of the project (https://github.com/chronis10/fossbot-hardware/tree/fossbot-v2/electronics_instructions), which also contains more detailed information regarding the circuit schematics.

11.4.2 3D-Printable Plastic Parts

All the 3D designs of the printable parts are freely available on Github (https://github.com/chronis10/fossbot-hardware/tree/main/designs).

Each .stl file in the designs folder is available for download. By clicking the download button on the right side, the .stl file will be downloaded. The downloaded file is an .stl file suitable for 3D printing. It represents geometric data of 3D object models as triangular meshes and contains information about the coordinates of each triangle and the direction of each surface (as can be seen in Fig. 11.3). The .stl file is one of the most popular files for 3D printing objects and can be used in design and

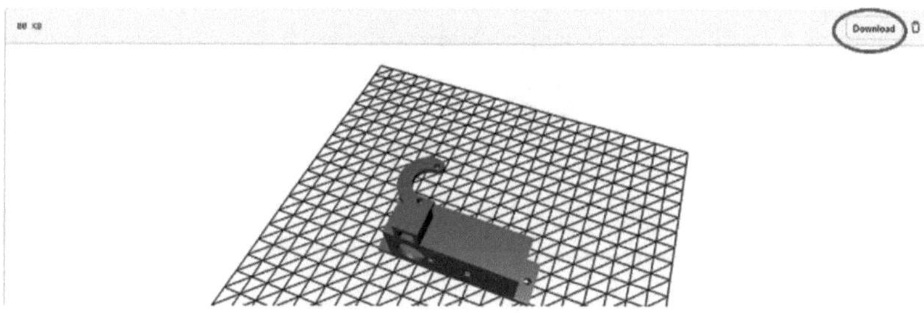

Fig. 11.3 Example of .stl file in Github

engineering programs such as AutoCAD, SolidWorks, Blender, and Tinkercad. To preview and modify the 3D-printable parts, you can use suitable software that is available for your operating system. When the 3D designs are final, they can be imported into a suitable program for slicing, such as the CURA software. The software takes the 3D model and converts it into G-code, which is the format needed to print with a 3D printer.

Before the printing of the plastic parts begins, the printing materials need to be inserted into the 3D printer. PETG is the material mentioned in the previous section, but any other suitable material can be chosen. Then, the printer needs to be set up. The temperature of the printing pin and the temperature of the bed, that is, the surface on which the object is going to be printed, must be set. The temperatures differ for each printing material as well as for different types of printing surfaces.

11.4.3 Software Stack

FOSSBot is based on a modular software stack, which allows for (1) the implementation of multiple operation/programming modes, (2) the orchestration of everything through its administration GUI, and (3) the control of the hardware in an easy way through a software library that plays the role of the FOSSBot operating system (Chronis & Varlamis, 2022).

The software stack of FOSSBot is presented in Fig. 11.4. It comprises the following interconnected components: Google Blockly (https://developers.google.com/blockly), Monaco Editor (https://microsoft.github.io/monaco-editor/), and Python Flask (https://flask.palletsprojects.com/en/2.1.x/). The core FOSSBot library is written in Python, controls the robot hardware, and is accessible by all other operation/

Fig. 11.4 Software stack

programming modules of the robot. The GUI consists of Blockly, Monaco, as well as the manual operation mode UI. Blockly is a block-based coding interface that allows the user to create full operational programs using coding blocks. Monaco is a fully featured code editor allowing you to write commands using Python. The manual operation mode UI provides a friendly interface for users to control the robot without any knowledge of programming.

11.4.4 Sensors

The sensors that FOSSBot supports are the following:

- Ultrasonic distance sensor
- Battery Sensor
- Accelerometer
- Gyroscope
- Odometers
- Light Sensors
- Line detection sensors/IR Obstacle sensor

The ultrasonic distance sensor is an electronic device that measures the distance between the robot and an object by emitting ultrasonic waves. Essentially, the ultrasonic sensor emits ultrasonic waves, which reflect on the surface of the object and then return to the ultrasonic sensor. The sensor has a transmitter, which emits the ultrasonic waves, and a receiver, which detects the reflected waves. From the waves that are detected, the sensor can determine both the type of defect, such as cuts, cracks, porosity, and other irregularities, as well as the size, shape, and position.

The battery sensor monitors the battery's voltage and temperature.

The accelerometer measures the acceleration with which it moves the robot at a specific instant. Acceleration is defined as the rate of change of speed and from the formula: $a = \Delta U / \Delta t$.

The gyroscope is a digital sensor, which detects the rotary motion on an axis, while it holds its orientation. Once it detects movement, it returns the number of degrees per second of rotation as well as a value (in degrees) that represents the total angle of rotation (angular velocity/ orientation).

Each odometer measures the distance that each wheel covers so that both wheels move with the same speed if the robot moves forward or backward. In case the robot needs to turn right or left, the speed of each wheel needs to be adjusted.

The light sensors understand whether or not there is a source of light close by

The line detection sensors use infrared light, which is electromagnetic radiation beyond the length of light visible to the human eye. Therefore, infrared light (or infrared rays) is not visible to the human eye. Longer wavelengths have a red (red) color, from where infrared ("beyond red") got its name. The sensor uses infrared light usually to avoid objects or detect black continuous lines. When there is no object, the infrared receiver receives no signals; when there is an object ahead that blocks and reflects the infrared light, the infrared receiver will receive signals. The sensor can be either analog or digital. In FOSSBot, the analog infrared reflective obstacle sensors are used and are placed at the bottom of the robot.

11.4.5 Examples

The sensors mentioned in the previous section can be used in various examples. In this section, some examples are presented to demonstrate the sensors' usage as well as to help educators understand the robot's features and inspire them to think of more complex scenarios.

11.4.5.1 Obstacle Avoidance

In this example, the robot uses its ultrasonic distance sensor to avoid obstacles. When the ultrasonic distance sensor locates an object, then it

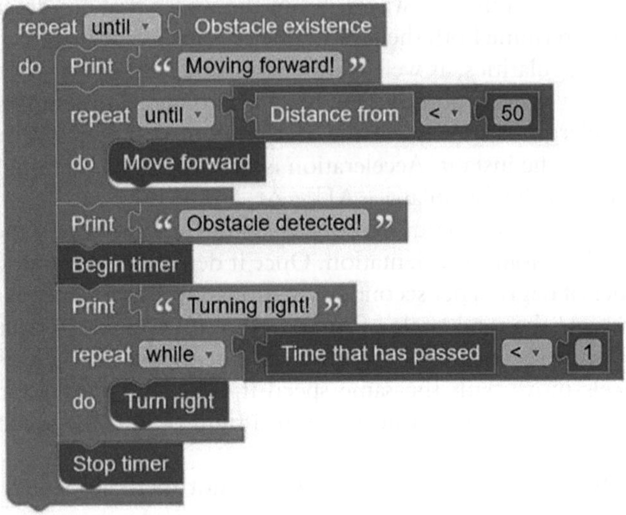

Fig. 11.5 Code for obstacle avoidance

checks the distance between the robot and the object found. In case the distance is greater than 50 cm, the robot continues to move forward. Otherwise, it turns right. In Fig. 11.5, the code for the obstacle avoidance example is presented. In Fig. 11.6, the scene used for obstacle avoidance is where the robot moves between the cones without falling on them.

11.4.5.2 Line Following

In this example, the robot uses its IR obstacle sensor for line following and the ultrasonic distance sensor for obstacle avoidance. While there is no obstacle in front of the robot, the code checks if any of the IR sensors on the bottom of the robot are above the black line. This ensures that the robot moves in the right direction. Figure 11.7 shows an example code that can be used for line following. Figure 11.8 shows the scene where the robot moves by following a single continuous line.

11.4.5.3 Light Detection

In this example, the robot uses its light detection sensor for the detection of a bright or dark source. Based on the detection of a bright or dark source, the robot turns right or left. The RGB color light changes

Fig. 11.6 Obstacle avoidance scene

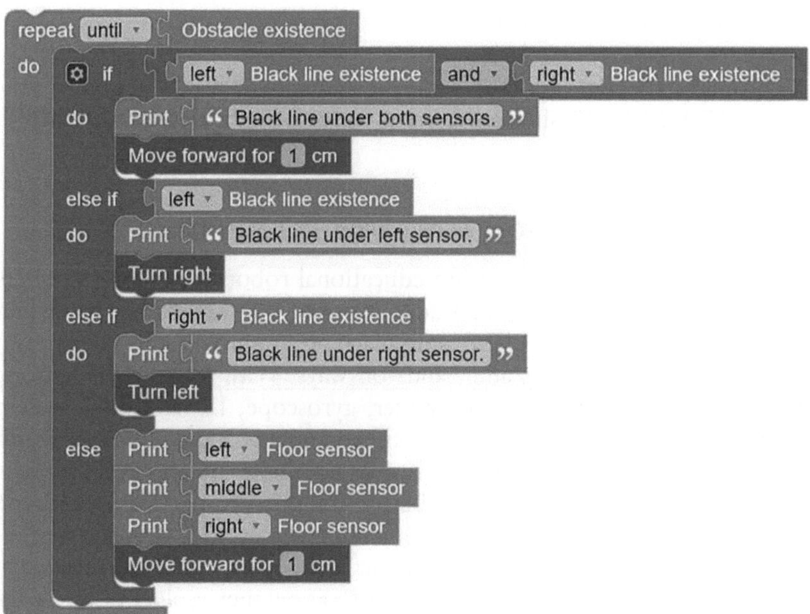

Fig. 11.7 Code for line following

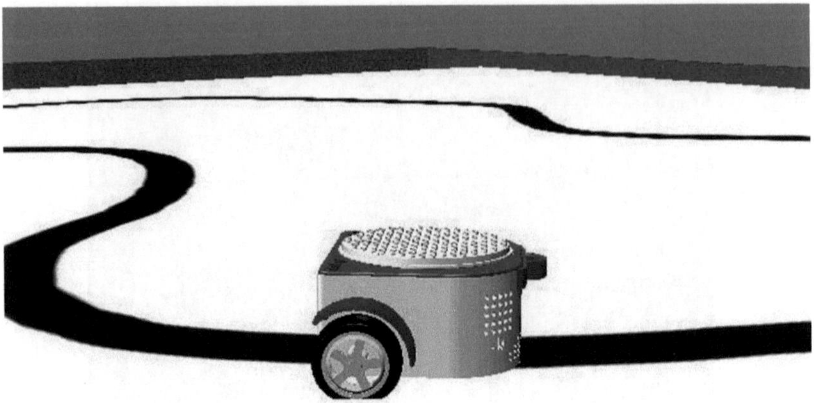

Fig. 11.8 Line following scene

accordingly. In Fig. 11.9, the code for detecting a bright source is presented. Figure 11.10 shows the scene used for light detection, where a bright source is included.

11.5 Customization for K-12 Education

An open-source and open-design educational robot like FOSSBot offers great flexibility to meet diverse academic needs in K-12 settings, providing educators with a versatile platform to teach programming, physics, or even mathematics in engaging and hands-on ways. With its array of sensors, including distance, line, accelerometer, gyroscope, IR, and light sensor, FOSSBot enables students to conduct experiments and collect real-world data that can be applied to mathematical and physics principles. For mathematics education, FOSSBot can be programmed to perform geometric calculations, such as measuring distances or angles, and visualize mathematical concepts through its movement and sensor data. Educators can design activities that explore concepts such as geometry, algebra, and trigonometry in a tangible and interactive manner, catering to the diverse learning styles and abilities of K-12 students.

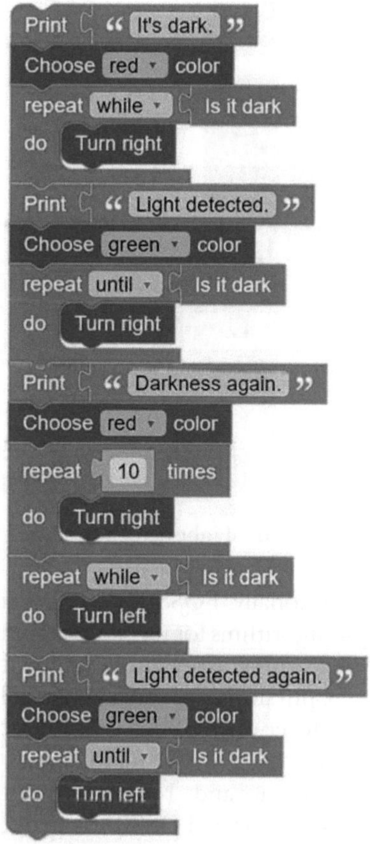

Fig. 11.9 Light detection scene

Students can experiment, observe, and collect results directly, search for solutions, and confirm or reject them, in the context of inquiry-based learning. In addition, FOSSBot is aligned with the new Greek school curricula, offering the possibility of creating an inclusive learning environment.

In the realm of physics, FOSSBot's sensors serve as invaluable tools for conducting experiments and collecting data on various physical phenomena. Students can design experiments to investigate concepts such as motion, force, and energy, using FOSSBot's sensors to measure parameters such as acceleration, velocity, and displacement. Through hands-on experimentation, students gain a deeper understanding of physics principles and develop essential skills such as data analysis and scientific inquiry.

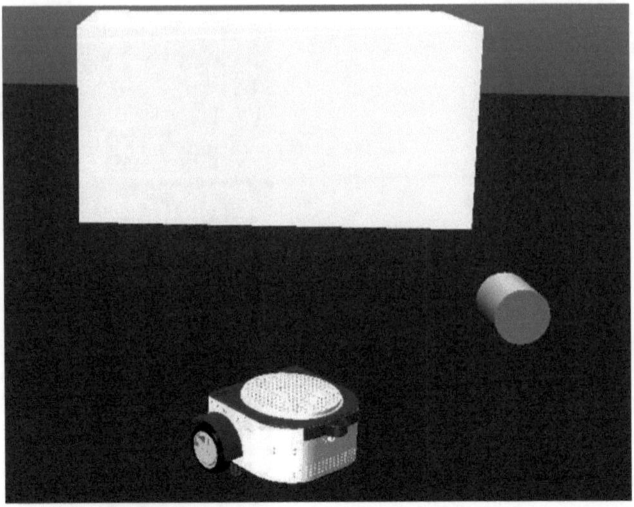

Fig. 11.10 Light detection scene

Additionally, FOSSBot's ability to simulate scenarios, such as implementing algorithms for motion planning or obstacle avoidance, further enhances the learning experience by allowing students to explore and visualize complex physical concepts in a virtual environment.

Moreover, FOSSBot presents a unique opportunity for interdisciplinary learning, where mathematics and physics concepts intersect with programming and robotics. By integrating mathematical and physical principles into FOSSBot's programming activities, students can explore concepts such as kinematics, dynamics, and control systems (González-García et al., 2020), while simultaneously developing programming skills and computational thinking. Through these interdisciplinary activities, students not only gain a deeper understanding of mathematical and physical concepts but also develop critical thinking, problem-solving, and teamwork skills that are essential for success in STEM fields.

The integration of visual blocks based on Google's Blockly architecture simplifies the learning process when programming FOSSBot. This intuitive interface replaces complex lines of code with graphical blocks that students can easily drag and drop to create sequences of commands. By abstracting the intricacies of traditional coding syntax, Blockly empowers learners of all ages and skill levels to engage with programming concepts without being hindered by syntax errors or the need for memorization. This approach enhances accessibility, particularly for younger students or those with limited prior programming experience, allowing them to focus on understanding fundamental concepts rather than syntax intricacies. Moreover, Blockly promotes a visual and interactive learning environment that fosters creativity and experimentation, encouraging students to explore and iterate on their ideas with immediate feedback. Overall, the use of visual blocks based on the Blockly architecture enhances the educational experience with FOSSBot, making programming more approachable and engaging for learners across diverse academic backgrounds.

In summary, FOSSBot's open-source and open-design nature, its intuitive no-code programming interface, coupled with its diverse sensors and capabilities, make it an ideal educational tool for teaching algorithms, mathematics, and physics concepts in K-12 settings. By tailoring activities to align with academic needs and leveraging FOSSBot's capabilities, educators can provide students with immersive and impactful learning experiences that foster curiosity, creativity, and a deeper appreciation for the mathematical and physical world around them. Through hands-on experimentation and interactive exploration with FOSSBot, students actively engage with abstract concepts, fostering deeper understanding and retention of key principles while developing critical problem-solving and analytical skills essential for success in STEM fields.

11.6 Case Studies and Practical Applications

The first step for putting an educational robot into practice is to make it easily accessible to anyone. FOSSBot is a DIY robot because anyone can download the designs for free, print the plastic parts using a 3D printer, and buy and assemble the electronics to have his/her own robot. The code library is also openly accessible and can be downloaded and installed on the robot using an SD card. Future updates are available through the application and can be downloaded and installed at the click of a button, thus making maintenance easy. However, FOSSBot is still a DIY project, which requires knowledge of 3D printing, electronics, and computers to create and operate. In order to support educators in gaining access to FOSSBot, we developed several manuals with step-by-step guidelines and videos that demonstrate how to assemble the robot, how to set up the software, and how to configure the robot for the first time and put it into practice. In an attempt to further facilitate educators in using the robot, we assembled 70 FOSSBots and sent them to primary schools around Greece.

The second step for expanding the user base of the robot and supporting educators is to create and organize printing and assembly instructions. More specifically, a complete course on 3D design and printing (https://elearn.ellak.gr/course/view.php?id=202), detailing how to print the plastic parts of the robot and how to modify them to your needs, has been developed. A snapshot of the electronics assembly instructions is provided in Fig. 11.11. The course begins with a general introduction to 3D printing and 3D design with Tinkercad but also contains several videos, step-by-step instructions, and demonstrations on how to modify the original

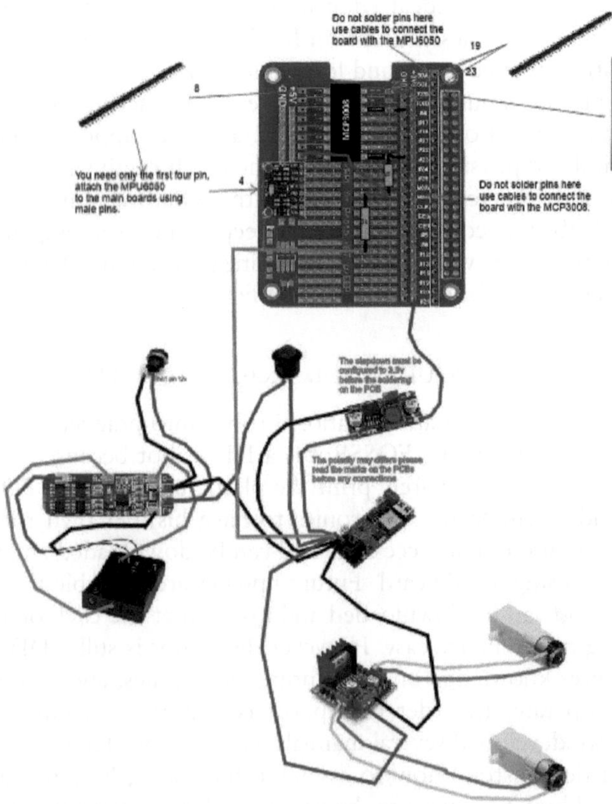

Fig. 11.11 Assembling instructions for the electronics

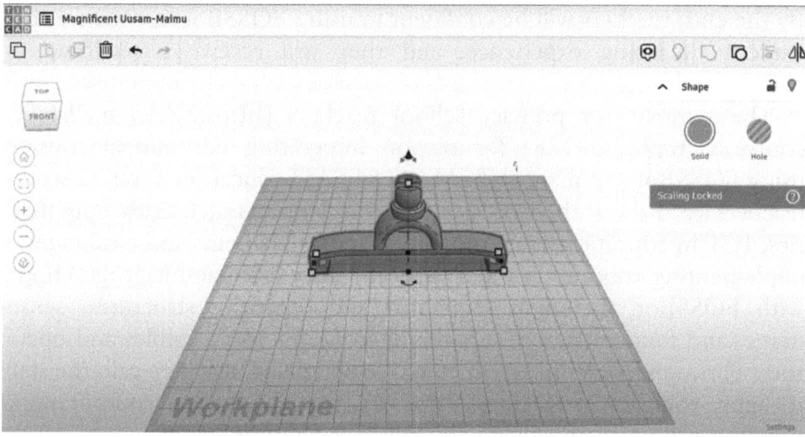

Fig. 11.12 A snapshot from the video that explains how to modify the robot's hook

design of the robot to make it appropriate for more use-case scenarios. As shown in Fig. 11.12, it illustrates how to modify a 3D-printed element, the robot's hook.

The third step is to provide training material and activities for educators who intend to use the educational robot so that they can support different courses. Since the robot is addressed to all school ages, from kindergarten to high school, with the help of educators, we developed material for two different educator groups: (1) kindergarten teachers focusing on the pedagogical utilization of the programmable robot FOSSBot in implementing collaborative activities and (2) teachers of all specialties focusing on the pedagogical utilization of educational robotics with the FOSSBot in integrating activities into lesson plans.

The seminar for kindergarten teachers (https://elearn.ellak.gr/course/view.php?id=135) on educational robotics with FOSSBot spans 12 weeks and equips educators with theoretical knowledge and practical skills to integrate robotics into their curriculum. Participants explore STEM/STEAM foundations, learning theories, and the role of robotics in education. Hands-on activities include programming with FOSSBot, creating math and science lesson plans, and developing interactive games. The seminar emphasizes sequential programming and simple command management, and encourages engagement through interactive learning.

By the end, teachers will be confident in using FOSSBot to enhance their students' learning experience, and they will receive a certificate of completion.

The seminar for primary school teachers (https://elearn.ellak.gr/course/view.php?id=138) focuses on integrating educational robotics using FOSSBot to enhance STEM and STEAM education. Over 12 weeks, teachers learn about theoretical and practical aspects such as learning theories, ICT in educational environments, and the benefits and challenges of implementing robotics in the classroom. They learn about programming with FOSSBot, including sequential and repetitive structures, sensor usage, and fundamental programming concepts like variables and operators. Through hands-on activities and lesson plans, teachers gain the skills to apply robotics in teaching various subjects, fostering computational thinking and problem-solving abilities in their students.

The fourth step is to provide synchronous training to the educators. An open call for training to educators through the Scientix (https://www.scientix.eu/) Network in Greece attracted the interest of more than 300 kindergarten, primary, and secondary school educators. With the assistance of the Greek Free Open Source Society and educators from the Scientix Network, we organized several seminar rounds and trained more than 300 teachers at all levels. In Fig. 11.13, a snapshot taken during a training session is presented, where the CoppeliaSim environment and the FOSSBot platform simulator are being used.

Finally, to engage students in using FOSSBot, we developed a simulated version of the robot in the CoppeliaSim simulator, which is also available for free download. CoppeliaSim is also free to download and using the guidelines provided on Github[1]; anyone interested in the robot can download and install it and gain access to the full functionality of the programming interface. By embedding CoppeliaSim in the same environment as the programming interface (see Fig. 11.2), we provide a unified user interface for working with the simulated robot.

The simulated version of the robot, which is presented in Fig. 11.14, comes with a predefined set of world environments, but it is possible to create more worlds in CoppeliaSim and upload them to the simulation environment.

Integrating FOSSBot into educational curricula has yielded numerous success stories and valuable lessons. We incorporated activity ideas

[1] https://github.com/chronis10/fossbot-app/

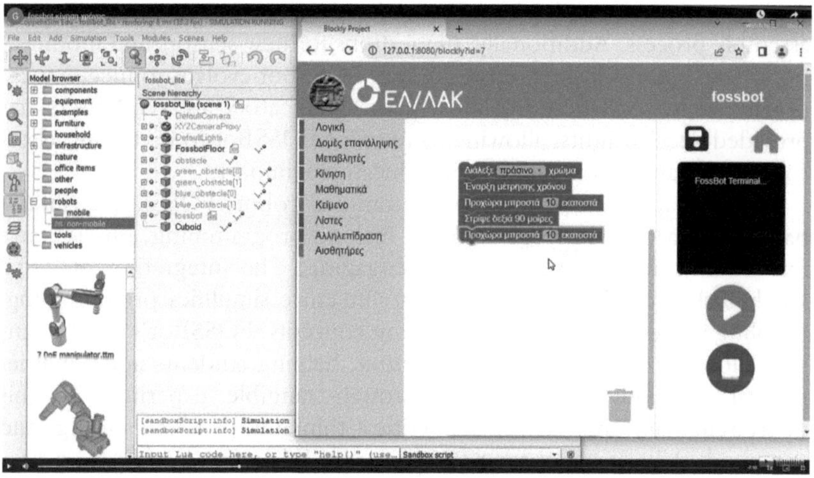

Fig. 11.13 Teaching in the Blockly environment using the simulated robot in CoppeliaSim

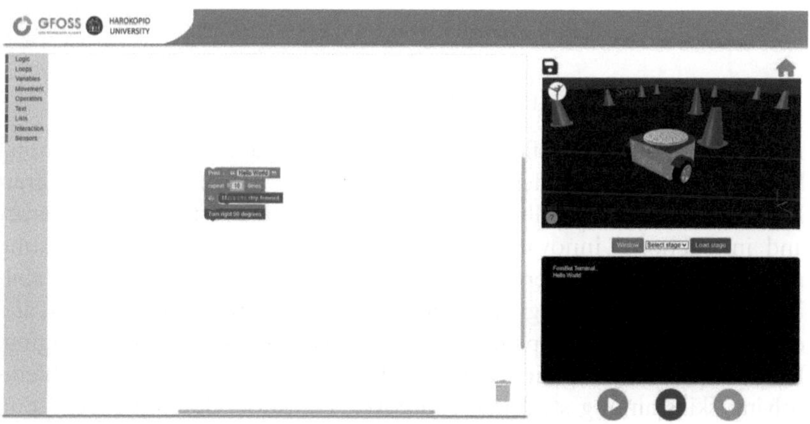

Fig. 11.14 The user interface of the simulated version of FOSSBot

gathered from seminars into our teaching material, enriching the educational experience. Feedback on assembling the robot, setting up the software, and using the environment has been instrumental in improving both the robot's design and the accompanying resources. In response to user feedback, a new, simpler board is under development to streamline the assembly process. Additionally, suggestions for new sensors and modifications have been collected, ensuring that FOSSBot evolves to meet educational needs. Collaborating closely with educators and end users has provided deep insights, allowing us to tailor FOSSBot to better suit classroom applications and enhance its educational impact.

As an adaptable open-source educational robot, FOSSBot facilitates hands-on learning in K-12 settings, making programming, physics, and mathematics more interactive and engaging. The integration of visual blocks based on Google's Blockly architecture simplifies programming, enabling students to easily grasp coding concepts. FOSSBot's diverse sensors allow for real-world data collection, helping students apply mathematical and physics principles through tangible experiments. This interdisciplinary approach fosters critical thinking, problem-solving, and teamwork skills essential for STEM success. Educators benefit from professional development, gaining insights into integrating technology into their curricula. Continuous feedback and collaboration with educators have led to improvements in FOSSBot's design and resources, ensuring it remains a valuable educational tool that meets diverse academic needs and enhances overall learning experiences.

11.7 Future Directions and Challenges

Educational robots significantly contribute to literacy development among both educators and students in K-12 education. For educators, integrating educational robots like FOSSBot into curricula fosters digital literacy and instructional innovation. Educators gain proficiency in navigating robot programming interfaces and utilizing sensors, enhancing their ability to integrate technology into subjects such as mathematics, physics, and computer science. This process empowers educators to design engaging, hands-on learning experiences that cultivate critical thinking and problem-solving skills among students. Moreover, educators develop expertise in adapting curriculum materials to leverage technological advancements, thereby enhancing their professional growth and instructional effectiveness.

Looking ahead, advancements in educational robot technology hold promise for further enhancing K-12 educational experiences. Future developments may include more advanced sensors for comprehensive data collection and analysis, expanding students' capabilities to conduct sophisticated experiments. Improvements in robot design, including easier assembly and maintenance, could enhance accessibility across diverse educational settings. Additionally, advancements in programming interfaces, such as AI-driven tools or augmented reality integration, could revolutionize student interactions with educational robots, providing intuitive and immersive learning experiences. These technological advancements have the potential to deepen student engagement and comprehension of complex concepts while preparing them for future STEM careers.

However, the widespread adoption of educational robots in K-12 education faces challenges and limitations. Issues such as cost constraints, technical support needs, and curriculum alignment complexities may hinder widespread deployment. Overcoming these challenges requires strategic initiatives, including targeted funding for schools, professional development programs for educators, and partnerships with industry to refine and scale educational robot technologies. Advocacy efforts to highlight the educational benefits of robots and their role in promoting STEM education are also critical for overcoming adoption barriers. By addressing these challenges proactively, stakeholders can pave the way for broader implementation of educational robots, ensuring equitable access to innovative learning opportunities for all students.

Extending the use of FOSSBot beyond K-12 education into higher levels, including high schools and universities, holds immense promise for advancing STEM education. In higher education, FOSSBot can serve as a platform for more advanced studies in robotics, AI, and interdisciplinary research projects (Mamatnabiyev et al., 2024b). Its open-source nature encourages customization and innovation, empowering students and researchers to explore cutting-edge technologies and tackle complex real-world challenges. By bridging the gap between theoretical knowledge and practical application, FOSSBot fosters a deeper understanding of robotics and AI concepts, preparing students for careers in emerging fields and contributing to ongoing advancements in technology and innovation.

11.8 Discussion

Among the challenges that educators face nowadays is to capture students' attention, motivate their problem-solving skills, and boost their creativity. Chatbots, as discussed in Chap. 12 by Maher (2025), aim to address this gap by providing personalized and interactive learning experiences, and FOSSBot similarly seeks to engage students through hands-on, creative exploration of robotics and AI concepts. However, educators face further challenges and limitations in the adoption of FOSSBot in K-12 education. First of all, not all educators have an engineering or programming background, so they are not familiar with sensors and complicated electronics, and, in some cases, they are not keen programmers. So, the first challenge is to provide them with basic electronics knowledge and help them get familiar with sensors, actuators, and their capabilities, as well as to provide them with some examples of their use. As a result, educators can start thinking about how to incorporate sensors and actuators in their teaching activities. The second challenge is to provide educators and students with easy interfaces for programming the robot. The development of no-code environments, which use blocks and the creation of some sample scripts and activities, can be beneficial in this direction. The establishment of a community of practice in which more experienced users share code, scripts, scenarios, and activities and introduce other users to the platform is also a step forward.

The integration of AI-driven tools in education must strike a balance between technological innovation and human interaction, a principle that is central to FOSSBot's design. This aligns with the argument by Klimova and Pikhart (2025), in Chap. 13, that technology should enhance, rather than replace, the human elements of teaching. While tools like Duolingo and ChatGPT focus on personalizing learning experiences, FOSSBot complements these approaches by fostering collaborative, hands-on environments that emphasize problem-solving, teamwork, and creativity. Additionally, proper technical support is needed when a robot malfunctions. Educators without an engineering or programming background may not be able to fix the robot, or others with an adequate background may need assistance. Last but not least, each classroom must be supplied with an adequate number of robotics tools. Insufficient assistance may make it difficult for teachers to incorporate all of FOSSBot's features into their classes, which could impede its integration into the curriculum. To help educators integrate FOSSBot into various academic disciplines, several actions can be implemented. The first solution is the provision of training, for example theoretical seminars as well as hands-on workshops, to get educators acquainted with the features that FOSSBot offers. Basic instructions on how to print, assemble, use, and program the robot should also be provided to all educators interested in using the robot. In order to support the community with more tools, a forum could be created where every educator can upload lesson plans per school grade so that others can use those if they find it difficult to create theirs or get inspired to create new ones. Supplementary content, such as videos, could also be created to explain further the provided lesson plans. Without proper support, many may struggle or others may not be interested enough to use FOSSBot in classes and therefore not contribute to the improvement of the quality of education offered in K-12.

While the development and implementation of FOSSBot represent a significant advancement in STEM education, it is essential to engage critically with the broader political economy, social justice, and sustainability implications of such technologies. We must interrogate not only the pedagogical benefits of FOSSBot but also the systemic inequities and ethical considerations it may inadvertently perpetuate or address. For instance, while FOSSBot's open-source and low-cost design democratizes access to robotics education, its reliance on 3D printing and electronic components may still exclude people without a technical background in electronics or communities lacking the necessary infrastructure. This raises questions about the equitable distribution of educational technologies and the need for systemic support to ensure that innovations like FOSSBot do not exacerbate existing educational disparities. In this direction, we are currently developing an online version of the programming environment (www.fossbot.gr), which allows people to start using FOSSBot in simulation, with zero cost and preparation, directly from their browsers. This is a step that we believe will further strengthen inclusion.

The integration of AI and robotics into K-12 education must also be examined through the lens of social justice and sustainability. While FOSSBot aims to foster AI literacy and prepare students for a technology-driven future, it is crucial to consider the ethical implications of normalizing AI technologies at an early age. Are students being equipped not only with technical skills but also with the critical awareness to question the societal, environmental, and ethical impacts of AI? For example, the extraction of rare earth metals for electronic components and the energy consumption associated with AI technologies pose significant sustainability challenges. Educators and policymakers must therefore balance the enthusiasm for technological innovation with a commitment to fostering responsible and sustainable practices among future generations. FOSSBot alternatives that use recyclable material instead of plastic, such as paper and wood (Mamatnabiyev et al., 2024b), are a step in this direction. By embedding these critical considerations into the discourse surrounding FOSSBot, we can ensure that its implementation aligns with broader goals of equity, justice, and environmental stewardship in education.

11.9 Conclusions

This chapter has underscored the profound impact of educational robots, exemplified by the open-source software and open design FOSSBot, on K-12 education. By enhancing educators' digital literacy and instructional strategies, FOSSBot empowers them to create dynamic learning activities that cultivate critical thinking and problem-solving skills among students. The integration of FOSSBot into STEM curricula not only enriches traditional subjects like mathematics and physics but also paves the way for interdisciplinary exploration of AI concepts. Its versatile sensors and intuitive programming interfaces facilitate hands-on experimentation, enabling students to engage deeply with AI principles and their applications across various academic disciplines.

Educational robots in STEM education can enhance K-12 educational experiences and become a part of the academic curriculum. Educators who integrate robots into their academic subjects become adept at navigating robot programming interfaces and using sensors as well as adapting curriculum materials to leverage technological advancements. This integration leads to designing engaging, hands-on learning experiences that cultivate critical thinking and problem-solving skills among students for a more effective learning process.

In future robot versions, more advanced sensors can be used so that students can conduct more complex experiments, including, to a greater extent, the capabilities that AI offers. For example, robots may be used not only in science courses but also in art or humanities classes, expanding their use in various academic fields and making more students be interested. Another example would be the cooperation between two or more robots, which would enhance the collaborative problem-solving principles among students. The development of educational robots means that the programming interfaces, such as AI-driven tools or augmented reality integration, advance too. Such tools can be used to detect the students' mental state, social skills, or learning style so that the robot can adapt to the students' needs for immersive learning experiences.

More specifically, FOSSBot can be a transformative tool for K-12 STEM education, offering opportunities to nurture AI literacy from an early age. By integrating AI concepts into educational activities, FOSSBot equips students with essential skills for future readiness in a technology-driven world. Its role in fostering creativity, collaboration, and innovation underscores its potential as a catalyst for educational transformation. Moving forward, continued research and development efforts are crucial to unlocking even greater educational potential and expanding access to FOSSBot's benefits across diverse educational settings. As enthusiasm grows for AI-driven learning experiences, the ongoing exploration and dissemination of FOSSBot's educational applications promise to inspire the next generation of learners and innovators worldwide.

Acknowledgments The authors would like to thank the Greek Free Open Source Software (GFOSS) society for the continuous support of the FOSSBot project. They would also like to thank Nikos Tzimopoulos, Maria Iossifidou, Paraskevi Foti, and Dimitris Tsiastoudis for the development of the training materials.

This work has been partially supported by the Erasmus+ project STEM Teaching and Education Platform for Students (S.T.E.P.S.). Agreement Number: 2023-1-FR01-KA220-HED-000165711

References

Afari, E., & Khine, M. S. (2017). Robotics as an educational tool: Impact of lego mindstorms. *International Journal of Information and Education Technology, 7*(6), 437–442. https://doi.org/10.18178/ijiet.2017.7.6.908

Alasbali, N., & Benatallah, B. (2015, October). Open source as an innovative approach in computer science education A systematic review of advantages and challenges. In *2015 IEEE 3rd International Conference on MOOCs, Innovation and Technology in Education (MITE)* (pp. 278–283). IEEE. https://doi.org/10.1109/MITE.2015.7375330

Bandura, A. (1977). *Social learning theory*. Englewood Cliffs.

Benitti, F. B. V. (2012). Exploring the educational potential of robotics in schools: A systematic review. *Computers & Education, 58*(3), 978–988. https://doi.org/10.1016/j.compedu.2011.10.006

Betancur-Vásquez, D., Mejia-Herrera, M., & Botero-Valencia, J. S. (2021). Open source and open hardware mobile robot for developing applications in education and research. *HardwareX, 10*, e00217. https://doi.org/10.1016/j.ohx.2021.e00217

Chatzichristofis, S. A. (2023). Recent advances in educational robotics. *Electronics, 12*(4), 925. https://doi.org/10.3390/electronics12040925

Chronis, C., & Varlamis, I. (2022). Fossbot: An open source and open design educational robot. *Electronics, 11*(16), 2606. https://doi.org/10.3390/electronics11162606

Fani, T., & Ghaemi, F. (2011). Implications of Vygotsky's zone of proximal development (ZPD) in teacher education: ZPTD and self-scaffolding. *Procedia, Social and Behavioral Sciences, 29*, 1549–1554. https://doi.org/10.1016/j.sbspro.2011.11.396

Fernando, S. Y., & Marikar, F. M. (2017). Constructivist Teaching/Learning Theory and participatory teaching methods. *Journal of Curriculum and Teaching, 6*(1), 110. https://doi.org/10.5430/jct.v6n1p110

González-García, S., Rodríguez-Arce, J., Loreto-Gómez, G., & Montaño-Serrano, V. M. (2020). Teaching forward kinematics in a robotics course using simulations: Transfer to a real-world context using LEGO mindstorms™. *International Journal on Interactive Design and Manufacturing (IJIDeM), 14*(3), 773–787. https://doi.org/10.1007/s12008-020-00670-z

Gucwa, K. J., & Cheng, H. H. (2018). RoboSim: A simulation environment for programming virtual robots. *Engineering with Computers, 34*, 475–485. https://doi.org/10.1007/s00366-017-0553-7

Heiss, G., Tiran Queney, E., Courbin, P., & Briere, A. (2023, April). esieabot: A Low-Cost, Open-Source, Modular Robot Platform Used in an Engineering Curriculum. In *International Conference on Robotics in Education (RiE)* (pp. 215–227). Springer Nature Switzerland. https://doi.org/10.1007/978-3-031-38454-7_19

Ivanitskaya, L., Clark, D., Montgomery, G., & Primeau, R. (2002). In Innovative higher education (27, 2, pp. 95–111). Springer Science and Business Media LLC. https://doi.org/10.1023/a:1021105309984

Kalaitzidou, M., & Pachidis, T. P. (2023). Recent robots in STEAM education. *Education Sciences, 13*(3), 272. https://doi.org/10.3390/educsci13030272

Khine, M. S. (Ed.). (2017). *Robotics in STEM education* (1st ed.). https://doi.org/10.1007/978-3-319-57786-9

Klimova, B., & Pikhart, M. (2025). Use of artificial intelligence in foreign language education in higher education settings. In S. Papadakis & G. Lampropoulos (Eds.), *Social robots and artificial intelligence in education: Integrating AI in K-12 and higher education* (pp. 329–353). Palgrave Macmillan.

Kolb, D. A. (1983). *Experiential learning*. Financial Times Prentice Hall.

Lapeyre, M., Rouanet, P., & Oudeyer, P. Y. (2013). The poppy humanoid robot: Leg design for biped locomotion. In *2013 IEEE/RSJ International Conference on Intelligent Robots and Systems* (pp. 349–356). IEEE. https://doi.org/10.1109/IROS.2013.6696375

Maher, D. (2025). Chatbots in schools: Opportunities and considerations. In S. Papadakis & G. Lampropoulos (Eds.), *Social robots and artificial intelligence in education: Integrating AI in K-12 and higher education* (pp. 305–328). Palgrave Macmillan.

Mamatnabiyev, Z., Chronis, C., Varlamis, I., Himeur, Y., & Zhaparov, M. (2024a). A holistic approach to use educational robots for supporting computer science courses. *Computers, 13*(4), 102. https://doi.org/10.3390/computers13040102

Mamatnabiyev, Z., Chronis, C., Varlamis, I., & Zhaparov, M. (2024b). Development of an educational robot for exploring the internet of things. *International Journal of Advanced Computer Science and Applications: IJACSA, 15*(4). https://doi.org/10.14569/ijacsa.2024.01504120

Moore, M. E., Vega, D. M., Wiens, K. M., & Caporale, N. (2020). Connecting theory to practice: Using self-determination theory to better understand inclusion in STEM. *Journal of Microbiology & Biology Education: JMBE, 21*(1), 05. https://doi.org/10.1128/jmbe.v21i1.1955

Papert, S. (1991). *Situating constructionism.* Constructionism/Ablex.

Pitonakova, L., Giuliani, M., Pipe, A., & Winfield, A. (2018). Feature and performance comparison of the V-REP, Gazebo and ARGoS robot simulators. In *Towards Autonomous Robotic Systems: 19th Annual Conference*, TAROS 2018, Bristol, UK July 25-27, 2018, Proceedings 19 (pp. 357–368). Springer International Publishing. https://doi.org/10.1007/978-3-319-96728-8_30

Rahman, S. M., Krishnan, V. J., & Kapila, V. (2017). Exploring the dynamic nature of TPACK framework in teaching STEM using robotics in middle school classrooms. In *2017 ASEE Annual Conference & Exposition.*

Rahman, S. M. M., Chacko, S., & Kapila, V. (2018). *Building trust in robots in robotics-focused STEM education under TPACK framework in middle schools. 2017 ASEE Annual Conference & Exposition Proceedings.* Presented at the 2017 ASEE Annual Conference & Exposition, Columbus, Ohio. https://doi.org/10.18260/1 2%2D%2D27990

Riedo, F., Chevalier, M., Magnenat, S., & Mondada, F. (2013, November). Thymio II, a robot that grows wiser with children. In *2013 IEEE Workshop on Advanced Robotics and its Social Impacts* (pp. 187–193). IEEE. https://doi.org/10.1109/ARSO.2013.6705527

Ryan, R. M., & Deci, E. L. (2000). Self-determination theory and the facilitation of intrinsic motivation, social development, and well-being. *American Psychologist, 55*(1), 68.

Stentoft, D. (2017). From saying to doing interdisciplinary learning: Is problem-based learning the answer? *Active Learning in Higher Education, 18*(1), 51–61. https://doi.org/10.1177/1469787417693510

Tsalmpouris, G., Tsinarakis, G., Gertsakis, N., Chatzichristofis, S. A., & Doitsidis, L. (2021). HYDRA: Introducing a low-cost framework for STEM education using open tools. *Electronics, 10*(24), 3056. https://doi.org/10.3390/electronics10243056

Tselegkaridis, S., & Sapounidis, T. (2021). Simulators in educational robotics: A review. *Education Sciences, 11*(1), 11. https://doi.org/10.3390/educsci11010011

Zhong, B., Zheng, J., & Zhan, Z. (2023). An exploration of combining virtual and physical robots in robotics education. *Interactive Learning Environments, 31*(1), 370–382. https://doi.org/10.1080/10494820.2020.1786409

CHAPTER 12

Chatbots in Schools: Opportunities and Considerations

Damian Maher

12.1 Introduction

As Sal Khan discussed in his Ted Talk (2023), educators are on the cusp of using artificial intelligence (AI) for possibly the biggest positive transformation that education has ever seen.

The way this will be achieved is that every student on the planet will be provided with their own personal tutor (or chatbot) and teachers around the world will also have their own teaching assistant (chatbot). He refers to data showing that personal tuition for students can have the effect of turning below average student into an above average student and an average student into an exceptional student.

Chatbots are computer programs designed to simulate human conversation through text or voice interactions. They use natural language processing (NLP) and artificial intelligence (AI) to understand and respond to user input, providing a conversational interface for various applications, from customer support to personal assistance. In the educational context,

D. Maher (✉)
University of Technology Sydney, Sydney, NSW, Australia
e-mail: damian.maher@uts.edu.au

S. Papadakis, G. Lampropoulos (eds.), *Social Robots and Artificial Intelligence in Education*, Digital Education and Learning,
https://doi.org/10.1007/978-3-032-03977-4_12

chatbots serve as virtual tutors, offering personalised learning experiences and support students in real-time. Educational chatbots can answer questions, provide feedback, and facilitate learning by guiding students through complex topics, offering practice exercises, and even supporting language learning. They can also assist teachers by handling routine tasks such as grading quizzes or offering classroom management support, ultimately enhancing the learning environment.

Chatbots are a technology application model that supports interpersonal communication and learning. They provide various types of information and knowledge through interactive methods and easy-to-operate interfaces (Hwang & Chang, 2023). Chatbots are now used across various sectors, including education. "Most of the latest intelligent AI chatbots are web-based platforms that adapt to the behaviors of both instructors and learners, enhancing the educational experience" (Labadze et al., 2023, p. 3). It should be noted that the use of chatbots is not new. The first recorded chatbot was invented in 1966 by Joseph Weizenbaum and was designed to emulate a psychologist (Weizenbaum, 1966). With the advent of ChatGPT and other AI platforms, the use of chatbots is becoming more widespread in schools, although the legal and ethical issues are meaning schools are having to carefully consider how they use chatbots.

Chatbots have the potential to take on some of the responsibility for providing the support for teachers and students where they take on roles of partners, assistants and even mentors in learning environments (Fidan & Gencel, 2022). Chatbots can take on this responsibility as they mimic human interactions, providing support for student learning (Yin et al., 2021) and teacher programming. Importantly, chatbots can simulate human conversations and provide a different learning experience (Chaves & Gerosa, 2021).

Chatbots offer a myriad of educational benefits, starting with providing access to an extensive array of learning resources (Kane, 2016). One way they facilitate learning is for language learning, including support for those students experiencing difficulties (Young & Maher, 2023). From grammar guides to vocabulary lists and interactive exercises, chatbots can act as virtual repositories, ensuring that students have a diverse range of materials at their fingertips to enhance language learning. Interactive language practice is another area where chatbots can support teachers.

This chapter will provide an overview of the ways chatbots are currently being incorporated into teaching and learning in schools drawing on the literature. To begin, the ways chatbots can support teacher's lesson planning is discussed followed feedback processes and then assessment. Next is discussion on how chatbots can support students by personalising learning. The chapter concludes with an examination of some considerations that need to be undertaken in using chatbots in schools, which include AI literacy, bias, and student well-being.

12.2 Chatbots in Education

12.2.1 Lesson Planning

In the realm of lesson planning, chatbots assist teachers in crafting effective lessons (Hashem et al., 2024). "*These tools have the capability to generate tailored educational content, alleviating time constraints while concurrently enhancing the quality of teaching*" (Kehoe, 2023, p. 172). Chatbots can assist educators in finding materials to support lesson plans. Through assessment of student data, chatbots can recommend tailored lesson plans and learning techniques to raise academic achievement. Additionally, by suggesting engaging activities, games, and exercises, these virtual assistants contribute to the creation of dynamic and student-centred lesson plans.

One of the benefits of using chatbots to support lesson planning that has been reported in the literature is time efficiency (Fitria, 2023). Chatbots provide this efficiency by automating routine administrative tasks like grading assignments or generating reports. "*This time efficiency allows educators to focus more on instructional activities, providing timely feedback, and engaging with students*" (Carr, 2023). Other aspects that AI can undertake include administrative tasks such as tracking student attendance and managing schedules.

12.2.2 Feedback

An important aspect of chatbots is that they can provide immediate personalised feedback (Cunningham-Nelson et al., 2019), which helps support the metacognitive processes of learning (Chang et al., 2022) and helps students understand their strengths and areas for improvement right away. This feedback can also limit a feeling of isolation for students (Maher, 2023). The importance of personalisation is also addressed in Klimova & Pikhart (2025 [Chap. 13]).

Immediate feedback has many educational benefits. One such benefit is that students are able to quickly identify their mistakes and understand the correct information or methodology they can use, thereby accelerating and enhancing the learning process. This real-time correction fosters a deeper understanding and retention of material (Shute, 2008). Having access to the feedback wherever and whenever they require gives students the flexibility to work at their own pace, which further enhances their learning. Providing feedback promptly can also boost learners' motivation. It encourages them to engage more deeply with the material and persist in their learning efforts, as they can immediately see the results of their efforts and understand how to improve (Hattie & Timperley, 2007).

A recommendation in the literature (Cheong-Trillo, 2023), and suggested by teachers in schools, is that chatbots can provide feedback that is part of a process that students use to help them shape their writing. In one school example, students write the first draft by themselves and then post this to their chatbot for feedback, and based on this feedback, they then produce a further version. The student submits all three versions to the teacher so they can see the development of the students' writing. Using chatbots for feedback can assist in writing becoming a process rather than a product and help students to understand that revision is an important part of the writing process.

Chatbots can also offer immediate feedback on students' progress, which can be recorded over time. By tracking their performance and highlighting areas of improvement, students gain a comprehensive view of their academic journey. The feedback provided by chatbots helps teachers understand student learning and can be supplemented by teachers' feedback. Research has shown that immediate feedback can lead to improved performance, as it allows learners to correct errors and refine their skills. This immediate correction helps prevent the reinforcement of incorrect practices and promotes the development of accurate skills and knowledge (Bangert-Drowns, et al., 1991).

An important consideration related to feedback is the age of the students. Young students require different types of feedback compared to senior students. At this point in time, chatbots such as ChatGPT can only be accessed by individuals over the age of 13, which in most countries, means only secondary school students can access chatbots in schools. However, trials by education departments are connecting primary school students with chatbots. An example of this is the New South Wales (NSW) Department of Educations who conducted a trial in the first two terms of

2024 including four primary schools. The department is using a purpose-built generative AI app designed called NSWEduChat.

In considering the age of students and feedback from AI, there are two aspects that are significant, which are digital literacy skills and forms of the feedback. Students require sophisticated digital literacy skills in order to be able to successfully learn with, and from chatbots. Some of these aspects related to AI literacy and safety are focused on later in this chapter.

In relation to the forms of feedback, young school students from kindergarten and grade 1 would need to receive feedback verbally as their skills in reading are limited. Currently, much of the mode of chatbot feedback is text-based, though there are some chatbots that can be verbally interacted with, such as the ChatGPT via Google Chrome and platforms like SchoolAI. There are other aspects relating to forms of feedback which need to be considered for young students. One of these aspects is non-verbal feedback, which includes hand gestures, touching, and facial expression (Phutela, 2015). Visual forms of feedback such as smiley faces would typically be a type of feedback that students in the junior years of primary school might receive. Older students receive much feedback through verbal and non-verbal means from teachers and it is important that chatbots can also provide this type of feedback. Currently, this type of feedback is limited for many chatbot platforms.

Another consideration relating to forms of feedback is the emotional tone used in giving feedback. The emotional tone conveyed through non-verbal communication can significantly affect students' emotional well-being and their sense of belonging (Pianta, 1999). Positive non-verbal feedback, such as a warm tone of voice can provide emotional support, reducing anxiety and encourage risk-taking in learning. As noted, the main mode used by chatbots currently is written text. Whilst animated chatbots are currently available, they are in their infancy although given the pace of change, it may be possible for animated chatbots to be part of schooling in the near future.

One other aspect related to emotional content is the 'relationship' between the chatbot and the student which is not the same as human relationships. However, chatbots can learn about the traits, skills, and interests of those it helps. This is done by the collecting and "*analyzing large amounts of personally identifiable information, such as our preferences, behaviors, and actions that are carried out online*" (Gupta et al., 2023, p. 185). For example, a chatbot could know that maths homework is due for Andi (example student) on a particular day, and that maths is an area of weakness for Andi. The chatbot would know that on Monday Andi has after school sports so sends out a 'friendly' reminder a few days before hand. In conversing with Andi, the chatbot is able to build up a profile which can be used to motivate Andi and to help them learn. In using this information, the chatbot could mimic a relationship. "*This is made possible as humans tend to personify objects, either treating, engaging with, or describing them to some degree as they would people*" (Zimmerman et al., 2023, p. 1). While this has benefits, as outlined here, there are also aspects that need to be considered which are discussed later in this chapter.

A key concern here, and in other parts of this chapter, is what aspects of feedback can chatbots assist in and what is essential for teachers to maintain? As Allagui (2023, p. 377) notes: "*From a social perspective, teacher written feedback is the method that teachers-as-readers use to interact with students.*" Teachers therefore need to carefully consider what they wish to provide feedback on and what they want to hand over to chatbots so that they maintain the important interactions with their students. One example is for the use of spelling and grammar, which has largely been undertaken by writing programs for several years. In moving towards AI-supported feedback, the teacher could identify with the student the area of feedback that is to be provided by them and by the chatbot.

12.2.3 Assessment

Recent studies speculate that chatbots may lead to fundamental changes in education and in assessment practices (Zhai 2022). One potential change is the facilitation of automated quizzing and testing where chatbots streamline the assessment process. Through a nuanced analysis of student performance and responses, chatbots can dynamically adapt tailoring assessments to individual strengths and weaknesses of students. This personalisation enhances the learning experience, ensuring that each student receives targeted and relevant assessments. One of the reasons touted for using AI for assessment is that it might reduce the teachers' workload as expressed in this statement: "*AI chatbots could act as an assistant to help teachers make dynamic assessments of each student (Jeon,* 2021*) which could reduce teachers' workload, burden and pressure*" (Wu and Yu, 2024).

Similar to the devolution associated with feedback, which was discussed earlier in this chapter, there is also careful consideration needed in using chatbots to support the assessment process. The outcome of assessment has many flow-on effects. As noted by Talanquer et al. (2015, p. 586):

"*Teacher judgments of student understanding and achievement play a central role in many important classroom decisions, including planning for instruction, providing feedback, adjusting instructional plans, and screening and placing students.*" In addition to these factors, the outcomes of assessment can dictate how much time and focus is given to a topic in the classroom. The groupings of students, and the resourcing of subjects, both at a class and school level are also important outcomes that are made based upon assessment of student work.

Other important aspects that form part of student assessment is whether assessment is based on verbal or written input, or some other form such as a movement piece or a musical presentation. The assessment may be summative or formative, which will impact on the type of feedback given and it may be individual or group-based assessment and be a formal or informal assessment. All these aspects form a cohesive whole that are used to continually assess students. Currently, chatbots do not have the capacity to undertake assessment considering all of these factors, which are necessary to support student learning.

The nature of assessment tasks, given the impact of generative AI, has been a topic of debate, particularly at the tertiary level. A concern is that "*students may use generative AI in assessment tasks, calling into question their personal learning attainment*" (Lodge et al., 2023, p. 1). The authors suggest that using multiple forms of assessment provides greater trustworthiness and allows for practices that are more inclusive. A strategy that is being used to support students' understanding of AI when completing assessment tasks is the traffic light system. Red light means AI cannot be used at all, orange light means permission is required from the teacher before using AI and green light means AI can be used at all levels in the assessment (Mormando, 2023).

12.3 Personalised Learning

"*Personalised learning refers to tailoring learning to cope with the needs, strengths and interests of each learner and to offer flexibility in what, how, when, and where to learn*" (Li & Wong, 2023, p. 423). One of the challenges facing teachers in personalising learning is how to differentiate the curriculum so that all students are engaged in their learning while having a chance to succeed. A challenge in differentiating learning is that inadequate time for planning and implementation (Gibbs, 2023). This is where the use of chatbots can be used to help teachers more effectively and efficiently personalise learning whilst reducing the amount of time required.

A pivotal aspect of chatbot utility lies in personalised learning paths (Tapalova & Zhiyenbayeva, 2022). Chatbots have the capability to analyse each student's unique learning habits and performance metrics. This data-driven analysis allows these systems to offer highly customised feedback and learning paths for students, ensuring that they can focus more efficiently on areas that require improvement. This personalised approach not only enhances learning outcomes by addressing students' specific needs but also promotes a more engaging and effective learning environment. In personalised learning paths, chatbots can adjust the difficulty level of tasks, suggest relevant resources, and even predict potential learning outcomes based on the students' performance history. This predictive capability ensures that students are neither under-challenged nor overwhelmed, maintaining an optimal learning curve.

One potential advantage in chatbots collecting data on student learning so that personalised learning can be supported is that it can be developed over time. Unlike current education systems, which tend to be compartmentalised by year and/or subject area, chatbots can build a sophisticated understanding of students' knowledge and understanding to support personalised learning. This understanding can potentially span from kindergarten through to tertiary levels of education. There is a lot of work to be undertaken by educators and chatbot designers if this is to be realised.

Another important aspect is the interplay between the role teachers undertake to personalise learning for their students and the role chatbots might play. Where chatbots interact directly with students to personalise their learning, teachers can draw on the student data to implement and augment their own teaching strategies. Teachers can also draw on chatbots, as their teaching assistant, to personalise their teaching plans and activities. In considering the interplay between human and AI support, Horver's and Molenaar's six levels of automation of personalised learning model (Molenaar, 2021b) is a useful model.

As can be seen in the model (Fig. 12.1), at the very left, at level 1, the teacher is completely in control. As the levels increase, the chatbot would increasingly take over control of the teaching and learning process. Currently, most learning in schools would be at level two in relation to the use of chatbots to support the personalisation process where computers are used. An example of chatbots taking over some of the responsibility of the writing process is where students or teachers would be using chatbots

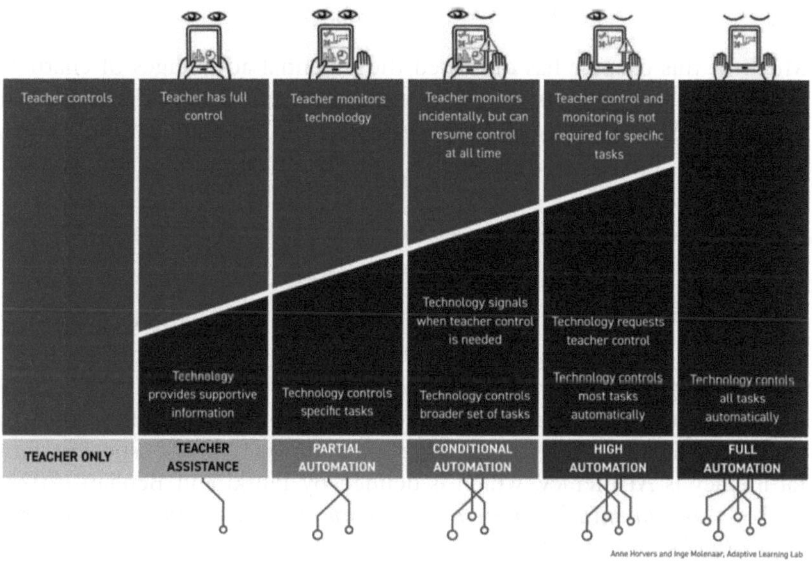

Fig. 12.1 Horver's and Molenaar's six levels of automation of personalised learning model (Molenaar, 2021b)

or automated systems to provide support on tasks such as spelling and grammar. Examples of personalised programs include Grammarly or Quillbot (Young & Maher, 2023). These systems have been available for some time now and most teachers and students see these as legitimate tools to help students with their writing. Moving to levels 3 and 4, automation becomes part of the process until at level 6, technology controls all tasks automatically.

As stated by Molennar: "*An important question is how viable the application of such technologies would be within the content of schools and what this would mean for teachers' responsibility and justification*" (2021a, p. 61). This is an important question and one that education systems will need to address before chatbots and other AI systems can sit within the educational infrastructure. It is not anticipated that we would see an education system where learning sits at level 6 of Horver's and Molenaar's model. One of the main reasons that this will most likely not happen is students' social need of emotional support for their well-being. This aspect is taken up later in the chapter.

12.4 Considerations

Much of this chapter has discussed the potential advantages of chatbots for school-aged students. In this section, some of the considerations that the use of AI in schools present is discussed. In focusing on this aspect, three areas are considered, which are AI literacy, bias, and student well-being.

12.4.1 AI Literacy

Broadly defined, digital literacy includes using digital tools to identify, evaluate, and analyse digital information (Falloon, 2020; Martin, 2005). This encompasses skills required to use both hardware and software as well as navigation skills and general literacy skills, which includes the ability to read and write (McBride, 2015). A recent addition to broad aspect of digital literacy is AI literacy, which is defined by Pinski and Benlian (2023, p. 168) as "*humans' socio-technical competence consisting of knowledge regarding human and AI actors in human-AI interaction, knowledge of the AI process steps, that is input, processing, and output, and experience in AI interaction*". According to Yi (2021, p. 356), "*AI literacy involves the acquisition of learner competencies required in the AI era based on learning functional, social, and technical literacy*". In conducting a review of articles published on AI literacy, Ng et al. (2021a) developed a framework with four aspects as set out in the Table 12.1.

As chatbots increasingly permeate education, understanding their basic functions and knowing how to use their applications is crucial for

Table 12.1 Coding framework of AI literacy

AI literacy	*Definition*
Know & understand AI	Know the basic functions of AI and how to use AI applications
Use & Apply AI	Applying AI knowledge, concepts and applications in different scenarios
Evaluate & create AI	Higher-order thinking skills (e.g., evaluate, appraise, predict, design) with AI applications
AI ethics	Human-centered considerations (e.g., fairness, accountability, transparency, ethics, safety)

Ng et al. (2021a, p. 4)

students. This knowledge equips students not only to engage competently with the technology but also to make informed decisions regarding its ethical and societal implications. Fostering AI literacy encourages critical thinking and problem-solving skills, making it essential to support student learning (Spector & Ma, 2019). For students in the early years of primary schools, it is important that they develop AI literacy learning about its limitations, ethical concerns, and basic understanding of the technology (Kong et al., 2021). Ng et al. (2021b) also suggest that it should be taught in K-2. AI-powered toys can also be used to provide students with a playful experience to learn and interact with robots and kits and develop their AI literacy (Su et al., 2023). In the early years, the use of AI would need to be heavily scaffolded, and the feedback students would receive from chatbots would be very minimal and would need to align with the practices of the classroom teacher.

One consideration, in relation to literacy and the use of chatbots is the effects of students' use of digital devices. One effect that has been reported in the literature is that students learn better via handwriting compared to writing on screens. It has been found that when students develop an automated personal handwritten script, that this contributes to writing success (Medwell & Wray, 2014; Schlagal, 2014). Additionally, the kinaesthetic process of writing is linked to writing achievement (Daffern, et al., 2017; Graham et al., 2000). An aspect related to writing is reading where it has been found that reading using traditional paper-based books provides for better learning outcomes than reading on digital screens. One meta-analysis concluded that paper-based reading yields better comprehension outcomes than digital-based reading although the authors claimed, "*comparing learning outcomes across printed and digital media are mixed, making conclusions difficult to reach*" (Delgado et al., 2018, p. 23).

As highlighted by the Mpu (2023), student AI literacy helps bridge the digital divide, ensuring that students from diverse backgrounds can equally participate in and benefit from the technological advancements. Through integrating AI into the curricula to prepare students for future challenges and opportunities, it can promote an inclusive educational environment. This approach not only prepares students for future job markets but also instils a strong foundation for lifelong learning and adaptation in the face of technological change. For students to benefit from the use of AI, it is important that they have access to the internet, regular electricity supply, and stable housing, which students from lower socio-economic backgrounds do not always have access to.

In relation to the use and application of AI, Ng et al. (2021a) reviewed 30 articles which emphasised the importance of educating learners to know how to apply AI concepts in different contexts and applications in everyday life. "*For example, Rodríguez-García et al. (2020) evaluated LearningML, a machine learning model builder, to educate citizens to understand AI applications and how it can affect our lives, as well as knowing the ethical issues regarding AI technologies*" (Ng et al., 2021a, p. 4).

The last aspect of Ng et al.'s framework relates to AI ethics, which they state "*measures an individual's perception of the social environment surrounding the behaviour, which is related to subjective norms*" (2020a, p. 4). This area is one that has caught the attention of many researchers, educators, and policy makers (e.g., Waltzer et al., 2023). One of the early ethical concerns that was raised with the introduction of ChatGPT related to students' use and academic integrity. The release of ChatGPT in late 2022 prompted widespread concern about the implications of AI for academic integrity (Waltzer et al., 2023). "*With its ability to provide specific answers to user questions, AI can be used to complete written assignments and examinations on behalf of students, leading to concerns about AI-assisted cheating*" (Lo, 2023, p. 15). In university courses, there has been a move away from this approach to a more nuanced ethical discussion (e.g., Kumar et al., 2024; McIntire et al., 2024).

It is important that students in schools are taught AI ethics and the expectations that the school has regarding how chatbots can be used to support the creative process. The inclusion of ethical consideration is important because AI has presented substantial ethical and socio-political challenges that call for thorough philosophical and ethical analysis (Zhang et al., 2023). There are resources being developed to support teachers' and students' understanding of AI ethics. One resource package produced is the AI Education Project, which consists of teacher-led lessons based on a project-based exploration of AI ethics and practicalities (https://www.aiedu.org/intro-to-ai). There are 35 lessons that are 45 minutes in duration and are aimed at grades 9 to 12. A summary of evidence-based resources to guide educators interested in understanding the AI tools available to them, the ethical implications of these tools, and developing best practices for using them in the classroom is presented at foundry10 (2023). This important consideration of ethical use of AI tools is also addressed in Wang & Watson (2025 [Chap. 14]).

12.4.2 Bias

One important consideration in providing students with access to chatbots relates to bias (Maher & Young, 2023). According to Akgun and Greenhow (2021, p. 432): *"The biggest risks of integrating these algorithms in education (particularly in K-12 contexts) are: (a) amplifying societies' existing systemic bias and discrimination, (b) perpetuating privacy, surveillance, and autonomy concerns for students from mostly marginalised and disadvantaged groups, and (c) amplifying exiting gender bias."*

One form of bias is algorithmic bias. If a chatbot's algorithms are trained on biased data sets, they can inadvertently perpetuate those biases. For example, if the training data primarily reflects a particular demographic, the chatbot might perform better for students from that demographic and worse for others, leading to unequal service quality.

Socioeconomic bias is defined as discriminating against an individual or group based on their socioeconomic standing or class (Schlosser, 2017). Chatbots can present ideas or information in the form of words or images that fail to consider the socioeconomic aspects of the students. For example, image-generating chatbots might default to depicting scenarios, landscapes, or people in ways that align with higher socioeconomic statuses, such as showing predominantly affluent urban settings, which might not reflect the everyday realities of students in lower-income or rural areas. There are tools than can help mitigate this socioeconomic bias such as Homework Helper, an AI app with content to support students with the help they need allowing them to practise their maths skills. In addition to bias, students from different socioeconomic backgrounds might have varying levels of access to the technology needed to interact effectively with chatbots. This can create a divide in which economically disadvantaged students benefit less from educational technologies, including chatbots. The issue of access to resources can be managed by schools to some degree so that economic disadvantage is minimised.

Related to socioeconomic bias is cultural bias. This is important to consider because of the psychological and emotional implications that may affect those who are subjected to this bias (Gupta et al., 2023), along with other bias discussed. A reason for cultural bias is that English is the main language used online, making up to 59% of the language used (Statista, 2024). Much of this English content comes from the United States. An example of bias provided by Shwartz, (2024) is: *"ChatGPT, upon learning about a couple dining in a restaurant in Madrid and tipping four per cent, suggested they were frugal, on a tight budget or didn't like the service. By default, ChatGPT followed the North American standard of a 15 to 25 per cent tip, ignoring the Spanish norm not to tip."*

The above example is one of many where cultural bias is built into AI chatbots. The dominance of the platforms which host chatbots such as ChatGPT pose challenges, particularly for non-native English speakers (Park, 2024). Chatbots might be limited in understanding or responding appropriately to dialects, slang, or cultural references that are specific to particular ethnic or regional groups. This can result in misunderstandings or a feeling of exclusion among students who do not communicate in the way the chatbot is optimised for.

Chatbots often struggle with languages and dialects that are less commonly used online or are underrepresented in training data. Students who speak in regional dialects or use non-standard language forms may find that chatbots do not understand them well or respond inappropriately, leading to a frustrating experience. In addition to this, the style of interaction preferred in different cultures can vary significantly. Some cultures may prefer formal and respectful interactions, while others might value informality and friendliness. A chatbot that is too casual or too formal can feel off-putting or disrespectful to students from particular backgrounds.

Gender is another aspect where students can be discriminated against because of inbuilt bias. The way AI systems are trained means they can discriminate against females with voice recognition that doesn't hear higher pitched (i.e., more feminine) voices (Tatman, 2017). Chatbots could develop gender biases based on the data they are trained on, potentially leading to different interactions based on perceived gender. This might involve reinforcing stereotypes or failing to address non-binary and transgender students appropriately. A final form of bias discussed here that can be entrenched through chatbot interactions is disability bias. If chatbots are not designed with accessibility in mind, they might exclude students with disabilities. For example, a chatbot that relies heavily on text-based information without auditory alternatives could be inaccessible to visually impaired students.

Some of the aspects discussed here related to bias can be resolved by AI architects and programmers who can train these models so that bias is minimised. An important way to help students so that bias is minimised is for teachers to help them understand bias of the chatbots they are using. "*Students need to learn how to use AI technologies judiciously, as well as to discriminate between ethical and unethical practices*" (Ng et al., 2021a, p. 2).

12.4.3 *Student Well-being*

Earlier in the chapter the aspect of relationships was raised. It seems inevitable that as chatbots take over some of the roles of the teacher that there will be less teacher–student interactions. This has implications for student well-being. The Melbourne Declaration characterises well-being in terms of its intellectual, physical, social, emotional, moral, spiritual, and aesthetic dimensions as set out by the Ministerial Council on Education, Employment, Training and Youth Affairs (MCEETYA, 2008). According to Roffey (2012), teachers bear the burden of youth mental health, of which well-being is a large part. Schonert-Reichl (2017) asserts that teacher–student relationships influence student well-being. In order to provide such support, it is important that teachers are able to identify factors that may be negatively impacting on students' well-being and put in place strategies to support students (Maher, 2022) and/or refer them to trained specialists such as student councillors.

Student well-being is currently falling. In New South Wales (NSW) for example, the Department of Education data revealed the lowest scores of student well-being since the well- being metric was first reported in 2015. Just 62% of high school students and 83% of primary school students reported a positive sense of belonging and an expectation of success in 2022 (NSW Teachers Federation, 2023). Similar figures are being reported in other countries. United States is in the midst of a student mental health crisis that has resulted in declarations of emergency from organisations like the American Academy of Paediatrics, the American Academy of Child and Adolescent Psychiatry, and the Children's Hospital Association (Posamentier, 2023). These figures are complicated by COVID-19, but they do underline a gradual downward trend in student well-being. This trend is highlighted by an Australian study that indicates mental health disorders in young Australians have increased by 47% over 15 years (ABS, 2023).

With student well-being on the decline, and the important role that teachers play in supporting well-being, handing over some of the opportunities for interactions to chatbots may see this decline continue. This is an area that requires further research. An understanding of this area is also complicated by a general teacher shortage across many countries, resulting in students missing out on opportunities. It may be that chatbots are able to fill some of the gaps created by teacher shortages, which may turn out to be beneficial. This is another area that will also require research.

One challenge related to the use of chatbots specifically, and more broadly to the use of computers, is the impact of screentime. Recommendations by the Australian Government are no more than one hour per day for children aged 2–5 years and no more than two hours of sedentary recreational screen time per day for children and young people aged 5–17 years (not including schoolwork) (Department of Health and Aged Care, n.d.). The recommendations are similar to other countries. The effects that excessive screentime can have on young people include impacts on their emotional, social, and cognitive development. It is important to note that these effects are strongest when "children use non-interactive, non-educational media instead of talking and playing with people and objects around them" (Wilkinson et al., 2021, p. 1). This is another area that requires further research.

12.4.4 *Broad Impacts*

AI is transforming various sectors, offering both opportunities and challenges. In the job market, there's a surge in demand for AI expertise, with nearly 25% of US tech job postings seeking AI skills. This trend spans multiple industries, including finance, retail, and healthcare, and offers premium pay and job security for those with advanced AI competencies (Rattner, 2025). This transformation has implications for the types of skills and knowledge young people are equipped with in schools. While schools and education systems were wary of initially introducing AI into the curriculum, this stance has now changed in many parts of the world. Chinese schools for example, will be required to provide at least eight hours of AI instruction per academic year (Sherman, 2025). This has resource implications as well as implications for teacher training, both at a pre-service level, and at an in-service level.

The environmental implications of AI have become increasingly significant, particularly concerning energy consumption, water usage, and electronic waste. Training and deploying large-scale AI models require substantial computational power, leading to heightened electricity demand and associated carbon emissions. For example, data centres consumed 21% of Ireland's national electricity in 2023, surpassing the energy usage of all urban homes combined, which stood at 18% (Varadarajan, 2024). Additionally, AI data centres necessitate vast amounts of water for cooling purposes. A study revealed that training Microsoft's GPT-3 language model in US data centres consumed 700,000 litres of water, underscoring the significant water footprint of AI technologies (Li et al., 2023). Additionally, the rapid development of AI infrastructure contributes to electronic waste, as servers typically have a short lifespan of around three years, leading to substantial e-waste accumulation. These environmental challenges necessitate a critical evaluation of AI's sustainability and the implementation of strategies to mitigate its ecological impact. These issues will need to be addressed to ensure that AI can be a sustainable aspect of education moving forward. Initiatives such as green energy can assist with this aspect but more needs to be done to ensure that agreements made at conferences, such as the United Nations Climate Change Conference (COP28), are adhered to.

12.5 Conclusions

The integration of chatbots in education presents significant opportunities, from streamlining lesson planning and providing personalised learning to automating feedback and assessment. However, the successful adoption of AI technologies in schools will depend on how well students and teachers embrace these tools and how effectively they are implemented within the educational infrastructure. While early signs indicate a growing acceptance, especially as students begin to experiment with AI at home, challenges remain, particularly in addressing concerns about academic integrity, digital literacy and teacher–student relationships.

The implications of chatbots in education extend beyond improving efficiency; they also raise critical questions about the ethical use of AI, the potential for bias, and the impact on student well-being. It is essential to consider how AI might inadvertently amplify existing disparities, whether through algorithmic, cultural, socioeconomic, or other biases. These concerns highlight the need for careful monitoring and the development of AI literacy programs that prepare students to engage with technology critically and responsibly.

Future research should focus on several key areas: understanding the long-term effects of AI on student well-being, particularly as chatbots take on more roles traditionally filled by teachers; exploring how AI can address the digital divide, ensuring equitable access to these technologies; and examining how chatbots can be designed to minimise bias and maximise inclusivity. Additionally, the impact of screen time on young learners, particularly in AI-driven educational environments, warrants further investigation. As AI continues to evolve, ongoing research will be crucial in guiding its ethical and effective integration into education.

References

ABS. (2023). *National study of mental health and wellbeing*. Australian Bureau of Statistics. https://www.abs.gov.au/statistics/health/mental-health/national-study-mental-health-and-wellbeing/latest-release.

Akgun, S., & Greenhow, C. (2021). Artificial intelligence in education: Addressing ethical challenges in K-12 settings. *AI and Ethics, 2*(3), 431–440. https://doi.org/10.1007/s43681-021-00096-7

Allagui, B. (2023, June). Chatbot feedback on students' writing: Typology of comments and effectiveness. In *International conference on computational science and its applications* (pp. 377–384). Springer Nature Switzerland. https://doi.org/10.1007/978-3-031-37129-5_31

Bangert-Drowns, R. L., Kulik, J. A., Kulik, C.-L. C., & Morgan, M. (1991). The instructional effect of feedback in test-like events. *Review of Educational Research, 61*(2), 213–238. https://doi.org/10.3102/00346543061002213

Carr, B. (2023). Revolutionizing education: Unleashing the power of Chat GPT/AI to empower educators. *Technology and the Curriculum: Summer 2023*. https://doi.org/10.21608/jstc.2024.361557

Chang, C.-Y., Kuo, S.-Y., & Hwang, G.-H. (2022). Chatbot-facilitated nursing education: Incorporating a knowledge-based chatbot system into a nursing training program. *Educational Technology & Society, 25*(1), 15–27.

Chaves, A. P., & Gerosa, M. A. (2021). How should my chatbot interact? A survey on social characteristics in human–chatbot interaction design. *International Journal of Human Computer Interaction, 37*(8), 729–758. https://doi.org/10.1080/10447318.2020.1841438

Cheong-Trillo, N. (2023). *Benefits and barriers of chatbot use in education*. Technology and the curriculum: Summer 2023.

Cunningham-Nelson, S., Boles, W., Trouton, L., & Margerison, E. (2019). A review of chatbots in education: Practical steps forward. In *In 30th annual conference for the Australasian association for engineering education (AAEE 2019): Educators becoming agents of change: innovate, integrate, motivate* (pp. 299–306). Engineers Australia.

Daffern, T., Mackenzie, N. M., & Hemmings, B. (2017). Predictors of writing success: How important are spelling, grammar and punctuation? *Australian Journal of Education, 61*(1), 75–87. https://doi.org/10.1177/0004944116685319

Delgado, P., Vargas, C., Ackerman, R., & Salmerón, L. (2018). Don't throw away your printed books: A meta-analysis on the effects of reading media on reading comprehension. *Educational Research Review, 25*, 23–38. https://doi.org/10.1016/j.edurev.2018.09.003

Department of Health and Aged Care. (n.d.). *Physical activity and exercise guidelines for all Australians*. Australian Government. https://www.health.gov.au/topics/physical-activity-andexercise/physical-activity-and-exercise-guidelines-for-all-australians?utm_source=health.gov.au&utm_medium=callout-auto-custom&utm_campaign=digital_transformation

foundry10 (2023) Navigating ethical AI: A professional development resource for K-12 educators, Medium. Retrieved September 15, 2024, from https://medium.com/the-foundry10-voice/navigating-ethical-ai-a-professional-development-resource-for-k-12-educators-c96a082a372f

Falloon, G. (2020). From digital literacy to digital competence: the teacher digital competency (TDC) framework. *Educational Technology Research and Development, 68*(5), 2449–2472. https://doi.org/10.1007/s11423-020-09767-4

Fidan, M., & Gencel, N. (2022). Supporting the instructional videos with chatbot and peer feedback mechanisms in online learning: The effects on learning performance and intrinsic motivation. *Journal of Educational Computing Research, 60*(7), 716–1741. https://doi.org/10.1177/07356331221077901

Fitria, T. N. (2023). Artificial intelligence (AI) technology in OpenAI ChatGPT application: A review of ChatGPT in writing English essay. Journal of English. *Language Teaching, 12*(1). https://doi.org/10.15294/elt.v12i1.64069

Gibbs, K. (2023). Voices in practice: challenges to implementing differentiated instruction by teachers and school leaders in an Australian mainstream secondary school. *The Australian Educational Researcher, 50*(4), 1217–1232. https://doi.org/10.1007/s13384-022-00551-2

Graham, S., Harris, K. R., & Fink, B. (2000). Is handwriting casually related to learning to write? Treatment of handwriting problems in beginning writers. *Journal of Educational Psychology, 92*(4), 620–626. https://doi.org/10.1037/0022-0663.92.4.620

Gupta, D., Singhal, A., Sharma, S., Hasan, A., & Raghuwanshi, S. (2023). Humans' emotional and mental well-being under the influence of artificial intelligence. *Journal for ReAttach Therapy and Developmental Diversities, 6*(6s), 184–197.

Hashem, R., Ali, N., El Zein, F., Fidalgo, P., & Khurma, O. A. (2024). AI to the rescue: Exploring the potential of ChatGPT as a teacher ally for workload relief and burnout prevention. *Research and Practice in Technology Enhanced Learning, 19*, 023–023. https://doi.org/10.58459/rptel.2024.19023

Hattie, J., & Timperley, H. (2007). The power of feedback. *Review of Educational Research, 77*(1), 81–112. https://doi.org/10.3102/003465430298487

Hwang, G. J., & Chang, C. Y. (2023). A review of opportunities and challenges of chatbots in education. *Interactive Learning Environments, 31*(7), 4099–4112.

Jeon, J. (2021). Chatbot-assisted dynamic assessment (CA-DA) for L2 vocabulary learning and diagnosis. *Computer Assisted Language Learning, 1–27*. https://doi.org/10.1080/09588221.2021.1987272

Kane, D. A. (2016). The role of chatbots in teaching and learning. *E-learning and the academic library: Essays on innovative initiatives*, 131.
Kehoe, F. (2023). Leveraging generative ai tools for enhanced lesson planning in initial teacher education at post primary. *Irish Journal of Technology Enhanced Learning, 7*(2), 172–182. https://doi.org/10.22554/ijtel.v7i2.124
Khan, S. (2023). How AI could save (not destroy) education. https://www.ted.com/talks/sal_khan_how_ai_could_save_not_destroy_education?language=en
Klimova, B., & Pikhart, M. (2025). Use of artificial intelligence in foreign language education in higher education settings. In S. Papadakis & G. Lampropoulos (Eds.), *Social robots and artificial intelligence in education: Integrating AI in K-12 and higher education* (pp. 329–353). Palgrave Macmillan.
Kong, S. C., Cheung, W. M. Y., & Zhang, G. (2021). Evaluation of an artificial intelligence literacy course for university students with diverse study backgrounds. *Computers and Education: Artificial Intelligence, 2*, 100026. https://doi.org/10.1016/j.caeai.2021.100026
Kumar, R., Eaton, S. E., Mindzak, M., & Morrison, R. (2024). Academic integrity and artificial intelligence: An overview. In *Second handbook of academic integrity*. Springer. https://doi.org/10.1007/978-3-031-54144-5_153
Labadze, L., Grigolia, M., & Machaidze, L. (2023). Role of AI chatbots in education: systematic literature review. *International Journal of Educational Technology in Higher Education, 20*(1), 56.
Li, K. C., & Wong, B. T. M. (2023). Artificial intelligence in personalised learning: A bibliometric analysis. *Interactive Technology and Smart Education, 20*(3), 422–445. https://doi.org/10.1007/978-3-031-54144-5_153
Li, P., Yang, J., Islam, M. A., & Ren, S. (2023). Making AI less 'thirsty': Uncovering and addressing the secret water footprint of AI models. arXiv preprint arXiv:2310.12345.
Lo, C. K. (2023). What is the impact of ChatGPT on education? A rapid review of the literature. *Education Sciences, 13*(4), 410. https://doi.org/10.3390/educsci13040410
Lodge, J. M., Howard, S., Bearman, M., Dawson, P., Agostinho, S., Buckingham Shum, S., et al. (2023). *Assessment reform for the age of artificial intelligence*. Australian Government.
Maher, D. (2022). In-service teacher professional learning in Australia: Lessons learnt from COVID-19. *International Journal for Research in Education (IJRE), 46*(2), 84–125. https://doi.org/10.36771/ijre.46.2.22-pp84-125
Maher, D. (2023). Blended learning for pre-service teachers. In J. Keengwe (Ed.), *Handbook of research on innovative frameworks and inclusive models for online learning*. IGI Global. https://doi.org/10.4018/978-1-6684-9072-3.ch013
Maher, D., & Young, K. (2023). Artificial Intelligence and literacy development in K-12 schools. In J. Keengwe (Ed.), *Creative AI tools and ethical implications in teaching and learning*. IGI Global. https://doi.org/10.4018/979-8-3693-0205-7.ch004

Martin, A. (2005). DigEuLit—A European framework for digital literacy: A progress report. *Journal of eLiteracy, 2*(2), 130–136. https://doi.org/10.18261/issn1891-943x-2006-02-06

McBride, C. (2015). *Children's literacy development: A cross-cultural perspective o nlearning to read and write.* Routledge. https://doi.org/10.4324/9781410613332-11

McIntire, A., Calvert, I., & Ashcraft, J. (2024). Pressure to plagiarize and the choice to cheat: Toward a pragmatic reframing of the ethics of academic integrity. *Education Sciences, 14*(3), 244. https://doi.org/10.3390/educsci14030244

Medwell, J., & Wray, D. (2014). Handwriting automaticity: The search for performance thresholds. *Language and Education, 28*(1), 34–51. https://doi.org/10.1080/09500782.2013.763819

Ministerial Council on Education, Employment, Training and Youth Affairs (MCEETYA). (2008). Melbourne declaration on educational goals for young Australians. Ministerial Council on Education, Employment, Training and Youth Affairs. http://www.curriculum.edu.au/verve/_resources/National_Declaration_on_the_Educational_Goals_for_Young_Australians.pdf

Molennar, I. (2021a). Personalisation of learning: Towards hybrid human-AI learning technologies. *Blockchain, and Robots,* 57–77. https://doi.org/10.1787/2cc25e37-en

Molennar, I. (2021b). Personalisation of learning: Towards hybrid human-AI learning Technologies. *In Organization for Economic Cooperation and Development OECD Digital Education Outlook 2021 Pushing the Frontiers with Artificial Intelligence, Blockchain and Robots.* OECD Publishing. https://doi.org/10.1787/2cc25e37-en

Mormando, S. (2023). *A stoplight model for guiding student AI usage.* Edutopia. https://www.edutopia.org/article/creating-ai-usage-guidelines-students/

Mpu, Y. (2023). Bridging the knowledge gap on special needs learner support: the use of artificial intelligence (AI) to combat digital divide post-covid-19 pandemic and beyond–a comprehensive literature review. In F. Altinay & Z. Altinay (Eds.), *Intellectual and learning disabilities—Inclusiveness and contemporary teaching environments.* IntechOpen. https://doi.org/10.5772/intechopen.113054

Ng, D. T. K., Leung, J. K. L., Chu, S. K. W., & Qiao, M. S. (2021a). Conceptualizing AI literacy: An exploratory review. *Computers and Education: Artificial Intelligence, 2,* 100041. https://doi.org/10.1016/j.caeai.2021.100041

Ng, D. T. K., Leung, J. K. L., Chu, K. W. S., & Qiao, M. S. (2021b). AI literacy: Definition, teaching, evaluation and ethical issues. *Proceedings of the Association for Information Science and Technology, 58*(1), 504–509. https://doi.org/10.1002/pra2.487

NSW Teachers Federation. (2023, June). Investment in student wellbeing. https://www.nswtf.org.au/news/2023/06/08/investment-in-student-wellbeing/

Park, S. (2024). AI chatbots and linguistic injustice. *Journal of Universal Language, 25*(1), 99–119. https://doi.org/10.22425/jul.2024.25.1.99

Phutela, D. (2015). The importance of non-verbal communication. *IUP Journal of Soft Skills, 9*(4), 43.

Pianta, R. C. (1999). *Enhancing relationships between children and teachers*. American Psychological Association. https://doi.org/10.1037/10314-000

Pinski, M., & Benlian, A. (2023). AI literacy-towards measuring human competency in artificial intelligence. *Proceedings of the 56th Hawaii International Conference on System Sciences*. https://doi.org/10.24251/hicss.2023.021

Posamentier, J. (2023). *The biggest 2023 policy trends affecting student well-being*. Committee for Children Blog. https://www.cfchildren.org/blog/2023/01/the-biggest-2023-policy-trends-affecting-student-well-being/

Rattner, N. (2025, March 10). How the AI talent race is reshaping the tech job market. *Wall Street Journal*

Rodríguez-García, J. D., Moreno-León, J., Román-González, M., & Robles, G. (2020). Introducing artificial intelligence fundamentals with LearningML: Artificial intelligence made easy. In *Eighth international conference on technological ecosystems for enhancing multiculturality* (pp. 18–20). https://doi.org/10.1145/3434780.3436705

Roffey, S. (2012). Pupil wellbeing—Teacher wellbeing: Two sides of the same coin? *Educational and Child Psychology, 29*(4), 8. https://doi.org/10.53841/bpsecp.2012.29.4.8

Schlagal, B. (2014). Best practices in spelling and handwriting. In S. Graham, C. MacArthur, & J. Fitzgerald (Eds.), *Best practices in writing instruction* (2nd ed., pp. 257–283). Guilford Press.

Schlosser, E. A. C. (2017). *Race, socioeconomic status, and implicit bias: Implications for closing the achievement gap* (Publication Number 10690797) [Ph.D., The University of Southern Mississippi]. ProQuest Dissertations & Theses Global; Publicly Available Content Database. Ann Arbor.

Schonert-Reichl, K. A. (2017). Social and emotional learning and teachers. *The Future of Children, 27*(1), 137–155. https://doi.org/10.1353/foc.2017.0007

Sherman, N. (2025, March 10). Beijing makes AI education mandatory for students.https://www.lawyer-monthly.com/2025/03/beijing-makes-ai-education-mandatory-for-students/

Shute, V. J. (2008). Focus on formative feedback. *Review of Educational Research, 78*(1), 153–189. https://doi.org/10.3102/0034654307313795

Shwartz, V. (2024). There is a growing need to address diversity in the datasets used to train artificial intelligence. *Conversation*. https://theconversation.

com/artificial-intelligence-needs-to-be-trained-on-culturally-diverse-datasets-to-avoid-bias-222811

Spector, J. M., & Ma, S. (2019). Inquiry and critical thinking skills for the next generation: From artificial intelligence back to human intelligence. *Smart Learning Environments, 6*(1), 1–11. https://doi.org/10.1186/s40561-019-0088-z

Statista. (2024). Languages most frequently used for web content as of January 2024, by share of websites. https://www.statista.com/statistics/262946/most-common-languages-on-the-internet/

Su, J., Ng, D. T. K., & Chu, S. K. W. (2023). Artificial intelligence (AI) literacy in early childhood education: The challenges and opportunities. *Computers and Education: Artificial Intelligence, 4*, 100124. https://doi.org/10.1016/j.caeai.2023.100124

Talanquer, V., Bolger, M., & Tomanek, D. (2015). Exploring prospective teachers' assessment practices: Noticing and interpreting student understanding in the assessment of written work. *Journal of Research in Science Teaching, 52*(5), 585–609. https://doi.org/10.1002/tea.21209

Tapalova, O., & Zhiyenbayeva, N. (2022). Artificial intelligence in education: AIEd for personalised learning pathways. Electronic Journal of. *e-Learning, 20*(5), 639–653. https://doi.org/10.34190/ejel.20.5.2597

Tatman, R. (2017). Gender and dialect bias in Youtube's automatic captions. In *Conference: Proceedings of the First ACL Workshop on Ethics in Natural Language Processing*. http://www.ethicsinnlp.org/workshop/pdf/EthNLP06.pdf

Varadarajan, S. (2024). Overcoming power and water challenges in data centers. https://www.linkedin.com/pulse/overcoming-power-water-challenges-data-centers-srinivas-varadarajannq7fc#:~:text=Energy%20Consumption%3A%20In%202023%2C%20data,rapid%20expansion%20of%20the%20sector

Waltzer, T., Cox, R. L., & Heyman, G. D. (2023). Testing the ability of teachers and students to differentiate between essays generated by ChatGPT and high school students. *Human Behavior and Emerging Technologies, 2023*(1), 1–9. https://doi.org/10.1155/2023/1923981

Wang, J., & Watson, S. (2025). Integrating generative AI in education: Affordances, issues, and directions. In S. Papadakis & G. Lampropoulos (Eds.), *Social robots and artificial intelligence in education: Integrating AI in K-12 and higher education* (pp. 355–379). Palgrave Macmillan.

Weizenbaum, J. (1966). ELIZA—A computer program for the study of natural language communication between man and machine. *Communications of the ACM, 9*(1), 36–45. https://doi.org/10.1145/365153.365168

Wilkinson, C., Low, F., & Gluckman, P. (2021). *Screen time: The effects on children's emotional, social, and cognitive development.* The Centre for Informed Futures.

Wu, R., & Yu, Z. (2024). Do AI chatbots improve students learning outcomes? Evidence from a meta-analysis. *British Journal of Educational Technology, 55*(1), 10–33. https://doi.org/10.1111/bjet.13334

Yi, Y. (2021). Establishing the concept of AI literacy. *Jahr–European Journal of Bioethics, 12*(2), 353–368. https://doi.org/10.21860/j.12.2.8

Yin, J., Goh, T. T., Yang, B., & Xiaobin, Y. (2021). Conversation technology with micro learning: The impact of chatbot-based learning on students' learning motivation and performance. *Journal of Educational Computing Research, 59*(1), 154–177. https://doi.org/10.1177/0735633120952067

Young, K., & Maher, D. (2023). Generative AI technology to support high school students experiencing writing challenges. In R. E. Ferdig, R. Hartshorne, E. Baumgartner, R. Kaplan-Rakowski, & C. Mouza (Eds.), *What Pre K-12 teachers should know about educational technology in 2023: A research-to-practice anthology*. AACE2023.

Zhai, X. (2022). ChatGPT user experience: Implications for education. *Available at SSRN 4312418*. https://doi.org/10.2139/ssrn.4312418

Zhang, H., Lee, I., Ali, S., DiPaola, D., Cheng, Y., & Breazeal, C. (2023). Integrating ethics and career futures with technical learning to promote AI literacy for middle school students: An exploratory study. *International Journal of Artificial Intelligence in Education, 33*(2), 290–324. https://doi.org/10.1007/s40593-022-00293-3

Zimmerman, A., Janhonen, J., & Beer, E. (2023). Human/AI relationships: challenges, downsides, and impacts on human/human relationships. *AI and Ethics*, 1–13. https://doi.org/10.1007/s43681-023-00348-8

CHAPTER 13

Use of Artificial Intelligence in Foreign Language Education in Higher Education Settings

Blanka Klimova and Marcel Pikhart

13.1 Introduction

In the contemporary era, technology has seamlessly woven itself into the fabric of nearly every human endeavor, prominently including the domain of foreign language education. This chapter examines the profound impact of technological advancements on enhancing students' proficiency in English as a Foreign Language (EFL), especially within the context of higher education (Godwin-Jones, 2023). It is a well-acknowledged fact that teachers of foreign languages are increasingly leveraging both formal and informal technological tools to facilitate language learning. The core objective of this chapter is to present a comprehensive discussion of the latest research findings on the incorporation of emerging technologies in EFL teaching and learning. This exploration extends to delineating

B. Klimova • M. Pikhart (✉)
Faculty of Informatics and Management, University of Hradec Kralove, Hradec Králové, Czech Republic
e-mail: blanka.klimova@uhk.cz; marcel.pikhart@uhk.cz

S. Papadakis, G. Lampropoulos (eds.), *Social Robots and Artificial Intelligence in Education*, Digital Education and Learning,
https://doi.org/10.1007/978-3-032-03977-4_13

specific pedagogical approaches that are most effective when utilizing various technological tools for teaching EFL. In an academic landscape increasingly dominated by digital innovation, the role of technology in language education has evolved from a mere adjunct to an integral component of the curriculum (Timotheou et al., 2023). The burgeoning field of artificial intelligence (AI) has introduced tools that have revolutionized traditional methodologies of teaching and learning.

This chapter not only addresses the pedagogical implications of employing these AI technologies in EFL classrooms but also provides a detailed analysis of their practical applications. The authors conduct a meticulous examination of some of the most prevalent AI applications in foreign language education at the university level, including but not limited to platforms, such as Duolingo, DeepL, and ChatGPT. These tools are discussed for their advantages and potential drawbacks in the context of language learning environments. The discussion extends to the pragmatic aspects of these technologies, assessing their role in enhancing specific language skills, such as writing, speaking, listening, reading, vocabulary, and grammar. This analysis is enriched through the inclusion of various case studies that illustrate the real-world implementation and efficacy of these tools in the educational sphere.

Additionally, this chapter offers a critical perspective on the use of AI technologies in language education, considering factors, such as accessibility, cultural relevance, and the balance between technology and human interaction in learning environments in the context of recent studies that are considered as case studies for this chapter. The discussion is not confined to the mere application of mentioned tools but also explores the broader implications of technology-assisted language learning on student engagement, motivation, and autonomy. The evolving landscape of AI in education prompts a need for continuous research and adaptation of pedagogical strategies. Therefore, the authors not only focus on the present state of affairs but also project into the future, proposing potential research directions in the field of EFL education and technology. This forward-looking perspective is crucial for educators, policymakers, and researchers in understanding the trajectory of language education in an increasingly digital world.

In conclusion, this chapter offers a holistic view of the intersection between technology and language education, providing educators with insights into effective practices and setting the stage for future innovations in the field. It stands as a testament to the transformative power of technology in shaping the future of foreign language education, particularly in the context of teaching English as a foreign language.

13.1.1 Current State of the Use of AI in Education

AI has rapidly emerged as a transformative force in education, particularly in the realm of language learning as it is also showcased in Maher (2025 [Chap. 12]) which deals with the topic of Chatbots. AI technologies are being increasingly integrated into educational environments, revolutionizing traditional methodologies. One of the primary advantages of AI in education is its ability to provide personalized feedback and adaptive learning experiences. AI systems, such as Neural Machine Translation (NMT) and chatbots like ChatGPT and Replika, offer individualized teaching materials and tests tailored to each student's needs and progress. These AI-driven tools enhance learner engagement and efficiency by delivering customized exercises and real-time feedback (Zawacki-Richter et al., 2019).

Moreover, AI-powered virtual reality (VR) and augmented reality (AR) technologies create immersive learning environments that simulate real-life language use scenarios (Kavanagh et al., 2017). These technologies are particularly effective for developing listening and speaking skills. By engaging learners in realistic contexts (Kyaw et al., 2019), VR and AR facilitate deeper language acquisition and retention (Fitria, 2021). Additionally, AI's role in natural language processing (NLP) aids in the evaluation of written and spoken texts (Litman, 2016; Shaik et al., 2023), providing automated assessments that help improve grammar and syntax (Son et al., 2023).

AI's integration into education extends beyond language learning. For instance, adaptive learning systems use AI to personalize educational content based on students' strengths and weaknesses (Gligorea et al., 2023). This approach ensures that learning experiences are tailored to individual needs, enhancing overall educational outcomes. Furthermore, AI can support teachers by automating administrative tasks, allowing them to focus more on instructional quality and student interaction.

In the context of EFL education, the integration of AI tools is not just a modern trend but is deeply rooted in a historical progression of technological advancements in education (Kamalov et al., 2023). Over the decades, the introduction of language labs, multimedia resources, and computer-assisted language learning (CALL) platforms laid the groundwork for the digital transformation we witness today (Han, 2020). As we trace the evolution of EFL teaching methods, it becomes clear that each technological milestone has brought us closer to more personalized and efficient learning experiences. The advent of AI is, therefore, not an isolated phenomenon but a continuation of this trajectory, pushing the boundaries of what is possible in language education (Jiang, 2022).

The influence of globalization on the expansion of EFL education cannot be overstated. As English solidified its role as the global lingua franca, the demand for effective and accessible language education has surged worldwide. In this context, AI technologies offer a promising solution to address the diverse needs of learners across different cultural and linguistic backgrounds. By providing tailored content that can adapt to various learning styles and paces, AI-driven tools like Duolingo and ChatGPT enable a more inclusive approach to language learning. This globalization of education, facilitated by AI, underscores the importance of understanding the broader sociocultural implications of these technologies. The ability to cater to a global audience while maintaining cultural sensitivity is a challenge that requires ongoing research and adaptation (Solak, 2024).

The intersection of AI with other disciplines further enriches the field of EFL education. The collaboration between linguists, computer scientists, and educators has led to the development of sophisticated AI systems capable of understanding and generating human language. For instance, the application of NLP in educational tools has significantly improved the accuracy and relevance of language learning content (Kurni et al., 2023). These interdisciplinary efforts highlight the potential for AI to bridge gaps between traditional language teaching methods and modern technological innovations. By fostering collaboration across fields, we can continue to refine AI-driven tools, making them more effective and aligned with educational goals (Hutson & Plate, 2023).

However, the rapid integration of AI in EFL education also raises important ethical considerations. The use of AI technologies in language learning brings with it concerns related to data privacy, the potential for bias in language models, and the digital divide (Patty, 2024). As these tools become more prevalent, it is crucial to address issues of equity and access to ensure that all learners benefit from technological advancements. Additionally, the reliance on AI raises questions about the role of teachers in the classroom. While AI can provide valuable support in terms of personalized feedback and adaptive learning, the irreplaceable value of human interaction in language learning must be recognized. Teachers play a critical role in guiding students, fostering critical thinking, and ensuring that technology enhances rather than diminishes the educational experience (Afzal et al., 2023; Darwin et al., 2023; Ghamrawi et al., 2024).

In examining the current landscape of AI in EFL education, it is essential to identify the research gaps that remain. Despite the significant progress made, there is still much to learn about the long-term effects of AI on language acquisition, particularly in terms of learner autonomy and motivation. Moreover, the impact of AI on less commonly taught languages and the effectiveness of AI in supporting language learners with special needs are areas that warrant further investigation (Law, 2024). As AI continues to evolve, so too must our understanding of its potential and limitations in the context of EFL education. Future research should focus on these areas to ensure that AI tools are developed and implemented in ways that truly enhance language learning outcomes.

Finally, the role of educators in this rapidly changing landscape cannot be overlooked. As AI tools become more sophisticated, teachers must adapt their pedagogical strategies to incorporate these technologies effectively (Kamalov et al., 2023). This requires ongoing professional development and a commitment to understanding the capabilities and limitations of AI in the classroom. By staying informed and engaged with the latest research, educators can ensure that they are not only keeping pace with technological advancements but also shaping the future of language education in meaningful ways.

In conclusion, the integration of AI in EFL education represents a significant shift in how languages are taught and learned. This chapter provides a comprehensive overview of the current state of AI in EFL education, highlighting both the opportunities and challenges presented by these technologies. By examining the historical context, the impact of globalization, interdisciplinary collaborations, ethical considerations, and the evolving role of teachers, we gain a deeper understanding of the complex dynamics at play. As we look to the future, it is clear that AI will continue to play a central role in shaping the landscape of language education. However, this will require a concerted effort to address the research gaps identified, ensuring that AI-driven tools are developed and used in ways that enhance learning outcomes and promote equity and inclusion. Through a careful balance of technological innovation and human insight, we can create a more effective and inclusive future for EFL education.

13.1.2 Enhancing Language Learning with AI

The application of AI in language learning has garnered significant interest due to its potential to revolutionize traditional teaching methods. AI technologies enable personalized and self-directed instruction, offering a more customized learning experience (Ng et al., 2023). This shift toward personalization is one of the most significant advantages of AI in education. For example, AI-driven language learning apps like Duolingo and Babbel use gamification to make learning engaging and fun, increasing motivation and retention. These apps adapt to individual learning paces, providing targeted exercises that address specific areas of improvement (Yuen & Schlote, 2024).

Furthermore, AI-powered chatbots and virtual assistants represent a cutting-edge technology with the potential to revolutionize language education. These tools simulate real-life conversations, offering learners opportunities to practice language skills in a safe and supportive environment. Chatbots provide instant feedback on user inputs, helping learners improve their fluency and accuracy. Despite their recent introduction, chatbots have shown promise in enhancing language acquisition by offering personalized language assistance and conversation practice (Zhang & Huang, 2024).

AI's role in machine translation (MT) is another area with significant implications for language learning. NMT technologies like DeepL and Google Translate have made substantial advancements, offering near-human translation quality (Moneus & Sahari, 2024). These tools can assist learners in understanding and translating foreign languages, making language learning more accessible and efficient. However, there is a need for clear and practical strategies to ensure that NMT applications are effectively integrated into the learning process. Further limitations are examined and discussed in Wang and Watson (2025 [Chap. 14]).

13.1.3 AI in Foreign Language Learning

AI technologies are revolutionizing language education by enabling personalized and self-directed instruction tailored to individual needs. Language learning apps like Duolingo and Babbel use AI to adapt to each learner's pace, offering targeted exercises that address specific areas for improvement (Yuen & Schlote, 2024). This customization significantly increases motivation and retention among learners, making the educational experience more effective and engaging.

Interactive tools such as AI-powered chatbots and virtual assistants represent a significant advancement in language education (Adiguzel et al., 2023; Labadze et al., 2023). These tools simulate real-life conversations, providing learners with opportunities to practice language skills in a supportive environment. Chatbots offer instant feedback, helping learners improve fluency and accuracy. Despite their recent introduction, these tools show great promise in enhancing language acquisition through personalized assistance and conversation practice (Poláková & Klímová, 2024).

Another area where AI has substantial implications for language learning is in machine translation. Technologies like NMT, exemplified by DeepL and Google Translate, have achieved near-human translation quality. These tools assist learners in understanding and translating foreign languages, making language learning more accessible and efficient. However, effective integration strategies are necessary to maximize the benefits of NMT applications in the learning process (Polakova & Klimova, 2023).

AI-powered VR and AR technologies create immersive learning environments that simulate real-life language use scenarios. These technologies are particularly effective for developing listening and speaking skills, facilitating deeper language acquisition and retention (Huang et al., 2021). The immersive nature of VR and AR makes language learning more engaging and realistic, helping learners practice in contexts similar to real-world situations.

Additionally, AI's capabilities in NLP aid in evaluating written and spoken texts, providing automated assessments that help improve grammar and syntax. This functionality streamlines the learning process and provides valuable feedback to learners, enabling them to refine their language skills more efficiently (Alqahtani et al., 2023).

In summary, AI in language education offers personalized, interactive, and immersive learning experiences. It enhances traditional teaching methods through tailored instruction, real-time feedback, advanced translation tools, and innovative VR and AR applications. These advancements make language learning more engaging, effective, and accessible, promising a brighter future for learners worldwide.

13.1.4 *Related Studies: Case Studies of the Use of AI in Foreign Language Education in Higher Education Settings*

In recent years, the integration of digital technologies into language learning has transformed educational practices and opened new avenues for enhancing second language acquisition (L2). Varlamis et al. (2025 [Chap. 11]) provided insights into the use of educational robots and their potential. This subchapter provides a comprehensive overview of several key studies examining the impact of various digital tools and methodologies on language learning. By analyzing systematic reviews and empirical studies, the following explores the effectiveness of VR, AR, digital corrective feedback, digital gaming, AI-driven chatbots, and the ethical considerations surrounding AI in education. The authors together with their colleagues, both at home institution and abroad, have been actively involved in research on the use of AI in foreign language education and they provide a summary of their pivotal research studies from 2022 to 2024 below. This collection of their studies emphasizes the transformative potential of digital technologies in language learning while also addressing the challenges and ethical considerations necessary for their successful integration into educational practices. Of course, there are also many other highly relevant studies that have to be taken into consideration to have a broader perspective of the developing field of the use of AI in education (Divekar et al., 2022; Kim, 2019; Schmidt & Strasser, 2022; Sharadgah & Sa'di, 2022; Sumakul et al. 2022).

Divekar et al. (2022) in their study explored Artificial Intelligence (AI) and Extended Reality (XR) and their use in language education. For this purpose, they developer the Cognitive Immersive Language Learning Environment (CILLE), which can interact with users in natural, multimodal conversations, and XR, which creates immersive, nonintrusive experiences. They conducted a seven-week study among ten university students learning Chinese as a foreign language. The results showed significant improvements in students' vocabulary, comprehension, and conversation skills, supported by positive student feedback and researcher observations.

Kim (2019) investigated the impact of utilizing AI chatbots on enhancing the English grammar proficiency of Korean university students. A total of 70 undergraduate students took part in this investigation, enrolled in a General English course provided by a Korean university. The study comprised two distinct groups: the chatbot group, consisting of 36 students, and the human group, comprising 34 students. Throughout a span of 16 weeks, the chatbot group engaged in ten chat sessions with an AI chatbot, while the human group conversed with a human chat partner. Pre- and post-assessments were conducted to assess alterations in the participants' grammar abilities over time. An independent t-test was then employed to compare the progress between the two groups. The primary results indicate that both sets of participants experienced a significant enhancement in their English grammar skills, demonstrating the positive impact of chat engagement. Moreover, a notable contrast in improvement was observed between the chatbot and human groups, highlighting the superior efficacy of chatbot utilization.

The research conducted by Schmidt and Strasser (2022) discusses the integration of AI in foreign language learning and teaching, emphasizing its potential to create adaptive and individualized learning environments. It explores how AI, particularly narrow AI, can enhance practice in language classrooms through tools like Intelligent Tutoring Systems (ITS), NLP, and Machine Learning. The authors highlight the importance of interdisciplinary collaboration to develop AI-driven systems that can provide personalized feedback, adapt to individual learning needs, and support communicative and task-based language learning. While current AI applications in language education show promise, many challenges remain, particularly in creating systems that can effectively analyze and respond to the complexities of human language. The article also outlines ongoing research projects aimed at developing these advanced learning environments and envisions a future where AI plays a central role in education, particularly in foreign language acquisition by 2040.

In 2022, Sharadgah and Sa'di (2022) provided a systematic review of research on the use of AI in English Language Teaching (ELT) from 2015 to 2021. They analyzed 64 articles, highlighting that AI in ELT is a promising yet nascent area, showing positive impacts on English language skills, translation, and assessment. However, they also revealed significant gaps, including insufficient research on integrating AI with ELT, a lack of detailed descriptions of AI-driven teaching materials, and the need for more interdisciplinary collaboration. The study underscores that while AI holds potential in enhancing ELT, much work remains to address the complexities of human language and to develop AI systems that can better simulate human-like interactions in language learning contexts.

Sindermann et al. (2021) evaluated the Attitude Towards Artificial Intelligence (ATAI scale) in the German, Chinese, and English languages among participants from Germany (N = 461; 345 females), China (N = 413; 145 females), and the UK (N = 84; 65 females). The results indicated that males generally showed higher acceptance and lower fear toward AI compared to females, with cultural differences also influencing attitudes. The scale was further validated by correlating participants' attitudes with their willingness to use specific AI products (i.e., self-driving cars, Siri, Alexa, the social robot Pepper, and the humanoid robot Erica), demonstrating its utility in cross-cultural research on human–AI interaction. However, the study also notes limitations, including skewed gender distributions and the need for further research to explore attitudes toward AI in various contexts.

Sumakul et al. (2022) explored the perceptions of English as a Foreign Language (EFL) teachers regarding the integration of Artificial Intelligence (AI) in their classrooms, specifically focusing on the use of AI-powered tools in Indonesian universities. The study, based on interviews with four EFL teachers, reveals that they generally view AI positively, recognizing its potential to assist in teaching and enhance student learning. Teachers appreciated how AI could reduce their workload and provide personalized learning experiences for students. However, they also highlighted the importance of considering students' motivation levels and the need for teachers to develop both technological and pedagogical competence when integrating AI into their teaching. The study concludes that while AI is seen as a beneficial tool, its successful implementation in EFL education requires careful consideration of these factors to ensure it effectively supports both teachers and learners.

The article by Klimova and Pikhart (2023) systematically reviewed the potential cognitive benefits of digital language learning. It explores whether digital language learning contributes to cognitive gains such as improved problem-solving skills, memory, and multitasking abilities. The review employs the Preferred Reporting Items for Systematic Reviews and Meta-Analyses (PRISMA) methodology, analyzing relevant research articles to answer two primary questions: whether digital language learning enhances cognitive gains and what pedagogical strategies can optimize these improvements.

The findings indicate that printed texts may be more effective for cognitive gains compared to electronic texts. However, incorporating multiple sensory modalities through digital education can also be beneficial. For instance, digital games have shown positive effects on vocabulary acquisition and engagement. Techniques such as the flipped classroom, which allows more in-depth classroom interactions, have been effective in improving writing and speaking skills. The review underscores the importance of aligning teaching strategies with students' needs and providing consistent feedback to maintain motivation. Overall, while digital tools offer various benefits, a balanced approach that includes traditional methods may be most effective for cognitive enhancement in language learning.

Another study by Klimova and Pikhart (2022) explored the perspectives of university-level foreign language teachers across Europe regarding the use of digital technologies in their teaching practices. The main findings highlight a generally positive attitude toward digital teaching tools, with many FL teachers recognizing the benefits of increased student engagement and enriched learning opportunities. However, the study also identifies several challenges, including technical issues, the time-consuming nature of integrating technology, and the lack of personal interaction, which is crucial for language learning. Teachers expressed a desire for more professional support and training from their institutions to effectively incorporate digital technologies. The data also revealed gender and experience-related differences, with female teachers and those with fewer years of teaching experience more likely to utilize gamification tools. Despite the positive overall outlook, the necessity for a well-defined strategy and ongoing support for digital teaching is emphasized. The findings suggest that while digital tools offer significant potential, their successful implementation requires careful planning and support from educational institutions.

The most recent study by Polakova and Klimova (2024) focused on the use of AI-driven chatbots for foreign language learning revealed significant findings on their impact and perceptions among students. The study involved 58 university students with B2 and C1 levels of English proficiency who engaged with the chatbot "Alex" over a month. The study generated both quantitative and qualitative results. The study showed a marked improvement in students' language skills. Both upper-intermediate and advanced groups exhibited significantly higher post-test scores compared to pre-test scores. For example, the upper-intermediate group's average scores increased from 59.28% to 74.62%, and the advanced group's scores improved from 79.98% to 89.58%. Therefore, the statistical analysis confirmed that the chatbot effectively helped students enhance their vocabulary and grammar.

The qualitative results then revealed that most students (85%) had never used a chatbot for language learning before this study. The chatbot sessions were generally well-received, with most interactions occurring at home. Furthermore, 64% of participants found the chatbot useful for practicing writing and conversational skills and acquiring new knowledge through interesting topics; 81% found the chatbot user-friendly, though some noted areas for improvement, such as sentence completion and text field size; 74% of students were satisfied with using the chatbot, primarily due to its novelty and the engaging nature of the interaction. Despite positive feedback, 79% preferred other methods of practicing English, such as speaking with native speakers or using other platforms like Duolingo.

Generally, the findings suggest that while chatbots like Alex can significantly aid in FL learning, their integration into educational practices should be carefully refined and personalized to meet diverse learner needs effectively. Maher (2025 [Chap. 12]) further explores the use of chatbots in education.

Similarly, another study (Klimova et al., 2024) explored the potential of ChatGPT for foreign language education at the university level to examine university students' attitudes and perceived usefulness of using ChatGPT for learning foreign languages. The research involved a qualitative approach, utilizing a questionnaire to gather data from students who had hands-on experience with ChatGPT. The findings revealed that students were generally fascinated, satisfied, and stimulated by the technology despite some reservations and perceived threats. Students appreciated ChatGPT's ability to assist with writing tasks, provide explanations, and generate creative content, which contributed to their learning efficiency and engagement.

The study also highlighted several pedagogical implications. While ChatGPT can significantly aid teachers by preparing materials and assessments, it challenges existing teaching methods and necessitates new approaches to enhance students' cognitive, creative, and critical thinking skills. Ethical considerations, such as data privacy and the potential for academic dishonesty, were also underscored. The research suggested the need for guidelines and protocols to ensure responsible use of AI technologies in education, emphasizing the importance of balancing the benefits with the potential risks to maximize the positive impact of AI in language learning. The authors identified key ethical concerns, such as data privacy, surveillance, and the potential for algorithmic bias. Their review highlighted the necessity of following four major ethical principles: beneficence, nonmaleficence, autonomy, and justice. One crucial concept introduced is "algorithm of vigilance," which involves continuous monitoring and assessment of AI algorithms to prevent adverse effects. In addition, the review emphasized the importance of interdisciplinary and multistakeholder collaboration in developing globally acceptable standards for the ethical use of AI in education. It stressed that ethical guidelines should cover the entire life cycle of AI systems, from their creation to their implementation and eventual decommissioning. Additionally, the need for further empirical research was emphasized to better understand and address the ethical challenges posed by AI in educational contexts. This study ultimately called for a balanced approach, integrating technical, social, and ethical perspectives to ensure the safe and beneficial use of AI technologies in education.

13.2 Drawbacks, Threats and Opportunities of AI in Education

Several drawbacks and opportunities related to the integration of AI and other technologies in EFL education have to be discussed. On the downside, one of the significant concerns highlighted is data privacy. The use of AI technologies raises crucial issues around the protection of students' personal information. As AI tools become more prevalent in educational settings, ensuring that data is not misused and that privacy is maintained becomes increasingly critical (Huang, 2023). Additionally, the manuscript addresses the potential for bias within AI language models. These models may inadvertently incorporate biases from the data they were trained on, leading to culturally insensitive or incorrect outputs that could affect the learning process.

Another challenge that must be noted is the digital divide, which refers to the disparity in access to AI tools and technologies. This issue is particularly concerning because students from underprivileged backgrounds may not have the same access to these advanced tools, potentially exacerbating existing educational inequalities. The manuscript also discusses the risk of overreliance on technology. While AI can significantly enhance the learning experience, there is a concern that it could reduce the essential human interaction in the classroom, potentially replacing the valuable role of teachers in guiding and mentoring students (Seo et al., 2021).

The rapid advancement of AI technologies also poses adaptation challenges for educators. Teachers are required to continuously update their skills and adapt to new tools, which can be a significant burden. This need for ongoing professional development highlights the complexities of integrating AI into the classroom. Furthermore, the manuscript touches on the impact of AI on less commonly taught languages. AI tools are often developed with a focus on widely spoken languages, which may limit their effectiveness for learners of languages that are not as commonly taught (Zawacki-Richter et al., 2019).

On the positive side, several opportunities that AI brings to EFL education can be explored. One of the most prominent advantages is the potential for personalized learning experiences. AI technologies, such as ChatGPT and Neural Machine Translation, offer tools that can tailor learning materials and feedback to individual students' needs and progress (Kasneci et al., 2023). This personalization enhances learner engagement and outcomes, providing a more effective educational experience.

Another opportunity is the use of adaptive learning systems driven by AI. These systems can adjust educational content based on each student's strengths and weaknesses, ensuring a more tailored and effective learning experience that can lead to improved educational outcomes. The manuscript also highlights the potential of AI-powered VR and AR technologies in creating immersive learning environments. These tools simulate real-life language scenarios, which are particularly effective for developing listening and speaking skills. By engaging learners in realistic contexts, VR and AR can deepen language acquisition and retention.

AI's role in NLP is also to be discussed, particularly in relation to automated assessments. AI enables the evaluation of written and spoken texts, providing real-time feedback that helps students improve their grammar, syntax, and overall language proficiency. This automated assessment capability represents a significant advancement in language education, offering students timely and precise feedback (Pang et al., 2024).

There is also the potential for AI to support teachers by reducing their administrative burden. By automating routine tasks, AI allows teachers to focus more on instructional quality and student interaction, leading to a more efficient and effective educational environment. Additionally, AI tools offer the opportunity to cater to diverse cultural and linguistic backgrounds, making EFL education more accessible and inclusive. The ability of AI to adapt content to different cultural contexts is a significant advantage in a globalized education landscape (Tafazoli, 2024).

The interdisciplinary collaboration encouraged by AI integration is another key opportunity. The convergence of linguistics, computer science, and education leads to the development of more sophisticated and effective AI tools that better meet the needs of learners. This interdisciplinary approach is essential for advancing the field of EFL education and more strategies related to the use of generative AI in education are presented in Wang and Watson (2025 [Chap. 14]).

13.3 Overcoming Challenges and Embracing Opportunities

Thus, as seen from the findings above, while AI offers numerous benefits in education, it also presents challenges that need to be addressed. One significant challenge is the cost associated with implementing and maintaining AI technologies. Schools and educational institutions must invest in infrastructure, training, and ongoing support to effectively integrate AI into their curricula. Additionally, there is a risk of overreliance on technology, which can lead to disruptions in learning if technical issues arise. Ensuring that both students and teachers have the necessary skills to utilize AI tools effectively is crucial for maximizing their benefits.

Another challenge is the ethical considerations surrounding AI in education. Privacy concerns, data security, and the potential for bias in AI algorithms are important issues that need to be carefully managed. Educators and policymakers must establish guidelines and best practices to address these ethical concerns and ensure that AI technologies are used responsibly and equitably.

Despite these challenges, the potential of AI in education is vast. AI-driven personalized learning experiences can cater to diverse learner needs, making education more inclusive and accessible. For instance, AI technologies can provide customized learning interfaces and assistive tools for students with disabilities, ensuring that they receive the support they need to succeed. Additionally, AI can enhance teacher roles by providing professional development opportunities and supporting them in delivering high-quality, technology-integrated instruction.

The integration of AI in higher education, particularly in the realm of English as a Foreign Language (EFL) instruction, brings forth several unique features that are reshaping the educational landscape. One of the most prominent features is the ability of AI to provide highly personalized learning experiences. Unlike traditional teaching methods that often follow a one-size-fits-all approach, AI technologies can analyze individual student data to tailor content, exercises, and feedback to the specific needs of each learner. For instance, AI-driven platforms like Duolingo and ChatGPT can adapt their teaching materials based on a student's proficiency level, learning speed, and areas of difficulty (Yuen & Schlote, 2024). This capability to facilitate immersive and interactive learning environments. AI-powered tools, such as VR and AR, allow students to experience language learning in simulated real-world contexts. These technologies can create virtual environments where students practice their language skills in scenarios that mimic everyday situations, such as ordering food in a restaurant or asking for directions. Such immersive experiences are particularly valuable in EFL education as they help students to develop practical communication skills and cultural understanding in a safe, controlled setting. Moreover, these tools can offer immediate feedback and allow learners to repeat exercises as needed, promoting deeper comprehension and retention of the language (Fitria, 2021; Klimova & Pikhart, 2022).

AI also significantly enhances the assessment and feedback mechanisms in higher education. Traditional assessment methods, which often involve delayed feedback and limited interaction, can be supplemented or even replaced by AI-driven evaluations that provide real-time, detailed feedback on both written and spoken language tasks. For example, AI tools equipped with NLP capabilities can automatically grade essays, correct grammar, and suggest improvements in syntax and vocabulary usage. These tools can also analyze spoken language for pronunciation accuracy and fluency. The immediacy and precision of AI-driven feedback empower students to make continuous improvements in their language skills, fostering a more dynamic and responsive learning process. Additionally, this feature reduces the administrative burden on educators, allowing them to focus more on personalized instruction and student mentorship.

13.4 Discussion

The integration of AI technologies into EFL education carries significant implications for both teaching and learning processes. One of the most critical implications is the potential transformation in the role of educators. Traditionally, language teachers have been the primary source of knowledge and feedback in the classroom. However, with the advent of AI-driven tools that can deliver personalized content and instant feedback, educators may find their roles shifting from that of a knowledge provider to a facilitator of learning experiences. This shift requires teachers to develop new skills in managing and integrating AI tools effectively, ensuring that these technologies complement rather than replace human instruction. It also highlights the need for ongoing professional development and training for educators to adapt to these emerging tools and maintain their relevance in the digital age (Faisal Rashid et al., 2024; Kamalov et al., 2023).

Furthermore, the use of AI in EFL education has significant implications for student engagement and autonomy. AI tools offer a level of interactivity and personalization that traditional methods cannot match, potentially leading to increased student motivation and involvement in the learning process. By providing students with tailored learning experiences and real-time feedback, AI encourages learners to take greater control of their education, fostering a sense of autonomy and self-directed learning. However, this also raises concerns about the balance between technology and human interaction (Kamalov et al., 2023; Malik et al., 2023). While AI can enhance engagement, there is a risk that overreliance on technology could diminish the importance of social interaction and cultural exchange, which are crucial elements of language learning. Educators must therefore strike a balance, ensuring that AI serves as a tool to support, rather than replace, the rich, human-centered aspects of language education (Klimova et al., 2024).

The widespread adoption of AI in EFL education also poses implications for equity and access. On the one hand, AI has the potential to democratize language learning by providing high-quality educational resources to a broader audience, including students in remote or underserved areas. AI-powered platforms can deliver consistent and personalized instruction regardless of geographical location, which can help bridge educational gaps and provide opportunities for learners who may not have access to traditional language learning resources. On the other hand, the implementation of AI in education may exacerbate existing inequalities if access to these technologies is uneven (Patty, 2024) Disparities in technological infrastructure, digital literacy, and financial resources could limit the benefits of AI for certain student populations (Kuhn et al., 2023). Addressing these challenges requires careful consideration of how AI tools are distributed and supported within educational systems to ensure equitable access for all learners.

In addition to educational implications, the integration of AI in EFL instruction raises important ethical considerations. The use of AI technologies in education involves the collection and analysis of large amounts of student data, which poses potential risks to privacy and security. Ensuring that student data is protected and used responsibly is a critical concern that educators and policymakers must address as AI becomes more prevalent in educational settings (Akgun & Greenhow, 2022). Moreover, the development and implementation of AI tools should be guided by ethical principles that prioritize student well-being and promote inclusive, fair, and transparent practices. This includes being mindful of potential biases in AI algorithms that could reinforce stereotypes or disadvantage certain groups of students. As AI continues to evolve, ongoing dialogue and research are needed to establish ethical frameworks that govern its use in education (Ferrara, 2024).

Furthermore, the integration of AI in foreign language education should be critically examined through the lens of educational technology, as discussed by Selwyn (2019). While AI promises advancements in personalized learning, it also raises concerns about the political economy, social justice, and sustainability. From a political economy perspective, the dominance of tech corporations in AI development risks prioritizing profit over pedagogy. In terms of social justice, AI could exacerbate educational inequalities by deepening the digital divide and reinforcing biases in AI models. Sustainability concerns also arise, as AI technologies can be resource-intensive and may lead to job displacement in education. A broader, critical perspective is essential to ensure AI's benefits are equitably distributed and its long-term implications responsibly addressed.

Finally, the integration of AI in EFL education invites reflection on the future directions of research and practice in the field. As AI technologies continue to advance, it is crucial for educators, researchers, and policymakers to remain proactive in exploring how these tools can be effectively and ethically integrated into language learning. Future research should focus on evaluating the long-term impacts of AI on student outcomes, the role of human–AI collaboration in education, and the development of best practices for AI-assisted teaching and learning. Additionally, there is a need to investigate how AI can be adapted to different cultural and linguistic contexts to ensure that its benefits are accessible and relevant to diverse populations. By addressing these areas, the educational community can harness the potential of AI to enhance language learning while also mitigating potential risks and challenges.

13.5 Conclusion

In summary, this manuscript has explored the transformative impact of artificial intelligence (AI) on English as a Foreign Language (EFL) education within the context of higher education. The integration of AI tools such as Duolingo, DeepL, and ChatGPT has been shown to significantly enhance language learning outcomes by providing personalized and adaptive learning experiences. These tools not only offer immediate feedback but also facilitate the development of critical language skills, including reading, writing, speaking, listening, vocabulary, and grammar. The analysis of case studies highlighted the practical applications and effectiveness of these technologies, illustrating their potential to revolutionize traditional language teaching methodologies.

However, the manuscript also underscores several challenges and considerations associated with the adoption of AI in EFL education. While AI can democratize access to high-quality language instruction, it may also widen existing inequalities if access to technology is uneven. Additionally, the increasing reliance on AI raises ethical concerns, particularly related to data privacy, security, and potential biases in AI algorithms. The manuscript emphasizes the importance of balancing technology with human interaction to preserve the social and cultural dimensions of language learning.

Furthermore, the discussion extends to the evolving role of educators, who must adapt to new pedagogical approaches that incorporate AI tools. This shift necessitates ongoing professional development to ensure that teachers are equipped to effectively integrate AI into their teaching practices. The manuscript also calls for continuous research to evaluate the long-term impacts of AI on language education and to explore how these technologies can be tailored to different cultural and linguistic contexts.

In conclusion, while AI presents remarkable opportunities to enhance EFL education, it also poses challenges that require careful consideration and strategic implementation. By addressing these challenges and leveraging the strengths of AI, educators and policymakers can create more effective, inclusive, and equitable language learning environments. This manuscript contributes to the ongoing discourse on the role of AI in education, offering insights into the potential benefits and limitations of these emerging technologies and providing a foundation for future research and innovation in the field.

The future of AI in education looks promising, with continued advancements expected to further transform teaching and learning practices. As AI technologies evolve, they will offer even more sophisticated and effective tools for language learning and beyond. The integration of AI, VR, AR, and other emerging technologies will create more immersive and engaging learning environments, providing students with authentic and interactive educational experiences.

AI's potential to personalize education will play a pivotal role in addressing the diverse needs of learners. By tailoring educational content to individual preferences, strengths, and learning styles, AI can help bridge the gap between traditional and modern teaching methodologies. This personalization will not only improve learning outcomes but also increase student motivation and engagement.

In conclusion, AI is poised to revolutionize education by offering innovative approaches to teaching and learning. Its ability to provide personalized feedback, create immersive learning environments, and support teachers in delivering high-quality instruction makes it a valuable tool in modern education. As we continue to explore and harness the potential of AI, it is essential to address the challenges and ethical considerations associated with its use. By doing so, we can ensure that AI technologies are integrated effectively and equitably, ultimately enhancing educational experiences and outcomes for all learners.

Future research in AI-enhanced English as a Foreign Language (EFL) education should address several key areas to optimize the integration and impact of these technologies. Long-term efficacy studies are essential to evaluate the sustained effectiveness of AI tools on language acquisition and retention, providing insights into the durability of learning gains over time. Additionally, research should focus on cultural and contextual adaptation, exploring how AI applications perform across diverse educational settings and how they can be tailored to various linguistic and cultural backgrounds. Ethical considerations, particularly regarding data privacy and algorithmic biases, also warrant further investigation to ensure that AI applications are used responsibly and equitably. Furthermore, exploring the integration of AI with traditional teaching methods can reveal how these tools can complement rather than replace human instruction, leading to more balanced pedagogical approaches. Finally, understanding the impact of professional development programs for educators on the effective use of AI tools is crucial for supporting teachers in adapting to these new technologies and enhancing their instructional practices. These research directions aim to refine the application of AI in EFL education, address existing gaps, and ensure that technological advancements positively contribute to the learning experience and further issues are discussed in Chap. 14 on the integrating generative AI in education.

References

Adiguzel, T., Kaya, M. H., & Cansu, F. K. (2023). Revolutionizing education with AI: Exploring the transformative potential of ChatGPT. *Contemporary Educational Technology, 15*(3), ep429.

Afzal, A., Kamran, F., & Naseem, A. (2023). The role of teachers in fostering critical thinking skills at the university level. *Qlantic Journal of Social Sciences and Humanities, 4*(3), 202–214.

Akgun, S., & Greenhow, C. (2022). Artificial intelligence in education: Addressing ethical challenges in K-12 settings. *AI and Ethics, 2*(3), 431–440. https://doi.org/10.1007/s43681-021-00096-7

Alqahtani, T., Badreldin, H. A., Alrashed, M., Alshaya, A. I., Alghamdi, S. S., Bin Saleh, K., Alowais, S. A., Alshaya, O. A., Rahman, I., Al Yami, M. S., & Albekairy, A. M. (2023). The emergent role of artificial intelligence, natural learning processing, and large language models in higher education and research. *Research in Social & Administrative Pharmacy: RSAP, 19*(8), 1236–1242. https://doi.org/10.1016/j.sapharm.2023.05.016

Darwin, Rusdin D., Mukminatien, N., Suryati, N., Laksmi, E. D., & Marzuki. (2023). Critical thinking in the AI era: An exploration of EFL students' perceptions, benefits, and limitations. *Cogent Education, 11*(1). https://doi.org/10.1080/2331186X.2023.2290342

Divekar, R. R., Drozdal, J., Chabot, S., Zhou, Y., Su, H., Chen, Y., et al. (2022). Foreign language acquisition via artificial intelligence and extended reality: Design and evaluation. *Computer Assisted Language Learning, 35*(9), 2332–2360. https://doi.org/10.1080/09588221.2021.1879162

Faisal Rashid, S., Duong-Trung, N., & Pinkwart, N. (2024). Generative AI in education: Technical foundations, applications, and challenges. *IntechOpen*. https://doi.org/10.5772/intechopen.1005402

Ferrara, E. (2024). Fairness and bias in artificial intelligence: A brief survey of sources, impacts, and mitigation strategies. *Science, 6*, 3. https://doi.org/10.3390/sci6010003

Fitria, T. N. (2021). Artificial intelligence (AI) in education: Using AI tools for teaching and learning process. *Prosiding Seminar Nasional & Call for Paper STIE AAS, 4*(1), 134–147. https://prosiding.stie-aas.ac.id/index.php/prosenas/article/view/106

Ghamrawi, N., Shal, T., & Ghamrawi, N. A. (2024). Exploring the impact of AI on teacher leadership: regressing or expanding? *Education and Information Technologies, 29*, 8415–8433. https://doi.org/10.1007/s10639-023-12174-w

Gligorea, I., Cioca, M., Oancea, R., Gorski, A.-T., Gorski, H., & Tudorache, P. (2023). Adaptive learning using artificial intelligence in e-learning: A literature review. *Educ. Sci., 13*, 1216. https://doi.org/10.3390/educsci13121216

Godwin-Jones, R. (2023). Emerging spaces for language learning: AI bots, ambient intelligence, and the metaverse. *Language Learning & Technology, 27*(2), 6–27.

Han, Y. (2020). Connecting the past to the future of Computer-assisted Language Learning: Theory, practice, and research. *Issues and Trends in Learning Technologies, 8*(1). https://doi.org/10.2458/azu_itlt_v8i1_han

Huang, L. (2023). Ethics of artificial intelligence in education: Student privacy and data protection. *Science Insights Education Frontiers, 16*(2), 2577–2587.

Huang, X., Zou, D., Cheng, G., & Xie, H. (2021). A systematic review of AR and VR enhanced language learning. *Sustainability, 13*, 4639. https://doi.org/10.3390/su13094639

Hutson, J., & Plate, D. (2023). Human-AI collaboration for smart education: Reframing applied learning to support metacognition. *IntechOpen*. https://doi.org/10.5772/intechopen.1001832

Jiang, R. (2022). How does artificial intelligence empower EFL teaching and learning nowadays? A review on artificial intelligence in the EFL context. *Frontiers in Psychology, 13*, 1049401.

Kamalov, F., Santandreu Calonge, D., & Gurrib, I. (2023). New era of artificial intelligence in education: Towards a sustainable multifaceted revolution. *Sustainability, 15*, 12451. https://doi.org/10.3390/su151612451

Kasneci, E., Seßler, K., Küchemann, S., Bannert, M., Dementieva, D., Fischer, F., … Kasneci, G. (2023, January 30). ChatGPT for good? *On opportunities and challenges of large language models for education*. https://doi.org/10.35542/osf.io/5er8f

Kavanagh, S., Luxton-Reilly, A., Wuensche, W., & Plimmer, B. (2017). A systematic review of Virtual Reality in education. *Themes in Science & Technology Education, 10*(2), 85–119.

Kim, N. Y. (2019). A study on the use of artificial intelligence chatbots for improving English grammar skills. *Journal of Digital Convergence, 17*(8), 37–46.

Klimova, B., & Pikhart, M. (2022). Application of corrective feedback using emerging technologies among L2 university students. *Cogent Education, 9*(1), 2132681.

Klimova, B., & Pikhart, M. (2023). Cognitive gain in digital foreign language learning. *Brain Sciences, 13*(7), 1074.

Klimova, B., Pikhart, M., & Al-Obaydi, L. H. (2024). Exploring the potential of ChatGPT for foreign language education at the university level. *Frontiers in Psychology, 15*, 1269319.

Kuhn, C., Khoo, S. M., Czerniewicz, L., et al. (2023). Understanding digital inequality: A theoretical kaleidoscope. *Postdigit Sci Educ, 5*, 894–932. https://doi.org/10.1007/s42438-023-00395-8

Kurni, M., Mohammed, M. S., & Srinivasa, K. G. (2023). Natural language processing for education. In *A beginner's guide to introduce artificial intelligence in teaching and learning*. Springer. https://doi.org/10.1007/978-3-031-32653-0_3

Kyaw, B. M., Saxena, N., Posadzki, P., Vseteckova, J., Nikolaou, C. K., George, P. P., Divakar, U., Masiello, I., Kononowicz, A. A., Zary, N., & Tudor Car, L. (2019). Virtual reality for health professions education: Systematic review and meta-analysis by the digital health education collaboration. *The Journal of Medical Internet Research, 21*(1), e12959. https://doi.org/10.2196/12959

Labadze, L., Grigolia, M., & Machaidze, L. (2023). Role of AI chatbots in education: systematic literature review. *International Journal of Educational Technology in Higher Education, 20*, 56. https://doi.org/10.1186/s41239-023-00426-1

Law, L. (2024). Application of generative artificial intelligence (GenAI) in language teaching and learning: A scoping literature review. *Computers and Education Open, 6*, 100174.

Litman, D. (2016). Natural language processing for enhancing teaching and learning. *Proceedings of the Thirtieth AAAI Conference on Artificial Intelligence, AAAI-16.*

Maher, D. (2025). Chatbots in schools: Opportunities and considerations. In S. Papadakis & G. Lampropoulos (Eds.), *Social robots and artificial intelligence in education: Integrating AI in K-12 and higher education* (pp. XX–XX). Palgrave Macmillan.

Malik, A.R., Pratiwi, Y., Andajani, K., Numertayasa, I.W., Suharti, S., Darwis, A., & Marzuki (2023). Exploring artificial intelligence in academic essay: Higher education student's perspective. *International Journal of Educational Research Open, 5*(2), 100296.

Moneus, A. M., & Sahari, Y. (2024). Artificial intelligence and human translation: A contrastive study based on legal texts. *Heliyon, 10*(6), e28106. https://doi.org/10.1016/j.heliyon.2024.e28106

Ng, D. T. K., Leung, J. K. L., Su, J., Ng, R. C. W., & Chu, S. K. W. (2023). Teachers' AI digital competencies and twenty-first century skills in the post-pandemic world. *Educational Technology Research and Development: ETR & D, 71*(1), 137–161. https://doi.org/10.1007/s11423-023-10203-6

Pang, T. Y., Kootsookos, A., & Ceng, C. T. (2024, April). Artificial intelligence use in feedback: A qualitative analysis. *Journal of University Teaching and Learning Practice, 21*, 06. https://doi.org/10.53761/40wmcj98

Patty, J. (2024). The use of AI in language learning: What you need to know. *Jurnal Review Pendidikan dan Pengajaran, 7*(1), 642–654.

Polakova, P., & Klimova, B. (2023). Using DeepL translator in learning English as an applied foreign language—An empirical pilot study. *Heliyon, 9*(8), e18595. https://doi.org/10.1016/j.heliyon.2023.e18595

Polakova, P., & Klimova, B. (2024). Implementation of AI-driven technology into education—A pilot study on the use of chatbots in foreign language learning. *Cogent Education, 11*(1), 2355385.

Schmidt, T., & Strasser, T. (2022). Artificial intelligence in foreign language learning and teaching: A CALL for intelligent practice. Anglistik: International Journal of. *English Studies, 33*(1), 165–184. https://doi.org/10.33675/ANGL/2022/1/14

Selwyn, N. (2019). *Should robots replace teachers? AI and the Future of Education* (1st ed.). Polity Press.

Seo, K., Tang, J., Roll, I., et al. (2021). The impact of artificial intelligence on learner–instructor interaction in online learning. *International Journal of Educational Technology in Higher Education, 18*, 54. https://doi.org/10.1186/s41239-021-00292-9

Shaik, T., Tao, X., Li, Y., Dann, C., Mcdonald, J., Redmond, P., & Galligan, L. (2023). A review of the trends and challenges in adopting natural language processing methods for education feedback analysis. *ArXiv.* https://doi.org/10.1109/ACCESS.2022.3177752

Sharadgah, T. A., & Sa'di, R. A. (2022). A systematic review of research on the use of artificial intelligence in English language teaching and learning (2015–2021): What are the current effects? *Journal of Information Technology Education: Research, 21*, 10.28945/4999.

Sindermann, C., Sha, P., Zhou, M., Wernicke, J., Schmitt, H. S., Li, M., et al. (2021). Assessing the attitude towards artificial intelligence: Introduction of a short measure in German, Chinese, and English language. *KI-Künstliche intelligenz, 35*(1), 109–118. https://doi.org/10.1007/s13218-020-00689-0

Solak, E. (2024). Revolutionizing language learning: How ChatGPT and AI are changing the way we learn languages. *International Journal of Technology in Education (IJTE), 7*(2), 353–372. https://doi.org/10.46328/ijte.732

Son, J., Ružić, N., & Philpott, A. (2023). Artificial intelligence technologies and applications for language learning and teaching. *Journal of China Computer-Assisted Language Learning.* https://doi.org/10.1515/jccall-2023-0015

Sumakul, D. T. Y., Hamied, F. A., & Sukyadi, D. (2022). Artificial intelligence in EFL classrooms: Friend or foe? *LEARN Journal: Language Education and Acquisition Research Network, 15*(1), 232–256.

Tafazoli, D. (2024). Exploring the potential of generative AI in democratizing English language education. *Computers & Education: Artificial Intelligence*, 100275.

Timotheou, S., Miliou, O., Dimitriadis, Y., Sobrino, S. V., Giannoutsou, N., Cachia, R., Monés, A. M., & Ioannou, A. (2023). Impacts of digital technologies on education and factors influencing schools' digital capacity and transformation: A literature review. *Education and Information Technologies, 28*(6), 6695–6726. https://doi.org/10.1007/s10639-022-11431-8

Varlamis, I., Chronis, C., Sofianopoulou, C., & Papageorgiou, E. (2025). An open-source and open-design robot for STEM education in K-12. In S. Papadakis & G. Lampropoulos (Eds.), *Social robots and artificial intelligence in education: Integrating AI in K-12 and higher education* (pp. 273–304). Palgrave Macmillan.

Wang, J., & Watson, S. (2025). Integrating generative AI in education: Affordances, issues, and directions. In S. Papadakis & G. Lampropoulos (Eds.), *Social robots and artificial intelligence in education: Integrating AI in K-12 and higher education* (pp. 355–379). Palgrave Macmillan.

Yuen, C. L., & Schlote, N. (2024). Learner experiences of mobile apps and artificial intelligence to support additional language learning in education. *Journal of Educational Technology Systems, 52*(4), 507–525. https://doi.org/10.1177/00472395241238693

Zawacki-Richter, O., Marín, V. I., Bond, M., et al. (2019). Systematic review of research on artificial intelligence applications in higher education—Where are the educators? *International Journal of Educational Technology in Higher Education, 16*, 39. https://doi.org/10.1186/s41239-019-0171-0

Zhang, Z., & Huang, X. (2024). The impact of chatbots based on large language models on second language vocabulary acquisition. *Heliyon, 10*(3), e25370.

CHAPTER 14

Integrating Generative AI in Education: Affordances, Issues, and Directions

Jiaqi Wang and Steven Watson

14.1 Introduction

Artificial intelligence (AI) has been called the "crude oil" of the twenty-first century (Holmes & Tuomi, 2022). If one were to give a clear definition of AI, its purpose would be to enable computers to perform tasks that the human brain can accomplish. Although AI requires a computer as a vehicle, it may be more accurate to regard it as a virtual machine, that is, an information-processing system that exists in people's minds as they use it (Boden, 2016). From the 1950s, when the phrase "artificial intelligence" was first introduced, this technology has continued to evolve. The origins of generative AI can also be traced back to this period when scientists began exploring the concept of using algorithms to create new data. However, it wasn't until landmark technological innovations brought generative AI into the public eye (Frey & Osborne, 2024). Since its release to the public by the US tech company OpenAI in November 2022, ChatGPT has gained worldwide attention. This generative AI tool has an

J. Wang (✉) • S. Watson
Faculty of Education, University of Cambridge, Cambridge, UK
e-mail: jw2319@cam.ac.uk; sw10014@cam.ac.uk

S. Papadakis, G. Lampropoulos (eds.), *Social Robots and Artificial Intelligence in Education*, Digital Education and Learning,
https://doi.org/10.1007/978-3-032-03977-4_14

incredibly sophisticated ability to perform complex tasks, in particular to create artificial relics by examining training examples and learning their patterns and distributions using existing digital content, including but not limited to video, images/graphics, text, audio, and video (Baïdoo-Anu & Ansah, 2023). As a result, there is an increasing urgency for research on the application of generative AI in various areas of society.

There has been a great deal of prediction and imagination about the potential of generative AI across various sectors of society, and the field of education is no exception (Fui-Hoon Nah et al., 2023). The emergent literature on integrating generative AI into education includes commentaries claiming or predicting a revolutionary transformation of education (Bahroun et al., 2023; Gao et al., 2024; Prather et al., 2023). A common theme in this literature, however, is the discussion of the dichotomy of opportunities and challenges presented by generative AI (Fengchun & Holmes, 2023; Michel-Villarreal et al., 2023; Yan et al., 2023). Overall, the current research base suggests that there are numerous possible applications for generative AI in education. At the same time, there are many concerns and barriers to its use educationally. This raise new research questions regarding how generative AI can be integrated into education to capitalise on its potential affordances while also addressing the challenges. The questions must now consider the practicalities of integration. By integration, we mean more formalised and strategic pedagogical and administrative implementation of generative AI in educational contexts.

In this section, we explore three main aspects. We begin this chapter by providing a global picture of the extent of generative AI integration in education. Second, we offer a conceptualisation of generative AI integration in terms of two components: Generative AI Literacy Education (GAILE) and Generative AI Education (GAIED). GAILE provides the necessary groundwork for individuals to understand generative AI at a conceptual level. This includes technical and ethical awareness, as well as critical thinking about the societal impacts of AI. We develop this by considering the technology from an interdisciplinary perspective, that is, both technically and linguistically, to demonstrate its capabilities in terms of meaning and semantics. Its *semantic transduction* capability—its capability to transform user input into another form while maintaining the input's meaning—is presented as central to GAILE. Semantic transduction, more than generative AI's capability to 'generate' new content, is central to understanding the affordances of the technology. **GAIED** builds on this foundation by exploring how generative AI can be practically implemented educationally. This builds on the idea of semantic transduction and provides an outline of how generative AI can be deployed as an assistive technology, providing alternative explanations and interpretations, summarising and explaining the text, as a writing assistant, supporting effective educational dialogue, supporting personalised formative assessment, assisting in multi-lingual learning environments, supporting educational inclusion (Nixon et al., 2024) and in assisting in administrative functions. Finally, we outline an approach to generative AI integration using design-based research methodologies, based on a contextual, co-designed, and iterative evaluation to develop GAIED, integrating GAILE approaches through semantic transduction.

14.2 The Current State of Generative AI Integration in Education

There is little research on the actual educational use of generative AI from a global and comparative perspective. Given the rapid introduction of generative AI, this is unsurprising. However, based on what follows, we conclude that despite the interest in generative AI and a great deal of speculation about its potential and limitations, its integration into educational practice has been relatively limited.

The following global integration landscape is somewhat speculative but is based on previous research and our extensive observations. This draws on our experiences in Europe and China, engagement with researchers and educators globally, and in consideration of surveys in the United Kingdom (UK) and research in the United States (US).

Surveys from the US and the UK show that it is used more by students than educators in higher education, and more by teachers than students in schools, but mainly for productivity rather than for teaching and learning (Hennessey, 2023; Laird et al., 2023; Shaw et al., 2023; Walton Family Foundation, 2023). It is also worth noting that, globally, university students appear to be making extensive use of technology in creative and appropriate ways. Klimova and Pikhart (2025 [Chap. 13]) provide examples of AI-driven tools, such as Duolingo, DeepL, and ChatGPT, that have been increasingly used in teaching foreign languages in higher education settings, transforming traditional teaching methods. Again, though, more rigorous research is necessary to understand the extent and nature of the use of generative AI by both teachers and students worldwide. A more recent study by RAND (Diliberti et al., 2024) surveyed 1020 K–12 public school teachers in the US between October and November 2023 on their use of AI tools and products but not specifically generative AI. This found that 18 percent of K–12 teachers reported using AI for teaching and another 15 percent have tried AI at least once. This investigation reflects our experiences and observations, highlighting the limited extent to which generative AI is integrated into education in Western societies, whereas the situation is somewhat different in China. The authors' recent extended visit during the summer of 2024 suggests that there are extensive government-led, strategically funded initiatives to integrate generative AI into education in all phases. However, this might not be reaching more rural parts of China.

We acknowledge the limitations of this impressionistic account of the extent of generative AI integration into education. However, this has some validity. Nonetheless, we argue that the limited use of generative AI is a result of two key factors. The first is related to the extent to which governments and institutions support integration through funding and guidance. The second factor can be attributed to educators' and educational leaders' fears and anxieties about the use of generative AI in education. This is especially the case in Western societies where these concerns translate to risk-based decision making, which can lead to reluctance and caution about the integration of generative AI. In China, it appears that a more pragmatic rather than risk-based approach is being taken, which has translated into an imperative to utilise the technology. However, the authors' engagement with schools and universities in China has revealed that teachers and academics do have many questions about how to integrate generative AI effectively, ethically, and safely into their teaching and existing educational processes.

Regardless of the accuracy of our account and the current level of integration, there is a clear need to understand how to integrate generative AI into education more effectively so that it can be used efficiently, safely, and ethically. This leads us to a discussion of generative AI literacy from a GAILE perspective and integration by considering GAIED approaches. We will develop specific explanations of these two concepts in subsequent sections.

It is essential to reflect on the concerns and reservations that educators have about the integration of generative AI into education, research on this issue is more developed. Concerns have centred on the potential for generative AI to undermine the integrity of the education system and the validity of the assessment process in both teaching and learning (Woolcock, 2023). First, generative AI has the potential to lead to plagiarism, that is, concerns that students can complete assignments by simply prompting the generative AI model (Grassini, 2023). Furthermore, the output produced by generative AI in response to user prompts is based on a complex, recursive, probabilistic model, which makes each response almost unique. Therefore, content generated by generative AI cannot be reliably detected by plagiarism detection systems (Chaka, 2023). Second, widespread concerns have also been raised about ethical issues. The extensive training sets used in developing generative AI inherently reflect structural biases. The dominant dataset of current AI technology is US and Western-centric (Hao, 2018), which has the potential to propagate limited cultural representations and perpetuate stereotypes from a Western standpoint. Third, regulatory issues related to copyright and attribution of original content creators in training datasets have not been resolved. Additionally, concerns regarding data privacy and personal data protection in training data and user prompts have also been raised (Kasneci et al., 2023). Indeed, when generative AI first emerged, some educational organisations in the West suggested that its use needed to be restricted and sometimes even banned (Johnson, 2023).

As an emerging technology, the challenges that AI poses to education are not limited to daily educational practices. Educational institutions have been experiencing digitisation over the past decade. According to West (2023), this trend was accelerated by the COVID-19 pandemic, which shifted education from schools to technologies at an unprecedented pace and scale. The over-reliance on various forms of digital products and electronic devices not only reduces the autonomy of schools but also exacerbates social inequalities. Digital education benefits well-resourced individuals while leaving long-standing educational challenges—such as resource shortages, unequal opportunities, and systemic deficiencies—largely unresolved (Selwyn et al., 2020). Just as the "digital divide" separates those with access to the Internet from those without, the advancement of AI further widens the gap due to its algorithmic principles (Yu, 2020). In addition, the integration of AI into education offers opportunities for economic growth. At present, the field of AI and education has become the focus of investment globally, and this sector is expected to see rapid commercial growth in the next few years (Holmes & Tuomi, 2022). However, the environmental impacts of this growth, including increased energy consumption and e-waste, are placing pressure on sustainable development. Therefore, abandoning the assumption of infinite technological growth and promoting ecologically focused digital technology use in education will be key to the future (Selwyn et al., 2020).

Much of the discussion about the potential role of and concerns about generative AI in education has taken place at a distance from practice and with seemingly minimal consideration of the capabilities of generative AI. We therefore argue that a fine-grained understanding of the educational use of generative AI is required to address the integration challenges presented by the imperative to make use of the technology, while also acknowledging its limitations. In the following section, we take a closer look at text-based Large Language Model (LLM) generative AI to explain its capability in terms of semantic transduction and as the foundation for GAILE. We follow this by considering what LLMs can be used for in education as the basis for GAIED, underpinned by GAILE principles.

14.3 Generative AI as a Language, Meaning, and Communication Technology: Towards a Framework for GAILE

The broad definition of generative AI is that it is an AI system that produces human-like content in response to user prompts in various forms, including text, images, audio, and video. However, this definition of generative AI does not fully capture the technology's capabilities. To comprehend this, it is necessary to consider the architecture and functioning of generative AI models, as well as their application from the perspectives of semantics, linguistics, and communication.

There are two main types of generative AI, all of which utilise deep learning technology based on neural networks. Text-based models, such as LLMs like ChatGPT, Gemini, Baidu's Ernie Bot, and Claude, employ the Transformer architecture, which utilises self-attention, enabling them to produce contextually specific and nuanced responses in text (Vaswani et al., 2017). Another type of generative AI is the Diffusion model, which is used for generating images, video, and audio applications. This model utilises bootstrapping techniques to enhance sample fidelity through sample diversity (Ramesh et al., 2022). Many generative AI technologies, such as ChatGPT, integrate both LLMs and Diffusion models giving them both text and image capabilities.

There are similarities in the overarching function of Transformer architecture and Diffusion models, but they are substantially different in the way in which they create content. In this chapter, we will focus on text-based LLMs that use the Transformer architecture. This is because text remains the dominant medium in educational settings, even though other multi-modal media have become and will continue to become increasingly important educationally. Furthermore, the overarching principles and capabilities that we explain in terms of LLMs and Transformers can be used to understand image, audio, and video generative AI based on Diffusion models.

14.3.1 A Simplified Overview of How LLM Generative Works: Tokens, Word Embeddings, and Transformers

14.3.1.1 Tokenisation and Word Embeddings

An LLM is initially trained on an extensive data set of publicly available text on the Internet. This allows it to quantify billions or even trillion parameters. These are weightings within the neural network architecture that enable it to identify relational features in the training data. This process begins with *tokenisation*, which differentiates the training data into elements. Usually, this is done at the word level, but can also be applied to phrases, syllables, characters, and symbols. The different approaches to tokenisation permit the model not only to deal with multiple languages but also different symbolic systems, such as computer code or mathematics.

The relational dependencies can be explained in terms of *word embeddings*. This approach predates LLMs and enables the model to establish and quantify relationships between words in a multidimensional vector space. These multidimensional relationships are also used with other tokenised symbolic systems. This, although very difficult to imagine in terms of the scale of its multidimensionality and vastness of the numbers of relationships, is relatively simple to comprehend when one considers an example.

Figure 14.1 illustrates a multidimensional vector space where words related to movement are positioned based on their semantic relationships, like how generative AI models, particularly those using word embeddings, organise language. In word embedding, the word 'jump' will be spatially close to similar words like 'leap', 'hop', 'bounce', 'skip', etc. Words that are similar but less close to 'jump' in the vector space of word embeddings

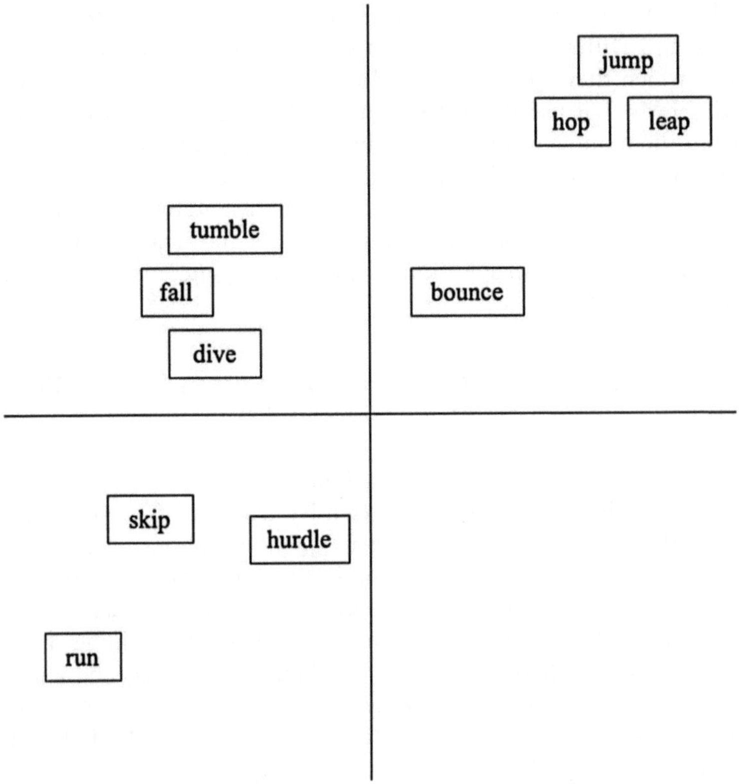

Fig. 14.1 Simplified illustration of word embeddings in 2-dimensions using proximity based on meaning

might include terms that share a broad theme of movement or transition but do not directly describe the action of jumping. These words may reflect related concepts or physical activities where jumping might be a component or associated action, such as 'run', 'climb', 'slide', 'glide', or 'fly'. This spatial arrangement reflects the varying degrees of similarity between these terms, with words closer together being more likely to share contexts in language use. The diagram illustrates how generative AI models comprehend and distinguish between words based on their meanings, utilising these relationships to generate contextually relevant content.

This is where it becomes difficult to visualise the vast multidimensional vector space of all possible relationships between words, with proximity and distance reflecting closeness or variance in meaning or syntactic function. This multidimensionality allows for words to have different meanings in various contexts. For a visualisation of the word embeddings, see http://projector.tensorflow.org. The possibility of imagining the complexity of this when extended to multiple languages and different symbolic systems is further stretched. But what is ultimately mind-bending is to consider that LLMs do this at the tokenisation level, where the multidimensional vector space represents parts of words, syllables, symbols, or characters, such as in Chinese. The approach to tokenisation is dependent on the model.

The process of training LLM generative can be understood as the identification of patterns of meaning within the text, in other words, what a specific word or token means in relation to all other words or tokens in the training data and multiple textual contexts. LLMs leverage this vast record of possible contextualised meanings and their relationships to semantic and syntactic structures to generate meaningful or plausible responses to user inputs. Unlike search engines that use keywords to find information, LLMs operate entirely based on meaning and syntactic relationships.

Meanings or semantics, derived from linguistic and human communication, are relational and context-dependent in relation to other words (Saussure, 1916/2011; Wittgenstein, 1953/1968). By itself, the word 'bank' is ambiguous and can refer to multiple unrelated concepts (e.g., a financial institution or the side of a river). The sentence 'She went to the bank to deposit some money' clarifies that 'bank' refers to a financial institution. Here, the surrounding words ('went to,' 'deposit,' and 'money') provide the relational context necessary to derive the correct meaning.

Through their training, LLMs identify diverse linguistic patterns and contexts and quantify them within their parameters that can be visualised in multidimensional vector space as described above. LLMs associate words with their relevant meanings based on surrounding text, making it possible for the technology to produce appropriate and coherent responses to user inputs.

Syntactic relationships are meaning-based rather than strictly rule-based, as illustrated by the sentences 'The cat chased the mouse' and 'The mouse was chased by the cat.' Both sentences convey the same event and meaning but have different syntactic structures. The first sentence follows a Subject-Verb-Object (SVO) structure, emphasising the agent ('the cat') performing the action. The second sentence follows an Object-Verb-Subject (OVS) structure, emphasising the recipient of the action ('the mouse'). Despite these structural differences, both sentences highlight the same semantic relationship: the cat is the agent, and the mouse is the recipient of the action. This flexibility in syntax, driven by the intended meaning or emphasis, demonstrates that syntactic relationships are fundamentally based on semantics.

A similar understanding can be applied to other language systems such as computer code, mathematics, and character-based languages such as Chinese. It is important to reflect on the fact that the function of generative AI is not rule-based but rather operates on principles of relational and contextualised meaning. This explains why, for example, an LLM can be valuable in explaining mathematical notation but has limitations in performing mathematical procedures. In summary, generative AI is a semantic system and not a logic-based system.

These ideas are central to understanding the appropriate use of generative AI in education. We argue that such understanding must be grounded in meaning and semantics. This underpins the GAILE approach that we outline further in this chapter.

However, tokenisation and word embeddings are just one part of the functioning of LLMs. The other key technologies are Transformers and self-attention, which enable LLMs to excel in Natural Language Processing (NLP).

14.3.1.2 Transformers and Natural Language Processing (NLP)

NLP enables models to interpret and transform user prompts given in natural language into various forms while preserving the original meaning. Natural language is a human language that is spoken, written, or signed and is used for everyday communication, encompassing complex grammar, vocabulary, and syntax. NLP allows the model to generate, for example, coherent paragraphs from a stream of thoughts, convert academic writing into poetry, explain steps in a mathematical proof, produce computer code from everyday language, or rephrase an angry email into a more neutral tone.

The core of this capability lies in the model's ability to process sequential data, such as text, where each element (word or phrase) contributes to the overall meaning. The advancement in NLP through the development of Transformer architecture, which uses self-attention, revolutionised this process. Unlike traditional sequential processing, self-attention enables the model to evaluate the importance of each element within the context of the entire text. This mechanism allows the model to weigh the contributions of different parts of the text to the overall meaning, leveraging its parameters or the multidimensional vector space of word embeddings. As a result, the model can generate responses that are coherent and aligned with the semantic structure of the user's input, effectively mimicking an understanding of the input text.

Consider the sentence, 'The cat that the dog chased ran up the tree.' In traditional models, words are processed sequentially, but Transformers use self-attention to evaluate the importance of each word in relation to others simultaneously. For example, the model assigns higher attention weights to the pairs 'cat-ran' and 'dog-chased' because of their strong syntactic and semantic connections. This context-aware approach allows the model to understand that 'The cat ran up the tree' is the primary action, while 'that the dog chased' provides additional information. By using contextually informed attention, the model can generate coherent and accurate responses, maintaining the overall meaning of the input.

14.3.2 *Meaning and Semantic Transduction as the Basis of GAILE*

Generative AI's power lies in its ability to transform and process contextualised meaning and semantics. This is central to GAILE and is based on an enhanced and interdisciplinary understanding of the technology as outlined above. This explanation is based on an understanding of the technical architecture of generative AI combined with philosophical aspects of linguistics. Thus, LLM generative AI, as a meaning machine, can be understood as a technology that can transform or transduce text from one form to another while maintaining the same meaning based on its training, word embeddings, transformers, and self-attention. Furthermore, it can change meaning in a way directed by the user while also transforming the form or genre of the text. We use the term *semantic transduction* to describe this capability which provides a more nuanced characterisation of the technology than something that is generative, that is, that simply generates content.

There is potentially a vast range of applications for generative AI when understood in terms of semantic transduction. For example, as an assistive technology it can provide exemplification, summarisation, and clarification of text, functions as a writing assistant, facilitates effective educational dialogue, supports personalised formative assessment, aids in multilingual learning environments, promotes educational inclusion, and assists in administrative tasks.

Moreover, we argue that the functionality of generative AI, understood in terms of meaning and semantic transduction, provides an improved understanding of how to deploy and utilise the technology in education and other contexts. Anecdotally, we have found that educators, most notably university students, who have engaged with technology and are using it to support teaching and learning, are using the technology intuitively in this way rather than simply using it to generate content, that is, to write their academic papers or generate an assignment for them. They are using it as an assistive language and communication tool in reading, writing, and refining their ideas.

This is not to say that generative AI is of no value as an information tool or in relation to knowledge. There is a clear overlap between what is considered plausible meaning and knowledge; in other words, statements like 'The sky is blue' are both meaningful and 'common' knowledge. However, understanding this relationship, as well as understanding generative AI in terms of its semantic transduction, is central to what we posit as GAILE and must underpin the integration of generative AI into educational contexts.

Furthermore, when used as a semantic transduction technology and not primarily for content generation, by introducing specific texts and user instructions about what is required, the possibility of generating misinformation or hallucinations, as well as biases, is reduced. Semantic transduction makes use of the models' training for transformational processes rather than generating content. This puts the user in more control of the technology, where this kind of use involves meaning as a reference rather than as a source of information or knowledge. Note here that we say that hallucinations, misinformation, and biases are reduced but not completely eradicated. It is therefore important that the user develops judgement through experiential learning, as we advocate in the GAILE and GAIED approaches.

We now turn to the issue of integration as GAIED, beginning with consideration of how generative AI might be deployed and an analysis of the types of integration that are emerging.

14.4 Three Categories of Generative AI Integration into Education: Toward an Understanding of GAIED

We categorise the applications of generative AI in education into three broad types based on our observation of the kinds of tools and applications that are being developed and deployed. These categories include Autonomous Generative AI Tools (GAITs), Integrated Generative AI Applications (IGAA), and Generative AI-based Automatic Decision-Making Systems (GADMS), see Table 14.1). Our aim here is to relate the ideas and principles of GAILE to GAIED and explain the affordances that each category of integration presents for GAILE.

Table 14.1 Three categories of generative AI integration

Category	*Advantages*	*Disadvantages*
Autonomous Generative AI tools (GAITS) e.g., ChatGPT, Gemini, & Claude	• Potential for personalised inclusion and access • Direct access to model • Opportunities to develop Generative AI literacy • Flexible and versatile	• Challenging to integrate • Need to develop high level of user knowledge and skill in Generative AI
Integrated Generative AI Applications (IGAA) e.g., Microsoft Copilot 365, Google Workspace	• Increased productivity and workflow • Some degree of personalisation • Some inclusion and access benefits	• Moderate integration challenges • Limited access to model • Limited development of user Generative AI literacy
Generative AI-based Automatic Decision-Making Systems (GADMS) e.g., intelligent or smart tutoring systems	• Easy to integrate • Improved efficiency in existing educational programmes and practices • Some inclusion and access benefits • Cost effective at scale	• Black box decision making (limited human-centricity) • Very limited development of generative AI literacy • Inflexible • Limited human/ user orientation • Personalisation in orthodox procedural knowledge and skill development.

The GAIT approach offers the potential for flexible inclusion and access, allowing learners to have direct interaction with the AI model and thereby enhancing learner orientation. The technology can translate curriculum content into forms meaningful to learners, including explanations of ideas in text, mathematics, or code. Advanced speech-to-text systems further extend this capability. For example, Akpan (2024) conducted second data research to compare LLM performance on educational benchmarks with Americans' average academic attainment and literacy levels. The results found that LLMs excelled in tasks such as undergraduate knowledge and advanced reading comprehension. Another study on the accuracy of LLMs in answering periodontology exam questions also indicates that LLMs perform well in terms of question comprehension (Ahmad et al., 2024). As Maher (2025 [Chap. 12]) discussed, chatbots, represented by ChatGPT, are increasingly being integrated into the school scenario by supporting lesson planning, providing instant feedback, and offering personalised learning experiences. However, there are fundamental differences between machine learning and human learning. While LLMs, as represented by ChatGPT, are capable of rapid information processing, they lack the nuanced understanding and creativity inherent in human knowledge. The integration of generative AI into education requires careful consideration of how it complements, rather than replaces, the development of human interaction and critical thinking skills (Wu, 2023). Therefore, the GAIT approach requires users to have a high level of generative AI literacy, necessitating substantial learning and experimentation to understand the technology's semantic transduction capability.

The IGAA approach integrates generative AI into existing applications, supporting users in word processing, email, and spreadsheet tasks through NLP. These tools are increasingly being adopted by organisations, promising significant benefits in terms of productivity and accessibility. The emergence of integrated generative AI represents the beginning of a new era where the way things are done is destined to change. Every role in an organisation will have new workflows, new goals and new challenges to meet (Khan, 2024). In the field of education, a representative example is found in Learning Management Systems (LMS) embedded with generative AI technology. Through the IGAA approach, teachers and students can teach and learn more efficiently through the digital platform. For example, generative AI can automatically generate response suggestions to help teachers quickly deal with student queries. Also, generative AI can automatically review and refine assignments submitted by students, providing personalised feedback. This not only improves work efficiency but also helps students understand and improve their learning outcomes more quickly (Arghir, 2024). While IGAA enhances productivity, it provides less flexibility and user orientation compared to GAIT and does not foster as deep an understanding of generative AI.

GADMS builds on existing adaptive learning systems, employing learning analytics and logic systems for automatic decision-making. These systems are effective in developing procedural and factual knowledge, identifying learner misconceptions through data analytics, and providing personalised responses (Tanweer & Ismail, 2024). While GADMS introduce greater personalisation than previous systems, they are limited in fostering deep conceptual understanding due to their logic-based decision-making processes, which are less exploratory and collaborative. GADMS are likely to become more prevalent in education, supplementing and potentially replacing traditional interactive educational experiences. They are efficient, scalable, and can generate significant revenue for technology companies. For example, the experiment of personalised generative libraries has obtained eye-catching experimental results (Štuikys & Burbaitė, 2024). However, the reliance of GADMS on logic systems for decision-making limits their human-centricity. Learners may blindly rely on these intelligent decision-making systems, making it challenging for them to develop a critical understanding of the outputs they produce (Chan & Hu, 2023).

We argue that the GAIT approach has the most educational value due to its flexibility, accessibility, and learner orientation. However, IGAA and GADMS are more likely to be adopted due to their economic advantages and lower integration costs. Education systems and policies must support the development of generative AI literacy, as understanding and using AI will be essential in the future.

14.5 Generative AI Literacy Education (GAILE) and Generative AI in Education (GAIED)

Although both GAILE and GAIED focus on the application and impact of AI on learning, teaching, and society, they are two sides of the same coin, interdependent and indispensable. The reason why we divide the integration of generative AI with education into GAILE and GAIED components is because they represent two dimensions of education: that is, principle learning and hands-on practice. In this section, we further explain these ideas.

GAILE focuses on imparting knowledge about generative AI, emphasising understanding what it is and how it works. This type of education aims to develop basic AI awareness and competencies among the public, enabling them to understand and critically evaluate AI technologies (Farrelly & Baker, 2023). The understanding of generative AI literacy is further developed from AI literacy, which is a concept that has only been introduced in recent years. AI literacy can be viewed as a set of competencies that enable individuals to evaluate AI technologies critically, effectively communicate and collaborate with AI, and use AI as a tool in a variety of scenarios (Long & Magerko, 2020).

As a subset of AI literacy, generative AI literacy encompasses a comprehensive set of knowledge and skills essential for understanding, interpreting, and engaging with generative AI technologies. It involves a technical awareness of generative AI, deep learning, LLMs, Transformer, and Diffusion models. Additionally, ethical awareness is a crucial component. This extends to issues related to data privacy, the potential biases in AI outputs due to biased training data, and the risks of misuse in creating deepfakes or spreading misinformation (Bozkurt, 2024).

Generative AI literacy also extends to the creative application of generative AI tools in various fields, such as writing, STEM, art, design, as well as interlingual and intralingual communication, centred on a comprehension of the semantic transduction capability of generative AI (Relmasira et al., 2023). Furthermore, critical evaluation skills are vital to enable individuals to understand the meaning of content generated by a hybrid of humans and technology, to assess the accuracy and relevance of AI responses, and to identify potential biases (Vartiainen & Tedre, 2023). Lastly, an awareness of the societal impact of generative AI, namely its potential to automate jobs, influence media and entertainment, and shape public opinion, must be integral to generative AI literacy (Nixon et al.,

2024). As the application of generative AI has lowered the barriers to using such technologies, educating the public is critical to the responsible use of these technologies and navigating the rapidly evolving digital and technological landscape (Farrelly & Baker, 2023). Moreover, this implies developing a philosophical perspective on technology, its societal role, and its relationships to individuals in society.

GAIED goes beyond the foundational knowledge provided by GAILE by focusing on a deep understanding of generative AI, including its application within educational settings. This approach aims to improve the efficiency and quality of education through the practical use of generative AI technologies, such as in the classroom.

Providing examples will make it easier to understand both concepts. For example, the GAILE programme could include introductory courses or workshops to teach participants the basics of generative AI, such as understanding how AI models, like ChatGPT, generate content and the ethical issues associated with their use (Sullivan et al., 2024). On the other hand, in the context of GAIED, schools may implement an AI-powered learning platform that assists teachers in creating personalised lesson plans, automating grading, and providing real-time feedback to students. Schools will provide training sessions for teachers and students on how to effectively use these AI tools, ensuring that the technology is seamlessly integrated into the educational process while improving learning outcomes. As such, combining knowledge from the GAILE project with practical applications from GAIED implementations are two integral parts of integrating generative AI into education.

Central to developing a deep and relational conceptual understanding of generative AI, specifically GAILE and GAIED, is the GAIT approach to integrating generative AI in education (Yilmaz & Yilmaz, 2023). While conceding in this, there is a high level of technological mediation even within GAIT approaches, educators and learners do have the opportunity to engage directly with generative AI models, and therefore, it offers a more human-centred approach to AI integration in education than IGAA and GADMS integration.

However, the major challenge of the GAIT approach is its integration into existing educational practice. This is mainly because generative AI is not well-understood by either educators or learners; therefore, approaches to research and development are required that are contextual and participatory, allowing for the development and dissemination of knowledge, skills, and new kinds of educational practice. In the remainder of this chapter, we will consider how to address the challenge of integrating generative AI into education, premised on the GAIT approach, and describe a research and development approach based on design-based research.

14.6 Participatory Co-design in Human-Centred Generative AI Integration in Education

In the previous section, we outlined the challenges in integrating human-centred generative AI in education within the GAIT framework. Regulatory compliance is essential to ensure the ethical use of AI tools like generative AI chatbots. Unethical use can lead to data breaches, privacy violations, and misuse of AI-generated content. Developing generative AI literacy is crucial for fully harnessing these technologies. This includes enhancing educators', decision-makers', and learners' understanding of the technology's capabilities, particularly in semantic transduction. However, challenges extend beyond learning the technology; they involve profound integration issues within the education system itself and its autonomous discourse in society (Watson & Shi, 2024).

Education is an autonomous discourse in society, interconnected with broader societal discourses, including science, politics, economics, media, and law. As a discursive system, education is subject to contingencies experienced by educators, learners, and other discursive systems, such as policymaking, law, media, and economics. Education provides a basis for meaning-making in programmes and practices, influencing individual actions and behaviours. The role individuals play and how they are affected by the education system structure are significant. Student performance is influenced by the curriculum, teaching methods, assessment methods, and resources, as well as family support, peer interactions, and extracurricular activities (Watson & Romic, 2024).

Education maintains its autonomy and integrity while interconnected with external contingencies through self-referential meaning-making in programs and practices. It operates as a generalised discursive medium for the learner and differentiates its operations based on learner progress. Integrity and autonomy depend on the continuity of programs and practices, such as curriculum, pedagogy, and assessment processes. Educational reforms and changes are challenging due to their systemic nature, requiring time and resources for meaningful adaptation. Reform initiatives cannot be easily anticipated or predicted regarding new forms of programs and practices (Watson & Romic, 2024).

This complexity was acknowledged by pragmatist philosophy and operationalised by Kurt Lewin post–Second World War. Lewin's Action Research approach introduced rigorous, contextualised, iterative participatory research and development with formative evaluation cycles, leading to emergent programmatic change and understanding of systemic change (Lewin, 1946).

Ann Brown further developed this principle in the 1980s through her work on design experiments in educational settings, which led to the emergence of design-based research (DBR) methodologies (Brown, 1992). DBR involves introducing new ideas or programmatic designs, including technological innovations, through contextualised and participatory formative evaluation cycles. This methodology allows innovations to be introduced into complex educational contexts, resulting in systemically coherent programs that can lead to sustained change and be adapted and scaled to similar contexts.

Despite challenges in complex research and development contexts and extensive data production (Dede, 2004), knowledge can be achieved through analytic generalisation. Transdisciplinary approaches to theory development are employed, drawing on a network of disciplinary concepts analysed abductively with data and in relation to the semantic structuration of their disciplinary origins (e.g., psychology, social psychology, anthropology, philosophy, and technology) (Marschall et al., 2024). For instance, psychologists can understand learner interactions with AI tools, philosophers address ethical considerations, and technologists ensure the development of robust AI systems. Integrating these diverse perspectives develops holistic solutions that are theoretically sound and practically viable.

Participatory co-design involves stakeholders, including educators, learners, and policymakers, in the design and implementation process of new technologies. This method ensures that the needs, perspectives, and experiences of all participants are considered, leading to more effective and human-centred outcomes.

Education systems, despite their inherent structure and stability, possess the flexibility to adapt to external changes and internal needs. This flexibility can be facilitated through participatory design and collaborative efforts, ensuring responsiveness. The primary goal of education systems is to meet learners' diverse needs. Adopting a human-centred approach allows these systems to incorporate individual and collective needs into their framework. Individuals, such as teachers, students, and policymakers, act as key components in the system's functioning, with their interactions and feedback being crucial. Teachers bring pedagogical insights into the classroom, influencing educational practice, while students contribute through their learning experiences and feedback, informing instructional strategies and curriculum development.

The GAIT approach advocates for integrating generative AI in education through participatory co-design and formative evaluation, involving learners, educators, and decision-makers in a structured and rigorous manner. Emergence is key, acknowledging that changes cannot be predicted linearly but through evolving and understanding new forms.

A participatory DBR approach provides robust frameworks for integrating generative AI in education. Future research should focus on longitudinal studies to assess the long-term impact of these technologies, exploring ways to scale successful interventions and continuously involving stakeholders to ensure the relevance and effectiveness of AI tools in diverse educational contexts.

14.7 Conclusion

In this chapter, we advocate for a human-centred approach to integrating generative AI into education, underpinned by the GAILE and GAIED principles. We offer a nuanced framework for understanding and utilising generative AI in education, emphasising its potential to enhance learning outcomes through semantic transduction. Central to this approach is the understanding of generative AI not merely as a content-generating tool but as a meaning-transducing technology. This approach entails utilising generative AI tools that enable direct interaction between educators and learners, as well as the AI models themselves, thereby fostering the development of generative AI literacy. By embracing this technology, educators and learners can effectively adapt to the educational and societal changes that its advancement has brought about. Additionally, we have outlined an approach to contextual and participatory research and development using DBR.

While generative AI holds great promise in transforming educational practices, we must acknowledge the limitations of our current understanding. First, generative AI integration is still in its infancy, particularly in formal educational settings. There is a lack of empirical studies that assess the long-term impacts of these tools on learning outcomes. Additionally, the ethical concerns surrounding bias, data privacy, and the potential misuse of AI-generated content need more extensive research and robust regulatory frameworks to ensure its safe deployment in education. Another limitation is that existing education systems and curricula are not designed to accommodate the iterative, participatory co-design approach we propose. This means that integrating generative AI with education requires reforming the education system, and it is then necessary to involve various stakeholders.

Given the sustained presence of generative AI, future research could address these gaps by focusing on longitudinal studies that evaluate the efficacy of generative AI tools across different educational contexts and learner demographics. Additionally, more attention should be given to co-designing AI tools with educators, learners, and policymakers to ensure their relevance and effectiveness. Research should also explore strategies to mitigate ethical risks and biases, ensuring that AI enhances rather than undermines educational equity and inclusion.

References

Ahmad, B., Saleh, K., Alharbi, S., Alqaderi, H., & Jeong, Y. N. (2024). Artificial intelligence in periodontology: Performance evaluation of ChatGPT, Claude, and Gemini on the in-service examination. medRxiv, 2024–05. https://doi.org/10.1101/2024.05.29.24308155.

Akpan, M. (2024). Have we reached AGI? Comparing ChatGPT, Claude, and Gemini to human literacy and education benchmarks. arXiv preprint arXiv:2407.09573. https://doi.org/10.48550/arXiv.2407.09573

Arghir, D. C. (2024). Implementation of learning management systems with generative artificial intelligence functions in the post-pandemic environment. *Information Technologies and Learning Tools, 100*(2), 217. https://doi.org/10.33407/itlt.v100i2.5518

Bahroun, Z., Anane, C., Ahmed, V., & Zacca, A. (2023). Transforming education: A comprehensive review of generative artificial intelligence in educational settings through bibliometric and content analysis. *Sustainability, 15*(17), Article 17. https://doi.org/10.3390/su151712983

Baídoo-Anu, D., & Ansah, L. O. (2023). Education in the era of generative artificial intelligence (AI): Understanding the potential benefits of ChatGPT in promoting teaching and learning. *Journal of AI, 7*(1), Article 1. https://doi.org/10.61969/jai.1337500

Boden, M. A. (2016). *AI: Its nature and future.* Oxford University Press.

Bozkurt, A. (2024). Why generative AI literacy, why now and why it matters in the educational landscape? Kings, Queens and GenAI Dragons. *Open Praxis, 16*(3), 283–290. https://doi.org/10.55982/openpraxis.16.3.739

Brown, A. L. (1992). Design experiments: Theoretical and methodological challenges in creating complex interventions in classroom settings. *Journal of the Learning Sciences, 2*(2), 141–178. https://doi.org/10.1207/s15327809jls0202_2

Chaka, C. (2023). Detecting AI content in responses generated by ChatGPT, YouChat, and Chatsonic: The case of five AI content detection tools. *Journal of Applied Learning and Teaching, 6*(2), Article 2. https://doi.org/10.37074/jalt.2023.6.2.12

Chan, C. K. Y., & Hu, W. (2023). Students' voices on generative AI: Perceptions, benefits, and challenges in higher education. *International Journal of Educational Technology in Higher Education, 20*(1), 43. https://doi.org/10.1186/s41239-023-00411-8

Dede, C. (2004). If design-based research is the answer, what is the question? A commentary on Collins, Joseph, and Bielaczyc; diSessa and Cobb; and Fishman, Marx, Blumenthal, Krajcik, and Soloway in the JLS Special Issue on design-based research. *Journal of the Learning Sciences, 13*(1), 105–114. https://doi.org/10.1207/s15327809jls1301_5

Diliberti, M. K., Schwartz, H. L., Doan, S., Shapiro, A., Rainey, L. R., & Lake, R. J. (2024). *Using Artificial intelligence tools in K–12 Classrooms.* RAND Corporation. https://www.rand.org/pubs/research_reports/RRA956-21.html

Farrelly, T., & Baker, N. (2023). Generative artificial intelligence: Implications and considerations for higher education practice. *Education Sciences, 13*(11), 1109. https://doi.org/10.3390/educsci13111109

Fengchun, M., & Holmes, W. (2023). *Guidance for generative AI in education and research.* UNESCO.

Frey, C. B., & Osborne, M. (2024). Generative AI and the future of work: A reappraisal. *Brown Journal of World Affairs, 30*, 161.

Fui-Hoon Nah, F., Zheng, R., Cai, J., Siau, K., & Chen, L. (2023). Generative AI and ChatGPT: Applications, challenges, and AI-human collaboration. *Journal of Information Technology Case and Application Research, 25*(3), 277–304. https://doi.org/10.1080/15228053.2023.2233814

Gao, L. (Xuehui), López-Pérez, M. E., Melero-Polo, I., & Trifu, A. (2024). Ask ChatGPT first! Transforming learning experiences in the age of artificial intelligence. Studies in Higher Education, 0(0), 1–25. https://doi.org/10.1080/03075079.2024.2323571

Grassini, S. (2023). Shaping the future of education: Exploring the potential and consequences of AI and ChatGPT in educational settings. *Education Sciences, 13*(7), Article 7. https://doi.org/10.3390/educsci13070692

Hao. (2018). AI has a culturally biased world view that Google has a plan to change. Retrieved March 26, 2024, from https://www.technologyreview.com/2018/12/02/138843/ai-has-a-culturally-biased-worldview-that-google-has-a-plan-to-change

Hennessey, M. (2023, April 21). Exclusive: Almost half of Cambridge students have used ChatGPT to complete university work. *Varsity Online*. https://www.varsity.co.uk/news/25463

Holmes, W., & Tuomi, I. (2022). State of the art and practice in AI in education. *European Journal of Education, 57*(4), 542–570. https://doi.org/10.1111/ejed.12533

Johnson, A. (2023, January 18). ChatGPT in schools: Here's where it's banned—And how it could potentially help students. *Forbes*. https://www.forbes.com/sites/ariannajohnson/2023/01/18/chatgpt-in-schools-heres-where-its-banned-and-how-it-could-potentially-help-students/

Kasneci, E., Sessler, K., Küchemann, S., Bannert, M., Dementieva, D., Fischer, F., Gasser, U., Groh, G., Günnemann, S., Hüllermeier, E., Krusche, S., Kutyniok, G., Michaeli, T., Nerdel, C., Pfeffer, J., Poquet, O., Sailer, M., Schmidt, A., Seidel, T., et al. (2023). ChatGPT for good? On opportunities and challenges of large language models for education. *Learning and Individual Differences, 103*, 102274. https://doi.org/10.1016/j.lindif.2023.102274

Khan, A. (2024). Copilot, Generative AI, and future of work. In *Introducing Microsoft Copilot for managers: Enhance your team's productivity and creativity with generative AI-Powered Assistant* (pp. 695–696). Apress. https://doi.org/10.1007/979-8-8688-0419-9_18

Klimova, B., & Pikhart, M. (2025). Use of artificial intelligence in foreign language education in higher education settings. In S. Papadakis & G. Lampropoulos (Eds.), *Social robots and artificial intelligence in education: Integrating AI in K-12 and higher education* (pp. 329–353). Palgrave Macmillan.

Laird, E., Dwyer, M., & Grant-Chapman, H. (2023). *OFF TASK! EdTech threats to student privacy and equity in the age of AI*. The Center for Democracy & Technology (CDT). https://cdt.org/wp-content/uploads/2023/09/091923-CDT-Off-Task-web.pdf

Lewin, K. (1946). Action research and minority problems. *Journal of Social Issues, 2*(4), 34–46. https://doi.org/10.1111/j.1540-4560.1946.tb02295.x

Long, D., & Magerko, B. (2020, April). What is AI literacy? Competencies and design considerations. In *Proceedings of the 2020 CHI conference on human factors in computing systems* (pp. 1–16). https://doi.org/10.1145/3313831.3376727

Maher, D. (2025). Chatbots in schools: Opportunities and considerations. In S. Papadakis & G. Lampropoulos (Eds.), *Social robots and artificial intelligence in education: Integrating AI in K-12 and higher education* (pp. 305–328). Palgrave Macmillan.

Marschall, G., Watson, S., Kimber, E., & Major, L. (2024). A transdisciplinary study of a novice mathematicsteacher's instructional decision-making. *Journal of Mathematics Teacher Education*, 1–26.

Michel-Villarreal, R., Vilalta-Perdomo, E., Salinas-Navarro, D. E., Thierry-Aguilera, R., & Gerardou, F. S. (2023). Challenges and opportunities of generative AI for higher education as explained by ChatGPT. *Education Sciences, 13*(9), Article 9. https://doi.org/10.3390/educsci13090856

Nixon, N., Lin, Y., & Snow, L. (2024). Catalyzing equity in STEM teams: Harnessing generative AI for inclusion and diversity. *Policy Insights from the Behavioral and Brain Sciences*, 23727322231220356. https://doi.org/10.1177/23727322231220356

Prather, J., Denny, P., Leinonen, J., Becker, B. A., Albluwi, I., Craig, M., Keuning, H., Kiesler, N., Kohn, T., Luxton-Reilly, A., MacNeil, S., Petersen, A., Pettit, R., Reeves, B. N., & Savelka, J. (2023). The robots are here: Navigating the generative AI revolution in computing education. In *Proceedings of the 2023 Working Group Reports on Innovation and Technology in Computer Science Education* (pp. 108–159). https://doi.org/10.1145/3623762.3633499.

Ramesh, A., Dhariwal, P., Nichol, A., Chu, C., & Chen, M. (2022). Hierarchical text-conditional image generation with CLIP Latents (arXiv:2204.06125). arXiv. https://doi.org/10.48550/arXiv.2204.06125.

Relmasira, S. C., Lai, Y. C., & Donaldson, J. P. (2023). Fostering AI literacy in elementary science, technology, engineering, art, and mathematics (STEAM) education in the age of generative AI. *Sustainability, 15*(18), 13595. https://doi.org/10.3390/su151813595

Saussure, F. de. (2011). *Course in general linguistics* (P. Meisel & H. Saussy, Eds.; W. Baskin, Trans.). Columbia University Press.. (Original work published 1916).

Selwyn, N., Hillman, T., Eynon, R., Ferreira, G., Knox, J., Macgilchrist, F., & Sancho-Gil, J. M. (2020). What's next for Ed-Tech? Critical hopes and concerns for the 2020s. *Learning, Media and Technology, 45*(1), 1–6.

Shaw, C., Yuan, L., Brennan, D., Martin, S., Janson, N., Fox, K., & Bryant, G. (2023). *Generative AI in higher education: Fall 2023 update of time for class study.* Technical Report. Tyton Partners, Boston, MA. Retrieved March 18, 2024, from https://tytonpartners.com/app/uploads/2023/10/GenAI-IN-HIGHER-EDUCATION-FALL-2023-UPDATE-TIME-FOR-CLASS-STUDY.pdf

Štuikys, V., & Burbaitė, R. (2024). Personal generative libraries for personalised learning: A case study. In *Evolution of STEM-driven computer science education: The perspective of big concepts* (pp. 109–134). Springer Nature Switzerland. https://doi.org/10.1007/978-3-031-48235-9_4

Sullivan, M., McAuley, M., Degiorgio, D., & McLaughlan, P. (2024). Improving students' generative AI literacy: A single workshop can improve confidence and understanding. *Journal of Applied Learning and Teaching, 7*(2). https://doi.org/10.37074/jalt.2024.7.2.7

Tanweer, M., & Ismail, A. (2024). Generative AI in curriculum development: A framework for adaptive, customized, and personalized learning. In *Impacts of generative AI on creativity in higher education* (pp. 193–226). IGI Global. https://doi.org/10.4018/979-8-3693-2418-9.ch008.

Vartiainen, H., & Tedre, M. (2023). Using artificial intelligence in craft education: Crafting with text-to-image generative models. *Digital Creativity, 34*(1), 1–21. https://doi.org/10.1080/14626268.2023.2174557

Vaswani, A., Shazeer, N., Parmar, N., Uszkoreit, J., Jones, L., Gomez, A. N., Kaiser, L., & Polosukhin, I. (2017). Attention is all you need (arXiv:1706.03762). arXiv. https://doi.org/10.48550/arXiv.1706.03762.

Walton Family Foundation. (2023, July 18). *Survey finds majority of teachers, parents report positive impact of ChatGPT on teaching and learning*. Walton Family Foundation. https://www.waltonfamilyfoundation.org/about-us/newsroom/survey-finds-majority-of-teachers-parents-report-positive-impact-of-chatgpt-on-teaching-and-learning

Watson, S., & Romic, J. (2024). ChatGPT and the entangled evolution of society, education, and technology: A systems theory perspective. *European Educational Research Journal, 14749041231221266*. https://doi.org/10.1177/14749041231221266

Watson, S., & Shi, S. (2024). Generative AI Integration in Education: Challenges and Approaches. *Artificial Intelligence in Education: The Intersection of Technology and Pedagogy*, 59–73.

West, M. (2023). *An Ed-Tech tragedy?: Educational technologies and school closures in the time of COVID-19*. UNESCO Publishing.

Wittgenstein, L. (1968). *Philosophical investigations*. Basil Blackwell. (Original work published 1953).

Woolcock, N. (2023, May 21). AI 'is clear and present danger to education'. *The Times*. https://www.thetimes.co.uk/article/ai-is-clear-and-present-danger-to-education-3sk09ftlf

Wu, Y. (2023). Integrating generative AI in education: How ChatGPT brings challenges for future learning and teaching. *Journal of Advanced Research in Education, 2*(4), 6–10. https://doi.org/10.56397/JARE.2023.07.02

Yan, L., Sha, L., Zhao, L., Li, Y., Martinez-Maldonado, R., Chen, G., Li, X., Jin, Y., & Gašević, D. (2023). Practical and ethical challenges of large language models in education: A systematic scoping review. *British Journal of Educational Technology, bjet.13370*. https://doi.org/10.1111/bjet.13370

Yilmaz, R., & Yilmaz, F. G. K. (2023). The effect of generative artificial intelligence (AI)-based tool use on students' computational thinking skills, programming self-efficacy and motivation. *Computers and Education: Artificial Intelligence, 4*, 100147. https://doi.org/10.1016/j.caeai.2023.100147

Yu, P. K. (2020). The algorithmic divide and equality in the age of artificial intelligence. *The Florida Law Review, 72*, 331.

Index

S. Papadakis, G. Lampropoulos (eds.), *Social Robots and Artificial Intelligence in Education*, Digital Education and Learning,
https://doi.org/10.1007/978-3-032-03977-4

MIX
Papier aus verantwortungsvollen Quellen
Paper from responsible sources
FSC® C105338

If you have any concerns about our products,
you can contact us on
ProductSafety@springernature.com

In case Publisher is established outside the EU,
the EU authorized representative is:
Springer Nature Customer Service Center GmbH
Europaplatz 3, 69115 Heidelberg, Germany

Printed by Libri Plureos GmbH
in Hamburg, Germany